Springer Undergraduate Mathematics Series

Advisory Board

P.J. Cameron *Queen Mary and Westfield College*
M.A.J. Chaplain *University of Dundee*
K. Erdmann *Oxford University*
L.C.G. Rogers *Cambridge University*
E. Süli *Oxford University*
J.F. Toland *University of Bath*

Other books in this series

A First Course in Discrete Mathematics *I. Anderson*
Analytic Methods for Partial Differential Equations *G. Evans, J. Blackledge, P. Yardley*
Applied Geometry for Computer Graphics and CAD, Second Edition *D. Marsh*
Basic Linear Algebra, Second Edition *T.S. Blyth and E.F. Robertson*
Basic Stochastic Processes *Z. Brzeźniak and T. Zastawniak*
Calculus of One Variable *K.E. Hirst*
Complex Analysis *J.M. Howie*
Elementary Differential Geometry *A. Pressley*
Elementary Number Theory *G.A. Jones and J.M. Jones*
Elements of Abstract Analysis *M. Ó Searcóid*
Elements of Logic via Numbers and Sets *D.L. Johnson*
Essential Mathematical Biology *N.F. Britton*
Essential Topology *M.D. Crossley*
Fields and Galois Theory *J.M. Howie*
Fields, Flows and Waves: An Introduction to Continuum Models *D.F. Parker*
Further Linear Algebra *T.S. Blyth and E.F. Robertson*
Geometry *R. Fenn*
Groups, Rings and Fields *D.A.R. Wallace*
Hyperbolic Geometry, Second Edition *J.W. Anderson*
Information and Coding Theory *G.A. Jones and J.M. Jones*
Introduction to Laplace Transforms and Fourier Series *P.P.G. Dyke*
Introduction to Ring Theory *P.M. Cohn*
Introductory Mathematics: Algebra and Analysis *G. Smith*
Linear Functional Analysis *B.P. Rynne and M.A. Youngson*
Mathematics for Finance: An Introduction to Financial Engineering *M. Capińksi and T. Zastawniak*
Matrix Groups: An Introduction to Lie Group Theory *A. Baker*
Measure, Integral and Probability, Second Edition *M. Capińksi and E. Kopp*
Multivariate Calculus and Geometry, Second Edition *S. Dineen*
Numerical Methods for Partial Differential Equations *G. Evans, J. Blackledge, P.Yardley*
Probability Models *J.Haigh*
Real Analysis *J.M. Howie*
Sets, Logic and Categories *P. Cameron*
Special Relativity *N.M.J. Woodhouse*
Symmetries *D.L. Johnson*
Topics in Group Theory *G. Smith and O. Tabachnikova*
Vector Calculus *P.C. Matthews*

Andrew Baker

Matrix Groups

An Introduction to Lie Group Theory

With 16 Figures

Springer

Andrew Baker, BSc, PhD
Department of Mathematics, University of Glasgow, Glasgow G12 8QW, UK

Cover illustration elements reproduced by kind permission of:
Aptech Systems, Inc., Publishers of the GAUSS Mathematical and Statistical System, 23804 S.E. Kent-Kangley Road, Maple Valley, WA 98038, USA. Tel: (206) 432 - 7855 Fax (206) 432 - 7832 email: info@aptech.com URL: www.aptech.com
American Statistical Association: Chance Vol 8 No 1, 1995 article by KS and KW Heiner 'Tree Rings of the Northern Shawangunks' page 32 fig 2
Springer-Verlag: Mathematica in Education and Research Vol 4 Issue 3 1995 article by Roman E Maeder, Beatrice Amrhein and Oliver Gloor 'Illustrated Mathematics: Visualization of Mathematical Objects' page 9 fig 11, originally published as a CD ROM 'Illustrated Mathematics' by TELOS: ISBN 0-387-14222-3, German edition by Birkhauser: ISBN 3-7643-5100-4.
Mathematica in Education and Research Vol 4 Issue 3 1995 article by Richard J Gaylord and Kazume Nishidate 'Traffic Engineering with Cellular Automata' page 35 fig 2. Mathematica in Education and Research Vol 5 Issue 2 1996 article by Michael Trott 'The Implicitization of a Trefoil Knot' page 14.
Mathematica in Education and Research Vol 5 Issue 2 1996 article by Lee de Cola 'Coins, Trees, Bars and Bells: Simulation of the Binomial Process' page 19 fig 3. Mathematica in Education and Research Vol 5 Issue 2 1996 article by Richard Gaylord and Kazume Nishidate 'Contagious Spreading' page 33 fig 1. Mathematica in Education and Research Vol 5 Issue 2 1996 article by Joe Buhler and Stan Wagon 'Secrets of the Madelung Constant' page 50 fig 1.

British Library Cataloguing in Publication Data
Baker, Andrew
Matrix groups : an introduction to Lie group theory.
(Springer undergraduate mathematics series)
1. Matrix groups 2. Lie groups
I. Title
512.5'5
ISBN 978-1-85233-470-3

Library of Congress Cataloging-in-Publication Data
Baker, Andrew, 1953-
Matrix groups : an introduction to Lie group theory / Andrew Baker.
p. cm. -- (Springer undergraduate mathematics series)
Includes bibliographical references and index.
ISBN 978-1-85233-470-3 ISBN 978-1-4471-0183-3 (eBook)
DOI 10.1007/978-1-4471-0183-3
1. Matrix groups. I. Title. II. Series.
QA174.2.B35 2001
512'.2—dc21 2001049261

Springer Undergraduate Mathematics Series ISSN 1615-2085
ISBN 978-1-85233-470-3

Additional material to this book can be downloaded from http://extra.springer.com.

springeronline.com

Originally published by Springer-Verlag London Limited in 2002

3rd printing 2006

Typesetting: Camera ready by the author

12/3830-5432 Printed on acid-free paper SPIN 11532026

Preface

This work provides a first taste of the theory of Lie groups accessible to advanced mathematics undergraduates and beginning graduate students, providing an appetiser for a more substantial further course. Although the formal prerequisites are kept as low level as possible, the subject matter is sophisticated and contains many of the key themes of the fully developed theory. We concentrate on *matrix groups*, *i.e.*, closed subgroups of real and complex general linear groups. One of the results proved is that every matrix group is in fact a Lie group, the proof following that in the expository paper of Howe [12]. Indeed, the latter, together with the book of Curtis [7], influenced our choice of goals for the present book and the course which it evolved from. As pointed out by Howe, Lie theoretic ideas lie at the heart of much of standard undergraduate linear algebra, and exposure to them can inform or motivate the study of the latter; we frequently describe such topics in enough detail to provide the necessary background for the benefit of readers unfamiliar with them.

Outline of the Chapters

Each chapter contains exercises designed to consolidate and deepen readers' understanding of the material covered. We also use these to explore related topics that may not be familiar to all readers but which should be in the toolkit of every well-educated mathematics graduate. Here is a brief synopsis of the chapters.

Chapter 1: The *general linear groups* $\mathrm{GL}_n(\Bbbk)$ for $\Bbbk = \mathbb{R}$ (the real numbers) and $\Bbbk = \mathbb{C}$ (the complex numbers) are introduced and studied both as groups and as topological spaces. *Matrix groups* are defined and a number of standard examples discussed, including *special linear groups* $\mathrm{SL}_n(\Bbbk)$, *orthogonal groups* $\mathrm{O}(n)$ and *special orthogonal groups* $\mathrm{SO}(n)$, *unitary groups* $\mathrm{U}(n)$ and *special unitary groups* $\mathrm{SU}(n)$, as well as more exotic examples such as *Lorentz groups* and *symplectic groups.* The relation of complex to real matrix groups is also studied. Along the way we discuss various algebraic, analytic and topological notions including *norms*, *metric spaces*, *compactness* and *continuous group actions.*

Chapter 2: The *exponential function* for matrices is introduced and *one-parameter subgroups* of matrix groups are studied. We show how these ideas can be used in the solution of certain types of differential equations.

Chapter 3: The idea of a *Lie algebra* is introduced and various algebraic properties are studied. *Tangent spaces* and Lie algebras of matrix groups are defined together with the *adjoint action.* The important special case of SU(2) and its relationship to SO(3) is studied in detail.

Chapters 4 and 5: *Finite dimensional algebras* over fields, especially $\mathbb{R}$ or $\mathbb{C}$, are defined and their units viewed as a source of matrix groups using the *reduced regular representation.* The *quaternions* and more generally the *real Clifford algebras* are defined and *spinor groups* constructed and shown to double cover the special orthogonal groups. The *quaternionic symplectic groups* $\mathrm{Sp}(n)$ are also defined, completing the list of compact connected classical groups and their universal covers. *Automorphism groups* of algebras are also shown to provide further examples of matrix groups.

Chapter 6: The geometry and linear algebra of *Lorentz groups* which are of importance in Relativity are studied. The relationship of $\mathrm{SL}_2(\mathbb{C})$ to the Lorentz group $\mathrm{Lor}(3,1)$ is discussed, extending the work on SU(2) and SO(3) in Chapter 3.

Chapter 7: The general notion of a *Lie group* is introduced and we show that all matrix groups are *Lie subgroups* of general linear groups. Along the way we introduce the basic ideas of *differentiable manifolds* and *smooth maps.* We show that not every Lie group can be realised as a matrix group by considering the simplest *Heisenberg group.*

Chapters 8 and 9: *Homogeneous spaces* of Lie groups are defined and we show how to recognise them as orbits of *smooth actions.* We discuss *connectivity* of Lie groups and use homogeneous spaces to prove that many familiar Lie groups are path connected. We also describe some important families of homogeneous spaces such as *projective spaces* and *Grassmannians*, as well as examples related to special factorisations of matrices such as *polar form.*

Chapters 10, 11 and 12: The basic theory of *compact connected Lie groups* and their *maximal tori* is studied and the relationship to some well-known matrix diagonalisation results highlighted. We continue this theme by describing the classification theory of compact connected *simple* Lie groups, showing how the families we meet in earlier chapters provide all but a finite number of the isomorphism types predicted. *Root systems*, *Weyl groups* and *Dynkin diagrams* are defined and many examples described.

Some suggestions for using this book

For an advanced undergraduate course of about 30 lectures to students already equipped with basic real and complex analysis, metric spaces, linear algebra, group and ring theory, the material of Chapters 1, 2, 3, 7 provide an introduction to matrix groups, while Chapters 4, 5, 6, 8, 9 supply extra material that might be quarried for further examples. A more ambitious course aimed at presenting the classical compact connected Lie groups might take in Chapters 4, 5 and perhaps lead on to some of the theory of compact connected Lie groups discussed in Chapters 10, 11, 12.

A reader (perhaps a graduate student) using the book on their own would find it useful to follow up some of the references [6, 8, 17, 18, 25, 29] to see more advanced approaches to the topics on differential geometry and topology covered in Chapters 7, 8, 9 and the classification theory of Chapters 10, 11, 12.

Each chapter has a set of Exercises of varying degrees of difficulty. Hints and solutions are provided for some of these, the more challenging questions being indicated by the symbols ⚠ or ⚠⚠ with the latter intended for readers wishing to pursue the material in greater depth.

Prerequisites and assumptions

The material in Chapters 1, 2, 3, 7 is intended to be accessible to a well-equipped advanced undergraduate, although many topics such as non-metric topological spaces, normed vector spaces and rings may be unfamiliar so we have given the relevant definitions. We do not assume much abstract algebra beyond standard notions of homomorphisms, subobjects, kernels and images and quotients; semi-direct products of groups are introduced, as are Lie algebras. A course on matrix groups is a good setting to learn algebra, and there are many significant algebraic topics in Chapters 4, 5, 11, 12. Good sources of background material are [5, 15, 16, 22, 28]

The more advanced parts of the theory which are described in Chapters 7, 8, 9, 10, 11, 12 should certainly challenge students and naturally point to more detailed studies of Differential Geometry and Lie Theory. Occasionally ideas from Algebraic Topology are touched upon (*e.g.*, the fundamental group and Lefschetz Fixed Point Theorem) and an interested reader might find it helpful to consult an introductory book on the subject such as [9, 20, 25].

Typesetting

This book was produced using LaTeX and the American Mathematical Society's `amsmath` package. Diagrams were produced using XY-pic. The symbol ⚠ was produced by my colleague J. Nimmo, who also provided other help with TeX and LaTeX.

Acknowledgements

I would like to thank the following: the Universität Bern for inviting me to visit and teach a course in the Spring of 2000; the students who spotted errors and obscurities in my notes and Z. Balogh who helped with the problem classes; the mathematicians of Glasgow University, especially I. Gordon, J. Nimmo, R. Odoni and J. Webb; the topologists and other mathematicians of Manchester University from whom I learnt a great deal over many years. Also thanks to Roger and Marliese Delaquis for providing a temporary home in the wonderful city of Bern. Finally, special thanks must go to Carole, Daniel and Laura for putting up with it all.

Remarks on the second printing

I have taken the opportunity to correct various errors found in the first printing and would like to thank R. Barraclough, R. Chapman, P. Eccles, S. Hendren, N. Pollock, R. Wickner and M. Yankelevitch for their helpful comments and error-spotting. The web page

`http://www.maths.gla.ac.uk/~ajb/MatrixGroups/`

contains an up to date list of known errors and corrections.

A reader wishing to pursue exceptional Lie groups and connections with Physics will find much of interest in the excellent recent survey paper of Baez [30], while representation theory is covered by Bröcker and tom Dieck [31] and Sternberg [27].

Contents

Part I

Basic Ideas and Examples

1
Real and Complex Matrix Groups

Throughout, $\Bbbk$ will denote a (commutative) field. Most of the time we will be interested in the cases of the fields $\Bbbk = \mathbb{R}$ (the real numbers) and $\Bbbk = \mathbb{C}$ (the complex numbers), however the general framework of this chapter is applicable to more general fields equipped with suitable norms in place of the absolute value. Indeed, as we will see in Chapter 4, much of it even applies to the case of a general *normed division algebra* or *skew field*, with the *quaternions* providing the most important non-commutative example.

1.1 Groups of Matrices

Let $\mathrm{M}_{m,n}(\Bbbk)$ be the set of $m \times n$ matrices whose entries are in $\Bbbk$. We will denote the (i, j) entry of an $m \times n$ matrix A by A_{ij} or a_{ij} and also write

$$A = [a_{ij}] = \begin{bmatrix} a_{11} & \cdots & a_{1n} \\ \vdots & \ddots & \vdots \\ a_{m1} & \cdots & a_{mn} \end{bmatrix}.$$

We will use the special notations

$$\mathrm{M}_n(\Bbbk) = \mathrm{M}_{n,n}(\Bbbk), \quad \Bbbk^n = \mathrm{M}_{n,1}(\Bbbk).$$

$\mathrm{M}_{m,n}(\Bbbk)$ is a $\Bbbk$-vector space with the operations of matrix addition and scalar multiplication. The zero vector is the $m \times n$ zero matrix $O_{m,n}$ which we will often denote O when the size is clear from the context. The matrices E^{rs} with $r = 1, \ldots, m$, $s = 1, \ldots, n$ and

$$(E^{rs})_{ij} = \delta_{ir}\delta_{js} = \begin{cases} 1 & \text{if } i = r \text{ and } j = s, \\ 0 & \text{otherwise,} \end{cases}$$

form a basis of $\mathrm{M}_{m,n}(\Bbbk)$, hence its dimension as a $\Bbbk$-vector space is

$$\dim_{\Bbbk} \mathrm{M}_{m,n}(\Bbbk) = mn. \tag{1.1}$$

When $n = 1$ we will denote the standard basis vectors of $\Bbbk^n = \mathrm{M}_{n,1}(\Bbbk)$ by

$$\mathbf{e}_r = E^{r1} \quad (r = 1, \ldots, m).$$

As well as being a $\Bbbk$-vector space of dimension n^2, $\mathrm{M}_n(\Bbbk)$ is also a ring with the usual addition and multiplication of square matrices, with zero $O_n = O_{n,n}$ and the $n \times n$ identity matrix I_n as its unity; $\mathrm{M}_n(\Bbbk)$ is not commutative except when $n = 1$. Later we will see that $\mathrm{M}_n(\Bbbk)$ is also an important example of a *finite dimensional* $\Bbbk$*-algebra* in the sense to be introduced in Chapter 4. The ring $\mathrm{M}_n(\Bbbk)$ acts on $\Bbbk^n$ by left multiplication, giving $\Bbbk^n$ the structure of a *left* $\mathrm{M}_n(\Bbbk)$*-module.*

Proposition 1.1

The determinant function $\det\colon \mathrm{M}_n(\Bbbk) \longrightarrow \Bbbk$ has the following properties.
i) For $A, B \in \mathrm{M}_n(\Bbbk)$, $\det(AB) = \det A \det B$.
ii) $\det I_n = 1$.
iii) $A \in \mathrm{M}_n(\Bbbk)$ is invertible if and only if $\det A \neq 0$.

We will use the notation

$$\mathrm{GL}_n(\Bbbk) = \{A \in \mathrm{M}_n(\Bbbk) : \det A \neq 0\}$$

for the set of invertible $n \times n$ matrices (also known as the set of units of the ring $\mathrm{M}_n(\Bbbk)$), and

$$\mathrm{SL}_n(\Bbbk) = \{A \in \mathrm{M}_n(\Bbbk) : \det A = 1\} \subseteq \mathrm{GL}_n(\Bbbk)$$

for the set of $n \times n$ *unimodular matrices.*

Theorem 1.2

The sets $\mathrm{GL}_n(\Bbbk)$, $\mathrm{SL}_n(\Bbbk)$ are groups under matrix multiplication. Furthermore, $\mathrm{SL}_n(\Bbbk) \leqslant \mathrm{GL}_n(\Bbbk)$, *i.e.*, $\mathrm{SL}_n(\Bbbk)$ is a subgroup of $\mathrm{GL}_n(\Bbbk)$.

Because of these group structures, $\mathrm{GL}_n(\Bbbk)$ is called the $n \times n$ *general linear group*, while $\mathrm{SL}_n(\Bbbk)$ is called the $n \times n$ *special linear* or *unimodular* group. When $\Bbbk = \mathbb{R}$ or $\Bbbk = \mathbb{C}$ we will refer to $\mathrm{GL}_n(\mathbb{R}), \mathrm{SL}_n(\mathbb{R})$ or $\mathrm{GL}_n(\mathbb{C}), \mathrm{SL}_n(\mathbb{C})$ as real or complex general linear groups. Of course, we can also consider subgroups of these groups, but before doing so we consider the topology of $\mathrm{M}_n(\mathbb{R})$ and $\mathrm{M}_n(\mathbb{C})$ as metric spaces.

1.2 Groups of Matrices as Metric Spaces

In this section we will always assume that $\Bbbk = \mathbb{R}$ or $\mathbb{C}$. Recall that $\mathrm{M}_n(\Bbbk)$ is a $\Bbbk$-vector space of dimension n^2. We will define a *norm* $\| \ \|$ on $\mathrm{M}_n(\Bbbk)$. It is worth remarking that we choose this particular norm mainly for the convenience of its multiplicative properties; in fact, as explained in Section 1.8, any other vector space norm would give an equivalent metric topology on $\mathrm{M}_n(\Bbbk)$. Other useful norms on $\mathrm{M}_n(\Bbbk)$ are discussed in Strang [28].

We begin with the usual notion of *length* for a vector $\mathbf{x} = \begin{bmatrix} x_1 \\ \vdots \\ x_n \end{bmatrix} \in \Bbbk^n$, namely

$$|\mathbf{x}| = \sqrt{|x_1|^2 + \cdots + |x_n|^2}.$$

This is an example of a *norm* on the vector space $\Bbbk^n$ as specified in Definition 1.51.

For $A \in \mathrm{M}_n(\Bbbk)$, consider the set

$$\mathcal{S}_A = \left\{ \frac{|A\mathbf{x}|}{|\mathbf{x}|} : \mathbf{0} \neq \mathbf{x} \in \Bbbk^n \right\}.$$

Then the subset

$$\mathcal{S}_A^1 = \{|A\mathbf{x}| : \mathbf{x} \in \Bbbk^n, \ |\mathbf{x}| = 1\} \subseteq \mathcal{S}_A$$

is actually equal to $\mathcal{S}_A$ since if $\mathbf{x} \neq \mathbf{0}$ we have

$$\frac{|A\mathbf{x}|}{|\mathbf{x}|} = |A\mathbf{x}'|,$$

where $\mathbf{x}' = (1/|\mathbf{x}|)\mathbf{x}$ has length $|\mathbf{x}'| = 1$. The subset

$$\{\mathbf{x} \in \Bbbk^n : |\mathbf{x}| = 1\} \subseteq \Bbbk^n$$

is closed and bounded and so is compact in the sense of Section 1.3, hence by Corollary 1.23, the real-valued function

$$\{\mathbf{x} \in \Bbbk^n : |\mathbf{x}| = 1\} \longrightarrow \mathbb{R}; \quad \mathbf{x} \longmapsto |A\mathbf{x}|$$

is bounded and attains its supremum

$$\sup \mathcal{S}_A = \sup \mathcal{S}_A^1 = \max \mathcal{S}_A^1 = \max \mathcal{S}_A.$$

This means that the real number

$$\|A\| = \max \mathcal{S}_A = \max \mathcal{S}_A^1$$

is defined. This *norm* function $\| \ \|\colon \mathrm{M}_n(\Bbbk) \longrightarrow \mathbb{R}$ is called the *operator* or *sup* (= *supremum*) *norm* on $\mathrm{M}_n(\Bbbk)$. For the general notion of a *norm* on a $\Bbbk$-*algebra* see Definition 4.31.

For a real matrix $A \in \mathrm{M}_n(\mathbb{R}) \subseteq \mathrm{M}_n(\mathbb{C})$, at first sight there appear to be two distinct norms of this type, namely

$$\|A\|_{\mathbb{R}} = \{|A\mathbf{x}| : \mathbf{x} \in \mathbb{R}^n,\ |\mathbf{x}| = 1\}, \quad \|A\|_{\mathbb{C}} = \{|A\mathbf{x}| : \mathbf{x} \in \mathbb{C}^n,\ |\mathbf{x}| = 1\}.$$

Lemma 1.3

If $A \in \mathrm{M}_n(\mathbb{R})$, then $\|A\|_{\mathbb{C}} = \|A\|_{\mathbb{R}}$.

Proof

It is obvious that $\|A\|_{\mathbb{R}} \leqslant \|A\|_{\mathbb{C}}$. Now for a vector $\mathbf{z} \in \mathbb{C}^n$ with $|\mathbf{z}| = 1$, write $\mathbf{z} = \mathbf{x} + i\mathbf{y}$ with $\mathbf{x}, \mathbf{y} \in \mathbb{R}^n$. Then $|\mathbf{x}|^2 + |\mathbf{y}|^2 = 1$ and

$$\begin{aligned} |A\mathbf{z}|^2 &\leqslant |A\mathbf{x}|^2 + |iA\mathbf{y}|^2 \\ &\leqslant |\mathbf{x}|^2\|A\|_{\mathbb{R}}^2 + |\mathbf{y}|^2\|A\|_{\mathbb{R}}^2 \\ &= (|\mathbf{x}|^2 + |\mathbf{y}|^2)\|A\|_{\mathbb{R}}^2 \\ &= \|A\|_{\mathbb{R}}^2, \end{aligned}$$

giving $|A\mathbf{z}| \leqslant \|A\|_{\mathbb{R}}$. Thus $\|A\|_{\mathbb{C}} \leqslant \|A\|_{\mathbb{R}}$ and hence $\|A\|_{\mathbb{R}} = \|A\|_{\mathbb{C}}$. □

Remark 1.4

There is a procedure for calculating $\|A\|$ which is important in numerical linear algebra. We describe this briefly; for further details see Strang [28].

All the eigenvalues of the positive hermitian matrix A^*A are non-negative real numbers, hence it has a largest non-negative real eigenvalue $\lambda_{\max}$. Then

$$\|A\| = \sqrt{\lambda_{\max}}.$$

In fact, for any unit eigenvector $\mathbf{v}$ of A^*A for the eigenvalue $\lambda_{\max}$, $\|A\| = |A\mathbf{v}|$. When A is real, $A^*A = A^TA$ is real positive symmetric and there are unit length eigenvectors $\mathbf{w} \in \mathbb{R}^n \subseteq \mathbb{C}^n$ of A^*A for the eigenvalue λ for which $\|A\| = |A\mathbf{w}|$. In particular, this also shows that $\|A\|$ is independent of whether A is viewed as a real or complex matrix.

The main properties of $\|\ \|$ are summarised in the next result and imply that $\|\ \|$ is a $\Bbbk$-*norm* on $\mathrm{M}_n(\Bbbk)$. General $\Bbbk$-norms are discussed in Section 4.2.

Proposition 1.5

The function $\|\ \|$ has the following properties.
i) If $t \in \Bbbk$, $A \in \mathrm{M}_n(\Bbbk)$, then $\|tA\| = |t|\,\|A\|$.
ii) If $A, B \in \mathrm{M}_n(\Bbbk)$, then $\|AB\| \leqslant \|A\|\,\|B\|$.
iii) If $A, B \in \mathrm{M}_n(\Bbbk)$, then $\|A + B\| \leqslant \|A\| + \|B\|$.
iv) If $A \in \mathrm{M}_n(\Bbbk)$, then $\|A\| = 0$ if and only if $A = 0$.
v) $\|I_n\| = 1$.

The norm $\|\ \|$ can be used to define a metric ρ on $\mathrm{M}_n(\Bbbk)$ by

$$\rho(A, B) = \|A - B\|$$

together with the associated metric topology on $\mathrm{M}_n(\Bbbk)$. Then a sequence $\{A_r\}_{r\geqslant 0}$ of elements in $\mathrm{M}_n(\Bbbk)$ converges to a limit $A \in \mathrm{M}_n(\Bbbk)$ if $\|A_r - A\| \to O$ as $r \to \infty$. We may also define continuous functions $\mathrm{M}_n(\Bbbk) \longrightarrow X$ into a topological space X.

For $A \in \mathrm{M}_n(\Bbbk)$ and $r > 0$, let

$$\mathrm{N}_{\mathrm{M}_n(\Bbbk)}(A; r) = \{B \in \mathrm{M}_n(\Bbbk) : \|B - A\| < r\},$$

which is the *open disc* of radius r in $\mathrm{M}_n(\Bbbk)$. Similarly, if $Y \subseteq \mathrm{M}_n(\Bbbk)$ and $A \in Y$, we set

$$\mathrm{N}_Y(A; r) = \{B \in Y : \|B - A\| < r\} = \mathrm{N}_{\mathrm{M}_n(\Bbbk)}(A; r) \cap Y.$$

Then a subset $V \subseteq Y$ is open in Y if and only if for every $A \in V$, there is a $\delta > 0$ such that $\mathrm{N}_Y(A; \delta) \subseteq V$.

Definition 1.6

Let $Y \subseteq \mathrm{M}_n(\Bbbk)$ and $(X, \mathcal{T})$ be a topological space. Then a function $f\colon Y \longrightarrow X$ is *continuous* or a *continuous map* if for every $A \in Y$ and $U \in \mathcal{T}$ such that $f(A) \in U$, there is a $\delta > 0$ for which

$$B \in \mathrm{N}_Y(A; \delta) \Longrightarrow f(B) \in U.$$

Equivalently, f is continuous if and only if for $U \in \mathcal{T}$, $f^{-1}U \subseteq Y$ is open in Y.

For a topological space $(X, \mathcal{T})$, a subset $W \subseteq X$ is *closed* if $X - W \subseteq X$ is open. For a metric space this is equivalent to requiring that whenever a sequence in W has a limit in X, the limit is in W. Yet another alternative formulation of the definition of continuity is that f is continuous if and only if for every closed subset $W \subseteq X$, $f^{-1}W \subseteq Y$ is closed in Y.

In particular, we may take $X = \mathbf{k}$ and $\mathcal{T}$ to be the natural metric space topology associated to the standard norm on $\mathbf{k}$ and consider continuous functions $Y \longrightarrow \mathbf{k}$.

Proposition 1.7

For $1 \leqslant r, s \leqslant n$, the *coordinate function*

$$\mathrm{coord}_{rs}\colon \mathrm{M}_n(\mathbf{k}) \longrightarrow \mathbf{k}; \quad \mathrm{coord}_{rs}(A) = A_{rs}$$

is continuous.

Proof

For the standard unit basis vectors $\mathbf{e}_i$ $(1 \leqslant i \leqslant n)$ of $\mathbf{k}^n$ we have

$$\begin{aligned} |A_{rs}| &\leqslant \sqrt{\sum_{i=1}^{n} |A_{is}|^2} \\ &= \left| \sum_{i=1}^{n} A_{is}\mathbf{e}_i \right| \\ &= |A\mathbf{e}_s| \\ &\leqslant \|A\|, \end{aligned}$$

hence for $A, A' \in \mathrm{M}_n(\mathbf{k})$,

$$|A'_{rs} - A_{rs}| \leqslant \|A' - A\|.$$

If $A \in \mathrm{M}_n(\mathbf{k})$ and $\varepsilon > 0$, then $\|A' - A\| < \varepsilon$ implies that $|A'_{rs} - A_{rs}| < \varepsilon$. This shows that coord_{rs} is continuous at every $A \in \mathrm{M}_n(\mathbf{k})$. □

Corollary 1.8

If $f\colon \mathbf{k}^{n^2} \longrightarrow \mathbf{k}$ is continuous, then the associated function

$$F\colon \mathrm{M}_n(\mathbf{k}) \longrightarrow \mathbf{k}; \quad F(A) = f((A_{ij})_{1 \leqslant i,j \leqslant n}),$$

is continuous.

Corollary 1.9

The determinant det: $\mathrm{M}_n(\Bbbk) \longrightarrow \Bbbk$ and trace tr: $\mathrm{M}_n(\Bbbk) \longrightarrow \Bbbk$ are continuous functions.

Proof

The determinant is obtained by composing a continuous function $\mathrm{M}_n(\Bbbk) \longrightarrow \Bbbk^{n^2}$ identifying $\mathrm{M}_n(\Bbbk)$ with $\Bbbk^{n^2}$ with a polynomial function $\Bbbk^{n^2} \longrightarrow \Bbbk$. Similarly,

$$\operatorname{tr} A = \sum_{k=1}^{n} A_{kk}$$

defines the trace as a polynomial function. □

There is a kind of converse to these results.

Proposition 1.10

For $A \in \mathrm{M}_n(\Bbbk)$,

$$\|A\| \leqslant \sum_{i,j=1}^{n} |A_{ij}|.$$

Proof

Let $\mathbf{x} = x_1\mathbf{e}_1 + \cdots + x_n\mathbf{e}_n$ with $|\mathbf{x}| = 1$. Since $|x_k| \leqslant 1$ for each k, we have

$$\begin{aligned} |A\mathbf{x}| &= |x_1 A\mathbf{e}_1 + \cdots + x_n A\mathbf{e}_n| \\ &\leqslant |x_1 A\mathbf{e}_1| + \cdots + |x_n A\mathbf{e}_n| \\ &\leqslant |A\mathbf{e}_1| + \cdots + |A\mathbf{e}_n| \\ &\leqslant \sqrt{\sum_{i=1}^{n} A_{i1}^2} + \cdots + \sqrt{\sum_{i=1}^{n} A_{in}^2} \\ &\leqslant \sum_{i,j=1}^{n} |A_{ij}|. \end{aligned}$$

As this is true for all vectors $\mathbf{x}$ with $|\mathbf{x}| = 1$,

$$\|A\| \leqslant \sum_{i,j=1}^{n} |A_{ij}|$$

by definition of $\|A\|$. □

Definition 1.11

A sequence $\{A_r\}_{r\geqslant 0}$ in $\mathrm{M}_n(\mathbb{k})$ is a *Cauchy sequence* if for every $\varepsilon > 0$, there is a natural number N such that $\|A_r - A_s\| < \varepsilon$ whenever $r, s > N$.

Theorem 1.12

For $\mathbb{k} = \mathbb{R}$ or $\mathbb{C}$, every Cauchy sequence $\{A_r\}_{r\geqslant 0}$ in $\mathrm{M}_n(\mathbb{k})$ has a unique limit $\lim\limits_{r\to\infty} A_r$ in $\mathrm{M}_n(\mathbb{k})$. Furthermore,

$$(\lim_{r\to\infty} A_r)_{ij} = \lim_{r\to\infty} (A_r)_{ij}. \tag{1.2}$$

Proof

It is standard that if such a limit exists it is unique so we need to show existence. By Proposition 1.7, the limit on the right-hand side of Equation (1.2) exists, so it is sufficient to show that the required limit is the matrix A for which

$$A_{ij} = \lim_{r\to\infty} (A_r)_{ij}.$$

For the sequence $\{A_r - A\}_{r\geqslant 0}$, as $r \to \infty$ we have

$$\|A_r - A\| \leqslant \sum_{i,j=1}^{n} |(A_r)_{ij} - A_{ij}| \to 0,$$

so $A_r \to A$ by Proposition 1.10. □

Because of this result, the metric space $(\mathrm{M}_n(\mathbb{k}), \| \ \|)$ is said to be *complete* with respect to the norm $\| \ \|$.

It can be shown that the metric topologies induced by $\| \ \|$ and the usual norm on $\mathbb{k}^{n^2}$ agree in the sense that they have the same open sets. Actually this is true for any two norms on $\mathbb{k}^{n^2}$; see Section 1.8 and [21, 22] for more on this. We summarise this as a useful criterion whose proof is left as an exercise.

Proposition 1.13

A function $F\colon \mathrm{M}_m(\mathbb{k}) \longrightarrow \mathrm{M}_n(\mathbb{k})$ is continuous with respect to the norms $\| \ \|$ if and only if each of the component functions $F_{rs}\colon \mathrm{M}_m(\mathbb{k}) \longrightarrow \mathbb{k}$ is continuous. In particular, a function $f\colon \mathrm{M}_m(\mathbb{k}) \longrightarrow \mathbb{k}$ is continuous with respect to the norm $\| \ \|$ and the usual metric on $\mathbb{k}$ if and only if it is continuous when viewed as a function $\mathbb{k}^{m^2} \longrightarrow \mathbb{k}$.

Next we consider the topology of some subsets of $\mathrm{M}_n(\Bbbk)$, in particular some groups of matrices.

Proposition 1.14

Let $\Bbbk = \mathbb{R}$ or $\mathbb{C}$. Then
i) $\mathrm{GL}_n(\Bbbk) \subseteq \mathrm{M}_n(\Bbbk)$ is an open subset;
ii) $\mathrm{SL}_n(\Bbbk) \subseteq \mathrm{M}_n(\Bbbk)$ is a closed subset.

Proof

We know that the function $\det\colon \mathrm{M}_n(\Bbbk) \longrightarrow \Bbbk$ is continuous. Then

$$\mathrm{GL}_n(\Bbbk) = \mathrm{M}_n(\Bbbk) - \det^{-1}\{0\},$$

which is open since $\{0\}$ is closed, hence (i) holds. Similarly,

$$\mathrm{SL}_n(\Bbbk) = \det^{-1}\{1\} \subseteq \mathrm{GL}_n(\Bbbk),$$

which is closed in $\mathrm{M}_n(\Bbbk)$ and $\mathrm{GL}_n(\Bbbk)$ since the singleton set $\{1\}$ is closed in $\Bbbk$, so (ii) is true. □

In the following we will make use of the *product topology* of two topological spaces X, Y; this is the topology on $X \times Y$ in which every open set is a union of sets of the form

$$U \times V \quad (U \subseteq X,\, V \subseteq Y \text{ open}).$$

We refer to $X \times Y$ as the product space if it has the product topology. If the topologies on X and Y come from metrics, it is possible to define a metric whose associated topology agrees with the product topology; this is discussed in the Exercises.

The addition and multiplication maps

$$\mathrm{add}\colon \mathrm{M}_n(\Bbbk) \times \mathrm{M}_n(\Bbbk) \longrightarrow \mathrm{M}_n(\Bbbk); \quad \mathrm{add}(X,Y) = X + Y,$$
$$\mathrm{mult}\colon \mathrm{M}_n(\Bbbk) \times \mathrm{M}_n(\Bbbk) \longrightarrow \mathrm{M}_n(\Bbbk); \quad \mathrm{mult}(X,Y) = XY,$$

are also continuous, where we take the product topology on the domain $\mathrm{M}_n(\Bbbk) \times \mathrm{M}_n(\Bbbk)$. Finally, the inverse map

$$\mathrm{inv}\colon \mathrm{GL}_n(\Bbbk) \longrightarrow \mathrm{GL}_n(\Bbbk); \quad \mathrm{inv}(A) = A^{-1},$$

is also continuous, since each entry of A^{-1} has the form

$$\frac{(\text{polynomial in the entries } A_{ij})}{\det A}$$

and as this is a continuous function of the entries of A it is a continuous function of A.

Definition 1.15

Let G be a topological space and view $G \times G$ as the product space. Suppose that G is also a group with multiplication map $\text{mult}: G \times G \longrightarrow G$ and inverse map $\text{inv}: G \longrightarrow G$. Then G is a *topological group* if mult and inv are continuous.

The simplest examples are obtained from arbitrary groups G given discrete topologies; in particular all finite groups can be viewed this way. Of course, the discussion above has already established the following.

Theorem 1.16

For $\Bbbk = \mathbb{R}$ or $\mathbb{C}$, each of the groups $\mathrm{GL}_n(\Bbbk)$, $\mathrm{SL}_n(\Bbbk)$ is a topological group with the evident multiplication and inverse maps and the subspace topologies inherited from $\mathrm{M}_n(\Bbbk)$.

1.3 Compactness

In this section we discuss the idea of compactness for topological spaces and explain its significance for subsets of $\Bbbk^n$ with the usual metric, where $\Bbbk = \mathbb{R}$ or $\mathbb{C}$. Many of the most useful results for continuous functions from a compact space into a metric space also apply more generally when the codomain is *Hausdorff* in the sense of Definition 1.24.

Definition 1.17

A subset $X \subseteq \Bbbk^m$ is *compact* if and only if it is closed and bounded.

Example 1.18

Identifying subsets of $\mathrm{M}_n(\Bbbk)$ with subsets of $\Bbbk^{n^2}$, we can consider compact subsets of $\mathrm{M}_n(\Bbbk)$. In particular, a subgroup $G \leqslant \mathrm{GL}_n(\Bbbk)$ is compact if it is compact as a subset of $\mathrm{GL}_n(\Bbbk)$, or equivalently of $\mathrm{M}_n(\Bbbk)$.

Our next result is standard for metric spaces.

Proposition 1.19

$X \subseteq \mathrm{M}_n(\Bbbk)$ is compact if and only if the following two conditions are satisfied:

- there is a $b \in \mathbb{R}^+$ such that for all $A \in X$, $\|A\| \leqslant b$;
- every sequence $\{C_n\}_{n \geqslant 0}$ in X which is convergent in $\mathrm{M}_n(\Bbbk)$ has a limit in X, *i.e.*, X is a closed subset of $\mathrm{M}_n(\Bbbk)$.

The following important characterisation of compact subsets of $\mathrm{M}_n(\Bbbk)$ leads to the general definition of *compact topological space.*

Theorem 1.20 (Heine–Borel Theorem)

$X \subseteq \mathrm{M}_n(\Bbbk)$ is compact if and only if every open cover $\{U_\alpha\}_{\alpha \in \Lambda}$ of X contains a finite subcover $\{U_{\alpha_1}, \ldots, U_{\alpha_k}\}$.

Definition 1.21

A topological space X is *compact* if and only if every open cover $\{U_\alpha\}_{\alpha \in \Lambda}$ of X contains a finite subcover $\{U_{\alpha_1}, \ldots, U_{\alpha_k}\}$.

Clearly our two notions of compactness coincide for a subset $X \subseteq \Bbbk^n$.

Proposition 1.22

Let X be a compact topological space and $f\colon X \longrightarrow Y$ be a continuous function. Then the image $fX \subseteq Y$ is a compact subspace of Y.

Proof

Let $\{V_\alpha\}_{\alpha \in \Lambda}$ be an open cover of fX. Then by definition of the subspace topology, there is a collection of open subsets $\{V'_\alpha\}_{\alpha \in \Lambda}$ for which $fX \cap V'_\alpha = V_\alpha$. For each $\alpha \in \lambda$,

$$f^{-1}V_\alpha = f^{-1}V'_\alpha,$$

so $\{f^{-1}V'_\alpha\}_{\alpha \in \Lambda}$ is an open covering of X. By compactness, there is a finite subcollection $\{f^{-1}V'_{\alpha_1}, \ldots, f^{-1}V'_{\alpha_k}\}$ which also covers X, hence $\{V'_{\alpha_1}, \ldots, V'_{\alpha_k}\}$ is a finite cover of fX. □

Corollary 1.23

Let X be a compact topological space and $f\colon X \longrightarrow \mathbb{R}$ be a continuous function. Then the image $fX \subseteq \mathbb{R}$ is a bounded subset and there are elements $x_+, x_- \in X$ for which

$$f(x_+) = \sup fX, \quad f(x_-) = \inf fX.$$

For later use we record some other useful results on continuous functions out of a compact space into a *Hausdorff* space. Omitted proofs can be found in books on point set topology. We start by introducing the notion of a *Hausdorff* space which provides a useful generalisation of the concept of a metric space particularly useful when dealing with quotient constructions such as the *homogeneous spaces* of Chapter 8.

Definition 1.24

A topological space X is *Hausdorff* if for every pair of points $u, v \in X$ with $x \neq y$, there are open subsets $U, V \subseteq X$ with $u \in U$, $v \in V$ and $U \cap V = \varnothing$.

Lemma 1.25

Every metric space is Hausdorff.

Proof

Let X be a metric space with metric ρ. If $u, v \in X$ are distinct, then $\rho(u, v) > 0$. If $r = \rho(u, v)/2$, the open discs $\mathrm{N}_X(u; r)$ and $\mathrm{N}_X(v; r)$ satisfy the conditions required for the open sets U and V. □

Proposition 1.26

Let X be a compact topological space. If $f: X \longrightarrow Y$ is a continuous function into a Hausdorff topological space Y, then $fX \subseteq Y$ is a closed subset. In particular, when $j: X \longrightarrow Y$ is the inclusion function for a subspace $X \subseteq Y$, we obtain that X is a closed subset of Y.

Our next definition provides a notion of equivalence of topological spaces.

Definition 1.27

A continuous bijection $f: X \longrightarrow Y$ between topological spaces X and Y is a *homeomorphism* if its inverse $f^{-1}: Y \longrightarrow X$ is continuous.

Proposition 1.28

Let X be a compact topological space and $f: X \longrightarrow Y$ a continuous bijection into a Hausdorff topological space Y. Then f is a homeomorphism.

Our final result will be useful when working with compact matrix groups.

Proposition 1.29

Let (X, ρ) be a compact metric space and let $\{s_n\}_{n\geqslant 1}$ be a sequence in X. Then there is a convergent subsequence $\{s_{\sigma(n)}\}_{n\geqslant 1}$.

Proof

Suppose that this is false. Then for every $x \in X$ which is not of the form $x = s_n$ for some n, there is an $r_x > 0$ for which

$$\rho(x, s_n) \geqslant r_x \quad (n \geqslant 1).$$

Also, for each $m \geqslant 1$, there is an $r_{s_m} > 0$ and an $n_m \geqslant m$ for which

$$\rho(s_m, s_n) \geqslant r_x \quad (n \geqslant n_m).$$

The open discs $\mathrm{N}_X(x; r_x)$ form an open cover of X, hence by compactness there are elements $x_1, \ldots, x_k$ for which

$$X = \mathrm{N}_X(x_1; r_{x_1}) \cup \cdots \cup \mathrm{N}_X(x_k; r_{x_k}).$$

But for sufficiently large n, s_n cannot be in any of the discs $\mathrm{N}_X(x_j; r_{x_j})$, so this provides a contradiction. □

1.4 Matrix Groups

Definition 1.30

A subgroup $G \leqslant \mathrm{GL}_n(\Bbbk)$ which is also a closed subspace is a *matrix group over* $\Bbbk$ or a $\Bbbk$*-matrix group*. In order to make the value of n explicit, we will sometimes say that G is a *matrix subgroup* of $\mathrm{GL}_n(\Bbbk)$.

Before considering some examples we record some useful general properties of matrix groups.

Proposition 1.31

Let $G \leqslant \mathrm{GL}_n(\Bbbk)$ be a matrix subgroup. Then a closed subgroup $H \leqslant G$ is a matrix subgroup H of $\mathrm{GL}_n(\Bbbk)$.

Proof

Every sequence $\{A_n\}_{n\geqslant 0}$ in H with a limit in $\mathrm{GL}_n(\Bbbk)$ actually has its limit in G since $A_n \in H \subseteq G$ for every n and G is closed in $\mathrm{GL}_n(\Bbbk)$. Since H is closed in G, this means that $\{A_n\}_{n\geqslant 0}$ has a limit in H. So H is closed in $\mathrm{GL}_n(\Bbbk)$ which shows that it is a matrix subgroup. □

This result suggests another definition.

Definition 1.32

A closed subgroup $H \leqslant G$ of a matrix group G is called a *matrix subgroup* of G.

Proposition 1.33

Let G be a matrix group and $H \leqslant K$, $K \leqslant G$ be matrix subgroups. Then H is a matrix subgroup of G.

Proof

This is a straightforward generalisation of Proposition 1.31. □

Example 1.34

$\mathrm{SL}_n(\Bbbk) \leqslant \mathrm{GL}_n(\Bbbk)$ is a matrix group over $\Bbbk$.

Proof

$\mathrm{SL}_n(\Bbbk)$ is closed in $\mathrm{M}_n(\Bbbk)$ by Proposition 1.14 and $\mathrm{SL}_n(\Bbbk) \leqslant \mathrm{GL}_n(\Bbbk)$. □

Example 1.35

We may consider $\mathrm{GL}_n(\Bbbk)$ as a subgroup of $\mathrm{GL}_{n+1}(\Bbbk)$ by identifying the $n \times n$ matrix $A = [a_{ij}]$ with

$$\begin{bmatrix} A & \mathbf{0} \\ \mathbf{0} & 1 \end{bmatrix} = \begin{bmatrix} a_{11} & \cdots & a_{1n} & 0 \\ \vdots & \ddots & \vdots & \vdots \\ a_{n1} & \cdots & a_{nn} & 0 \\ 0 & \cdots & 0 & 1 \end{bmatrix}.$$

Using the restrictions of the coordinate functions of Proposition 1.7 to continuous functions $\mathrm{coord}_{r,s}\colon \mathrm{GL}_{n+1}(\Bbbk) \longrightarrow \Bbbk$ we have

$$\mathrm{GL}_n(\Bbbk) = \left(\bigcap_{j=1}^{n} \mathrm{coord}_{n+1,j}^{-1}\{0\}\right) \cap \left(\bigcap_{i=1}^{n} \mathrm{coord}_{i,n+1}^{-1}\{0\}\right) \cap \mathrm{coord}_{n+1,n+1}^{-1}\{1\},$$

which is a closed subset of $\mathrm{GL}_{n+1}(\Bbbk)$ since it is the intersection of inverse images of closed sets under a continuous function. Hence $\mathrm{GL}_n(\Bbbk)$ is a matrix subgroup of $\mathrm{GL}_{n+1}(\Bbbk)$.

We can also restrict this to an embedding of $\mathrm{SL}_n(\Bbbk)$ which then appears as a closed subgroup of $\mathrm{SL}_{n+1}(\Bbbk) \leqslant \mathrm{GL}_{n+1}(\Bbbk)$. So $\mathrm{SL}_n(\Bbbk)$ is a matrix subgroup of $\mathrm{SL}_{n+1}(\Bbbk)$. More generally, with the aid of this embedding, any matrix subgroup of $\mathrm{GL}_n(\Bbbk)$ can also be viewed as a matrix subgroup of $\mathrm{GL}_{n+1}(\Bbbk)$.

Given a matrix subgroup $G \leqslant \mathrm{GL}_n(\Bbbk)$, it is often useful to restrict the determinant on $\mathrm{GL}_n(\Bbbk)$ to a function

$$\det_G\colon G \longrightarrow \Bbbk^{\times}; \quad \det_G A = \det A.$$

We usually write this function as det when no ambiguity is likely to result. Of course, $\det_G$ is always a continuous group homomorphism.

When $\Bbbk = \mathbb{R}$, we set

$$\mathbb{R}^+ = \{t \in \mathbb{R} : t > 0\}, \quad \mathbb{R}^- = \{t \in \mathbb{R} : t < 0\}, \quad \mathbb{R}^\times = \mathbb{R}^+ \cup \mathbb{R}^-.$$

Notice that $\mathbb{R}^+$ is a subgroup of $\mathrm{GL}_1(\mathbb{R}) = \mathbb{R}^\times$ which is both closed and open as a subset, while $\mathbb{R}^-$ is an open subset; thus $\mathbb{R}^+$ and $\mathbb{R}^-$ are *clopen* subsets, *i.e.*, both closed and open. For $G \leqslant \mathrm{GL}_n(\mathbb{R})$,

$$\det_G^{-1} \mathbb{R}^+ = G \cap \det^{-1} \mathbb{R}^+,$$

and also

$$G = \det_G^{-1} \mathbb{R}^+ \cup \det_G^{-1} \mathbb{R}^-.$$

Hence G is a disjoint union of the clopen subsets

$$G^+ = \det_G^{-1} \mathbb{R}^+, \quad G^- = \det_G^{-1} \mathbb{R}^-.$$

Since $I_n \in G^+ = \det_G^{-1} \mathbb{R}^+$, the component G^+ is never empty. Indeed, G^+ is a closed subgroup of G, hence it is a matrix subgroup of $\mathrm{GL}_n(\mathbb{R})$. When $G^- \neq \varnothing$, the space G is not connected since it is the union of two disjoint open subsets. When $G^- = \varnothing$, $G = G^+$ may or may not be connected.

1.5 Some Important Examples

We now discuss some important examples of real and complex matrix groups.

Groups of Upper Triangular Matrices

For $n \geqslant 1$, an $n \times n$ matrix $A = [a_{ij}]$ is *upper triangular* if it has the form

$$\begin{bmatrix} a_{11} & a_{12} & \cdots & \cdots & \cdots & a_{1n} \\ 0 & a_{21} & \ddots & \ddots & \ddots & a_{2n} \\ 0 & 0 & \ddots & \ddots & \ddots & \vdots \\ \vdots & \vdots & \ddots & \ddots & \ddots & \vdots \\ \vdots & \vdots & \ddots & 0 & a_{n-1\,n-1} & a_{n-1\,n} \\ 0 & 0 & \cdots & 0 & 0 & a_{nn} \end{bmatrix},$$

i.e., $a_{ij} = 0$ if $i < j$. A matrix is *unipotent* if it is upper triangular and also has all diagonal entries equal to 1, *i.e.*, $a_{ij} = 0$ if $i < j$ and $a_{ii} = 1$.

The *upper triangular subgroup* or *Borel subgroup* of $\mathrm{GL}_n(\Bbbk)$ is

$$\mathrm{UT}_n(\Bbbk) = \{A \in \mathrm{GL}_n(\Bbbk) : A \text{ is upper triangular}\},$$

while the *unipotent subgroup* of $\mathrm{GL}_n(\Bbbk)$ is

$$\mathrm{SUT}_n(\Bbbk) = \{A \in \mathrm{GL}_n(\Bbbk) : A \text{ is unipotent}\}.$$

It is easy to verify that $\mathrm{UT}_n(\Bbbk)$ and $\mathrm{SUT}_n(\Bbbk)$ are closed subgroups of $\mathrm{GL}_n(\Bbbk)$. Notice also that $\mathrm{SUT}_n(\Bbbk) \leqslant \mathrm{UT}_n(\Bbbk)$ and is a closed subgroup.

For the case

$$\mathrm{SUT}_2(\Bbbk) = \left\{ \begin{bmatrix} 1 & t \\ 0 & 1 \end{bmatrix} \in \mathrm{GL}_2(\Bbbk) : t \in \Bbbk \right\} \leqslant \mathrm{GL}_2(\Bbbk),$$

the function

$$\theta \colon \Bbbk \longrightarrow \mathrm{SUT}_2(\Bbbk); \quad \theta(t) = \begin{bmatrix} 1 & t \\ 0 & 1 \end{bmatrix},$$

is a continuous group homomorphism which is an isomorphism with continuous inverse. This allows us to view $\Bbbk$ as a matrix group.

Affine Groups

The *n-dimensional affine group* over $\Bbbk$ is

$$\mathrm{Aff}_n(\Bbbk) = \left\{ \begin{bmatrix} A & \mathbf{t} \\ 0 & 1 \end{bmatrix} : A \in \mathrm{GL}_n(\Bbbk),\ \mathbf{t} \in \Bbbk^n \right\} \leqslant \mathrm{GL}_{n+1}(\Bbbk).$$

This is clearly a closed subgroup of $\mathrm{GL}_{n+1}(\Bbbk)$. If we identify $\mathbf{x} \in \Bbbk^n$ with $\begin{bmatrix} \mathbf{x} \\ 1 \end{bmatrix} \in \Bbbk^{n+1}$, then as a consequence of the formula

$$\begin{bmatrix} A & \mathbf{t} \\ 0 & 1 \end{bmatrix} \begin{bmatrix} \mathbf{x} \\ 1 \end{bmatrix} = \begin{bmatrix} A\mathbf{x} + \mathbf{t} \\ 1 \end{bmatrix},$$

we obtain an action of $\mathrm{Aff}_n(\Bbbk)$ on $\Bbbk^n$. Transformations of $\Bbbk^n$ with the form $\mathbf{x} \mapsto A\mathbf{x}+\mathbf{t}$ with A invertible are called *affine transformations* and they preserve lines, *i.e.*, translates of 1-dimensional subspaces of the $\Bbbk$-vector space $\Bbbk^n$. The associated geometry is *affine geometry* and it has $\mathrm{Aff}_n(\Bbbk)$ as its symmetry group. The vector space $\Bbbk^n$ itself can be viewed as the *translation subgroup* of $\mathrm{Aff}_n(\Bbbk)$,

$$\mathrm{Trans}_n(\Bbbk) = \left\{ \begin{bmatrix} I_n & \mathbf{t} \\ 0 & 1 \end{bmatrix} : \mathbf{t} \in \Bbbk^n \right\} \leqslant \mathrm{Aff}_n(\Bbbk),$$

and this is a closed subgroup. There is also the closed subgroup

$$\left\{ \begin{bmatrix} A & 0 \\ 0 & 1 \end{bmatrix} : A \in \mathrm{GL}_n(\Bbbk) \right\} \leqslant \mathrm{Aff}_n(\Bbbk)$$

which we will identify with $\mathrm{GL}_n(\Bbbk)$. The following is a standard notion in group theory.

Definition 1.36

Let G be a group with $H \leqslant G$ and $N \lhd G$. Then G is the *semi-direct product* of H and N if $G = HN$ and $H \cap N = \{1\}$; this is often denoted by $G = H \ltimes N$ or $G = N \rtimes H$.

When $G = H \ltimes N$, there is a group isomorphism $q\colon G/N \longrightarrow H$ as well as the inclusion homomorphism $j\colon H \longrightarrow G$ and these satisfy $q \circ j = \mathrm{Id}_H$. Notice that H acts on N by conjugation since N is normal in G. The simplest kind of semi-direct product is the *direct product* $H \times N$, where the conjugation action of H on N is trivial.

Proposition 1.37

$\mathrm{Trans}_n(\Bbbk)$ is a normal subgroup of $\mathrm{Aff}_n(\Bbbk)$ and $\mathrm{Aff}_n(\Bbbk)$ can be expressed as the semi-direct product of $\mathrm{Trans}_n(\Bbbk)$ and $\mathrm{GL}_n(\Bbbk)$,

$$\mathrm{Aff}_n(\Bbbk) = \mathrm{GL}_n(\Bbbk) \ltimes \mathrm{Trans}_n(\Bbbk) = \{AT : A \in \mathrm{GL}_n(\Bbbk),\ T \in \mathrm{Trans}_n(\Bbbk)\},$$

with $\mathrm{Trans}_n(\Bbbk) \cap \mathrm{GL}_n(\Bbbk) = \{I_{n+1}\}$.

Proof

To see that $\mathrm{Trans}_n(\Bbbk) \lhd \mathrm{Aff}_n(\Bbbk)$, notice that if

$$\begin{bmatrix} A & 0 \\ 0 & 1 \end{bmatrix} \in \mathrm{GL}_n(\Bbbk), \quad \begin{bmatrix} I & \mathbf{t} \\ 0 & 1 \end{bmatrix} \in \mathrm{Trans}_n(\Bbbk),$$

then

$$\begin{aligned} \begin{bmatrix} A & 0 \\ 0 & 1 \end{bmatrix} \begin{bmatrix} I & \mathbf{t} \\ 0 & 1 \end{bmatrix} \begin{bmatrix} A & 0 \\ 0 & 1 \end{bmatrix}^{-1} &= \begin{bmatrix} A & A\mathbf{t} \\ 0 & 1 \end{bmatrix} \begin{bmatrix} A^{-1} & 0 \\ 0 & 1 \end{bmatrix} \\ &= \begin{bmatrix} I & A\mathbf{t} \\ 0 & 1 \end{bmatrix}, \end{aligned}$$

which is in $\mathrm{Trans}_n(\Bbbk)$. The equality

$$\mathrm{Trans}_n(\Bbbk) \cap \mathrm{GL}_n(\Bbbk) = \{I_{n+1}\}$$

follows from the fact that non-trivial translations do not fix $\mathbf{0}$ while all elements of $\mathrm{GL}_n(\Bbbk)$ do. □

Orthogonal and Isometry Groups

For $n \geqslant 1$, an $n \times n$ real matrix A for which $A^T A = I_n$ is called an *orthogonal matrix*; here A^T is the *transpose* of $A = [a_{ij}]$, whose entries are given by

$$(A^T)_{ij} = a_{ji}.$$

Such an orthogonal matrix has an inverse, namely A^T, and the product of two orthogonal matrices A, B is orthogonal since

$$(AB)^T(AB) = B^T A^T AB = BI_nB^T = BB^T = I_n.$$

Notice also that $I_n \in \mathrm{O}(n)$. So the subset

$$\mathrm{O}(n) = \{A \in \mathrm{GL}_n(\mathbb{R}) : A^T A = I_n\} \subseteq \mathrm{M}_n(\mathbb{R})$$

is a subgroup of $\mathrm{GL}_n(\mathbb{R})$ and is called the $n \times n$ *(real) orthogonal group*. The single matrix equation $A^T A = I_n$ is equivalent to the n^2 polynomial equations

$$\sum_{k=1}^{n} a_{ki} a_{kj} = \delta_{ij} \tag{1.3}$$

for the n^2 real numbers a_{ij}, where the *Kronecker symbol* δ_{ij} is defined by

$$\delta_{ij} = \begin{cases} 1 & \text{if } i = j, \\ 0 & \text{if } i \neq j. \end{cases}$$

Since polynomial functions are continuous, in similar fashion to Example 1.35 we can express $\mathrm{O}(n)$ as a closed subset of $\mathrm{GL}_n(\mathbb{R}) \subseteq \mathrm{M}_n(\mathbb{R})$. Hence $\mathrm{O}(n)$ is a matrix subgroup of $\mathrm{GL}_n(\mathbb{R})$.

Consider the determinant function restricted to $\mathrm{O}(n)$, $\det\colon \mathrm{O}(n) \longrightarrow \mathbb{R}^{\times}$. For $A \in \mathrm{O}(n)$,

$$(\det A)^2 = \det A^T \det A = \det(A^T A) = \det I_n = 1,$$

which implies that $\det A = \pm 1$. Thus we have

$$\mathrm{O}(n) = \mathrm{O}(n)^+ \cup \mathrm{O}(n)^-,$$

where

$$\mathrm{O}(n)^+ = \{A \in \mathrm{O}(n) : \det A = 1\}, \quad \mathrm{O}(n)^- = \{A \in \mathrm{O}(n) : \det A = -1\}.$$

Notice that

$$\mathrm{O}(n)^+ \cap \mathrm{O}(n)^- = \varnothing,$$

so $\mathrm{O}(n)$ is the *disjoint union* of the subsets $\mathrm{O}(n)^+$ and $\mathrm{O}(n)^-$. The important subgroup

$$\mathrm{SO}(n) = \mathrm{O}(n)^+ \leqslant \mathrm{O}(n)$$

is the $n \times n$ *special orthogonal group*.

One of the main reasons for the study of the orthogonal groups $\mathrm{O}(n)$ and $\mathrm{SO}(n)$ is their relationship with *isometries*, where an *isometry* of $\mathbb{R}^n$ is a distance-preserving bijection $f\colon \mathbb{R}^n \longrightarrow \mathbb{R}^n$, *i.e.*,

$$|f(\mathbf{x}) - f(\mathbf{y})| = |\mathbf{x} - \mathbf{y}| \quad (\mathbf{x}, \mathbf{y} \in \mathbb{R}^n).$$

If such an isometry fixes the origin $\mathbf{0}$ then it is actually a linear transformation, often referred to as a *linear isometry*, and so with respect to the standard basis it corresponds to a matrix $A \in \mathrm{GL}_n(\mathbb{R})$. Our next result summarizes the properties of such matrices.

Proposition 1.38

If $A \in \mathrm{GL}_n(\mathbb{R})$, then the following conditions are equivalent.

- A is a linear isometry.
- $A\mathbf{x} \cdot A\mathbf{y} = \mathbf{x} \cdot \mathbf{y}$ for all vectors $\mathbf{x}, \mathbf{y} \in \mathbb{R}^n$.
- $A^T A = I_n$, *i.e.*, A is orthogonal.

Proof

If A is a linear isometry then for every $\mathbf{v} \in \mathbb{R}^n$, $|A\mathbf{v}| = |\mathbf{v}|$. Now for every pair of vectors $\mathbf{x}, \mathbf{y} \in \mathbb{R}^n$,

$$\begin{aligned} |A(\mathbf{x} - \mathbf{y})|^2 &= (A\mathbf{x} - A\mathbf{y}) \cdot (A\mathbf{x} - A\mathbf{y}) \\ &= |A\mathbf{x}|^2 + |A\mathbf{y}|^2 - 2A\mathbf{x} \cdot A\mathbf{y} \\ &= |\mathbf{x}|^2 + |\mathbf{y}|^2 - 2A\mathbf{x} \cdot A\mathbf{y}, \end{aligned}$$

and similarly,

$$\begin{aligned} |A(\mathbf{x} - \mathbf{y})|^2 &= |\mathbf{x} - \mathbf{y}|^2 \\ &= (\mathbf{x} - \mathbf{y}) \cdot (\mathbf{x} - \mathbf{y}) \\ &= |\mathbf{x}|^2 + |\mathbf{y}|^2 - 2\mathbf{x} \cdot \mathbf{y}. \end{aligned}$$

Hence,

$$A\mathbf{x} \cdot A\mathbf{y} = \mathbf{x} \cdot \mathbf{y}.$$

For $\mathbf{u}, \mathbf{v} \in \mathbb{R}^n$, $\mathbf{u} \cdot \mathbf{v} = \mathbf{u}^T \mathbf{v}$ and $(A\mathbf{u}) \cdot (A\mathbf{v}) = \mathbf{u}^T A^T A\mathbf{v}$. For $i, j = 1, \ldots, n$,

$$\mathbf{e}_i^T A^T A\mathbf{e}_j = (i, j) \text{ entry of } A^T A$$

and $\mathbf{e}_i^T \mathbf{e}_j = \delta_{ij}$. Thus $A^T A = I_n$.

Finally, if $A^T A = I_n$ then for each $\mathbf{w} \in \mathbb{R}^n$,

$$\begin{aligned} |A\mathbf{w}|^2 &= (A\mathbf{w}) \cdot (A\mathbf{w}) \\ &= \mathbf{w}^T A^T A\mathbf{w} \\ &= \mathbf{w}^T \mathbf{w} \\ &= |\mathbf{w}|^2, \end{aligned}$$

showing that A is a linear isometry. □

Elements of $SO(n)$ are often called *direct isometries* or *rotations*, while elements of $O(n)^-$ are sometimes called *indirect isometries*. We can also define the full *isometry group* of $\mathbb{R}^n$,

$$\mathrm{Isom}_n(\mathbb{R}) = \{f\colon \mathbb{R}^n \longrightarrow \mathbb{R}^n : f \text{ is an isometry}\},$$

which clearly contains the subgroup of translations. In fact, $\mathrm{Isom}_n(\mathbb{R}) \leqslant \mathrm{Aff}_n(\mathbb{R})$ and it is also a closed subset, hence it is a matrix subgroup. There is also a semi-direct product decomposition

$$\mathrm{Isom}_n(\mathbb{R}) = \left\{ \begin{bmatrix} A & \mathbf{t} \\ \mathbf{0} & 1 \end{bmatrix} : A \in O(n),\ \mathbf{t} \in \mathbb{R}^n \right\}.$$

Proposition 1.39

$\mathrm{Trans}_n(\mathbb{R})$ is a normal subgroup of $\mathrm{Isom}_n(\mathbb{R})$ and $\mathrm{Isom}_n(\mathbb{R})$ can be expressed as the semi-direct product of $\mathrm{Trans}_n(\mathbb{R})$ and $O(n)$,

$$\mathrm{Isom}_n(\mathbb{R}) = O(n) \ltimes \mathrm{Trans}_n(\mathbb{R}) = \{AT : A \in O(n),\ T \in \mathrm{Trans}_n(\mathbb{R})\},$$

with $\mathrm{Trans}_n(\mathbb{R}) \cap O(n) = \{I_{n+1}\}$.

Proof

This is proved in a similar fashion to Proposition 1.37. □

For later use, we record some important ideas about elements of $O(n)$. First we recall that a subspace $H \subseteq \mathbb{R}^n$ of dimension $\dim H = (n-1)$ is called a *hyperplane* in $\mathbb{R}^n$. Associated with such a hyperplane is a linear transformation $\theta_H\colon \mathbb{R}^n \longrightarrow \mathbb{R}^n$ called *reflection in the hyperplane H*. To define θ_H, observe that every element $\mathbf{x} \in \mathbb{R}^n$ can be uniquely expressed as $\mathbf{x} = \mathbf{x}_H + \mathbf{x}'_H$ with $\mathbf{x}_H \in H$ and $\mathbf{y} \cdot \mathbf{x}'_H = 0$ for every $\mathbf{y} \in H$. Then

$$\theta_H\colon \mathbb{R}^n \longrightarrow \mathbb{R}^n; \quad \theta_H(\mathbf{x}) = \mathbf{x}_H - \mathbf{x}'_H. \tag{1.4}$$

Lemma 1.40

For a hyperplane $H \subseteq \mathbb{R}^n$, the hyperplane reflection in H is in an indirect isometry of $\mathbb{R}^n$, $\theta_H \in O(n)$.

Proof

This is proved by observing that on choosing an orthonormal basis for H and adjoining a unit vector orthogonal to H, the matrix of θ_H with respect to this basis has the block form

$$\begin{bmatrix} I_{n-1} & O_{(n-1)\times 1} \\ O_{1\times(n-1)} & -1 \end{bmatrix} = \operatorname{diag}(1,\dots,1,-1),$$

which clearly has determinant -1. □

We will refer to an element of $\mathrm{O}(n)$ as a hyperplane reflection if it represents a hyperplane reflection with respect to the standard basis of $\mathbb{R}^n$, hence it has the form

$$P\operatorname{diag}(1,\dots,1,-1)P^T$$

for some $P \in \mathrm{O}(n)$.

Proposition 1.41

Every element $A \in \mathrm{O}(n)$ is a product of hyperplane reflections. The number of these is even if $A \in \mathrm{SO}(n)$ and odd if $A \in \mathrm{O}(n)^-$.

Proof

We sketch a direct proof, noting that the result also follows from the Principal Axis Theorem 10.13.

We proceed by induction on n. When $n = 1$, $A = \pm 1$ and as $1 = (-1)^2$, the result is true. Now assume that it holds for all $B \in \mathrm{O}(k)$ with $k < n$. Suppose that $A \in \mathrm{O}(n)$.

If ± 1 is an eigenvalue of A with corresponding unit eigenvector $\mathbf{u}$, then taking

$$H = \{\mathbf{v} \in \mathbb{R}^n : \mathbf{v}\cdot\mathbf{u} = 0\},$$

we find that for $\mathbf{v} \in H$,

$$(A\mathbf{v})\cdot\mathbf{u} = \mathbf{v}\cdot A^T\mathbf{u} = \pm\mathbf{v}\cdot\mathbf{u} = 0,$$

where we have used the fact that

$$A^T\mathbf{u} = A^{-1}\mathbf{u} = \pm\mathbf{u}.$$

Using an orthonormal basis for H extended by $\mathbf{u}$ we see that

$$A = P\begin{bmatrix} A_{n-1} & O_{(n-1)\times 1} \\ O_{1\times(n-1)} & \pm 1 \end{bmatrix}P^T$$

for some $P \in \mathrm{O}(n)$ and $A_{n-1} \in \mathrm{O}(n-1)$. By induction, A_{n-1} is a product of hyperplane reflections and so also is A.

We must still consider the case where all the eigenvalues of A have the form $e^{\pm\alpha i}$ with $0 < \alpha < \pi$. Given a unit eigenvector $\mathbf{v} \in \mathbb{C}^n$ for the eigenvalue $e^{\alpha i}$, $\overline{\mathbf{v}}$ is a unit eigenvector for the eigenvalue $e^{-\alpha i}$. With respect to the usual hermitian inner product on $\mathbb{C}^n$, $\mathbf{v}$ and $\overline{\mathbf{v}}$ are orthogonal. Then

$$\mathbf{v}' = \frac{1}{\sqrt{2}}(\mathbf{v} + \overline{\mathbf{v}}), \quad \mathbf{v}'' = \frac{i}{\sqrt{2}}(\mathbf{v} - \overline{\mathbf{v}})$$

are real, orthogonal unit vectors which also satisfy the equations

$$A\mathbf{v}' = \cos\alpha\,\mathbf{v}' + \sin\alpha\,\mathbf{v}'', \quad A\mathbf{v}'' = -\sin\alpha\,\mathbf{v}' + \cos\alpha\,\mathbf{v}''.$$

Consider the subspace

$$H' = \{\mathbf{w} \in \mathbb{R}^n : \mathbf{w}\cdot\mathbf{v}' = 0 = \mathbf{w}\cdot\mathbf{v}''\} \subseteq \mathbb{R}^n.$$

An orthonormal basis of H' can be extended to one for $\mathbb{R}^n$ by adding $\mathbf{v}', \mathbf{v}''$. Then with respect to this basis, the matrix of the linear transformation A has the form

$$\begin{bmatrix} A_{n-2} & O_{(n-2)\times 2} \\ O_{2\times(n-2)} & R_2(\alpha) \end{bmatrix},$$

where

$$A_{n-2} \in \mathrm{O}(n-2), \quad R_2(\alpha) = \begin{bmatrix} \cos\alpha & -\sin\alpha \\ \sin\alpha & \cos\alpha \end{bmatrix} \in \mathrm{SO}(2).$$

So for some $Q \in \mathrm{O}(n)$,

$$A = Q\begin{bmatrix} A_{n-2} & O_{(n-2)\times 2} \\ O_{2\times(n-2)} & R_2(\alpha) \end{bmatrix} Q^T.$$

By the inductive assumption, A_{n-2} is a product of hyperplane reflections in $\mathbb{R}^{n-2}$, so it only remains to show that the matrix $R_2(\alpha)$ is a product of hyperplane reflections in $\mathbb{R}^2$. By direct calculation we obtain

$$R_2(\alpha) = \begin{bmatrix} \cos\alpha & \sin\alpha \\ \sin\alpha & -\cos\alpha \end{bmatrix} \begin{bmatrix} 1 & 0 \\ 0 & -1 \end{bmatrix},$$

where the first factor represents reflection in the line

$$\cos(\alpha/2)\,y = \sin(\alpha/2)\,x$$

and the second represents reflection in the x-axis, $y = 0$.

The other statements follow from the fact that a hyperplane reflection is an indirect isometry. □

Generalised Orthogonal Groups

A more general situation is associated with an $n \times n$ real symmetric matrix Q. Then there is an analogue of the orthogonal group,

$$\mathrm{O}_Q = \{A \in \mathrm{GL}_n(\mathbb{R}) : A^T Q A = Q\}.$$

It is easy to see that this is a closed subgroup of $\mathrm{GL}_n(\mathbb{R})$, *i.e.*, a matrix group. Moreover, if $\det Q \neq 0$, then for $A \in \mathrm{O}_Q$ we have $\det A = \pm 1$. We can also define

$$\mathrm{O}_Q^+ = \det^{-1} \mathbb{R}^+, \quad \mathrm{O}_Q^- = \det^{-1} \mathbb{R}^-$$

and can write O_Q as a disjoint union of clopen subsets $\mathrm{O}_Q = \mathrm{O}_Q^+ \cup \mathrm{O}_Q^-$ where O_Q^+ is a subgroup.

An important example of this occurs in relativity where $n = 4$ and

$$Q = \begin{bmatrix} 1 & 0 & 0 & 0 \\ 0 & 1 & 0 & 0 \\ 0 & 0 & 1 & 0 \\ 0 & 0 & 0 & -1 \end{bmatrix}.$$

The *Lorentz group* Lor is the closed subgroup of $\mathrm{O}_Q^+ \cap \mathrm{SL}_2(\mathbb{R})$ which preserves each of the two connected components of the hyperboloid

$$x_1^2 + x_2^2 + x_3^2 - x_4^2 = -1.$$

We will study this example in greater detail in Chapter 6.

Symplectic Groups

Symplectic geometry has become an actively studied mathematical topic and is the geometry associated with Hamiltonian Mechanics and therefore with Quantum Mechanics; it is also important as an area of differential geometry and in the study of 4-dimensional manifolds. Symplectic groups are the natural symmetry groups for such geometries. We will discuss the related notion of a *symplectic form* in Section 8.8.

Similar considerations to those in the last section apply to an $n \times n$ real *skew symmetric* matrix S, *i.e.*, one for which $S^T = -S$. For such a matrix,

$$\det S^T = \det(-S) = (-1)^n \det S,$$

giving

$$\det S = (-1)^n \det S. \tag{1.5}$$

The most interesting case occurs if $\det S \neq 0$ when n must be even and we then write $n = 2m$. The standard example of this is built up using the 2×2 block

$$J = \begin{bmatrix} 0 & 1 \\ -1 & 0 \end{bmatrix}.$$

If $m \geqslant 1$ we have the non-degenerate skew symmetric matrix

$$J_{2m} = \begin{bmatrix} J & O_2 & \cdots & O_2 \\ O_2 & J & \cdots & O_2 \\ \vdots & \vdots & \ddots & \vdots \\ O_2 & O_2 & \cdots & J \end{bmatrix}.$$

The matrix group

$$\mathrm{Symp}_{2m}(\mathbb{R}) = \{A \in \mathrm{GL}_{2m}(\mathbb{R}) : A^T J_{2m} A = J_{2m}\} \leqslant \mathrm{GL}_{2m}(\mathbb{R}),$$

is called the $2m \times 2m$ *(real) symplectic group.*

There is an alternative version of the symplectic group defined using the skew symmetric matrix

$$J'_{2m} = \begin{bmatrix} O_m & I_n \\ -I_n & O_m \end{bmatrix}$$

in place of J_{2m}. For the precise relationship see the discussion preceding Corollary 8.25. Then we define

$$\mathrm{Symp}'_{2m}(\mathbb{R}) = \{A \in \mathrm{GL}_{2m}(\mathbb{R}) : A^T J'_{2m} A = J'_{2m}\} \leqslant \mathrm{GL}_{2m}(\mathbb{R}).$$

There is a simple relationship between $\mathrm{Symp}'_{2m}(\mathbb{R})$ and $\mathrm{Symp}_{2m}(\mathbb{R})$, see the discussion at the end of Section 8.8 for further explanation.

Proposition 1.42

Let $P \in \mathrm{GL}_{2m}(\mathbb{R})$ be any matrix for which $J'_{2m} = P^T J_{2m} P$. Then

$$\mathrm{Symp}'_{2m}(\mathbb{R}) = P^{-1}\,\mathrm{Symp}_{2m}(\mathbb{R}) P = \{P^{-1}AP : A \in \mathrm{Symp}_{2m}(\mathbb{R})\}.$$

Remark 1.43

It is straightforward to see that

$$\mathrm{Symp}_2(\mathbb{R}) = \mathrm{Symp}'_2(\mathbb{R}) = \mathrm{SL}_2(\mathbb{R}),$$

but in general

$$\mathrm{Symp}_{2m}(\mathbb{R}) \neq \mathrm{SL}_{2m}(\mathbb{R}), \quad \mathrm{Symp}'_{2m}(\mathbb{R}) \neq \mathrm{SL}_{2m}(\mathbb{R}).$$

It is also easy to show that $\det A = \pm 1$ if $A \in \mathrm{Symp}_{2m}(\mathbb{R})$ or $A \in \mathrm{Symp}'_{2m}(\mathbb{R})$. In fact the methods of Chapters 8 and 9 can be used to prove that $\det A = 1$, so

$$\mathrm{Symp}_{2m}(\mathbb{R}) \leqslant \mathrm{SL}_{2m}(\mathbb{R}), \quad \mathrm{Symp}'_{2m}(\mathbb{R}) \leqslant \mathrm{SL}_{2m}(\mathbb{R}).$$

Unitary Groups

For $A = [a_{ij}] \in \mathrm{M}_n(\mathbb{C})$,

$$A^* = (\overline{A})^T = \overline{(A^T)},$$

is the *hermitian conjugate* of A, *i.e.*, $(A^*)_{ij} = \overline{a_{ji}}$. The $n \times n$ *unitary group* is the subgroup

$$\mathrm{U}(n) = \{A \in \mathrm{GL}_n(\mathbb{C}) : A^*A = I\} \leqslant \mathrm{GL}_n(\mathbb{C}).$$

Again the unitary condition amounts to n^2 equations for the n^2 complex numbers a_{ij} (compare Equation (1.3)),

$$\sum_{k=1}^{n} \overline{a}_{ki} a_{kj} = \delta_{ij}. \tag{1.6}$$

By taking real and imaginary parts, these equations actually give $2n^2$ bilinear equations in the $2n^2$ real and imaginary parts of the a_{ij}, although there is some redundancy.

The $n \times n$ *special unitary group* is

$$\mathrm{SU}(n) = \{A \in \mathrm{GL}_n(\mathbb{C}) : A^*A = I \text{ and } \det A = 1\} \leqslant \mathrm{U}(n).$$

Again we can specify that a matrix is special unitary by requiring that its entries satisfy the $(n^2 + 1)$ equations

$$\begin{cases} \displaystyle\sum_{k=1}^{n} \overline{a}_{ki} a_{kj} = \delta_{ij} \quad (1 \leqslant i, j \leqslant n), \\ \det A = 1. \end{cases} \tag{1.7}$$

Of course, $\det A$ is a polynomial in the entries a_{ij}. Notice that $\mathrm{SU}(n)$ is a normal subgroup of $\mathrm{U}(n)$, $\mathrm{SU}(n) \lhd \mathrm{U}(n)$.

The dot product on $\mathbb{R}^n$ can be extended to $\mathbb{C}^n$ by setting

$$\mathbf{x} \cdot \mathbf{y} = \mathbf{x}^* \mathbf{y} = \sum_{k=1}^{n} \overline{x}_k y_k,$$

where $\mathbf{x} = \begin{bmatrix} x_1 \\ \vdots \\ x_n \end{bmatrix}$ and $\mathbf{y} = \begin{bmatrix} y_1 \\ \vdots \\ y_n \end{bmatrix}$. Note that $\cdot$ is not $\mathbb{C}$-linear but for $u, v \in \mathbb{C}$ it satisfies

$$(u\mathbf{x}) \cdot (v\mathbf{y}) = \overline{u}v(\mathbf{x} \cdot \mathbf{y}).$$

This dot product allows us to define the length of a complex vector by

$$|\mathbf{x}| = \sqrt{\mathbf{x} \cdot \mathbf{x}}$$

since $\mathbf{x} \cdot \mathbf{x}$ is a non-negative real number which is zero only when $\mathbf{x} = \mathbf{0}$. Then a matrix $A \in \mathrm{M}_n(\mathbb{C})$ is unitary if and only if

$$A\mathbf{x} \cdot A\mathbf{y} = \mathbf{x} \cdot \mathbf{y} \quad (\mathbf{x}, \mathbf{y} \in \mathbb{C}^n).$$

1.6 Complex Matrices as Real Matrices

Recall that the complex numbers can be viewed as a 2-dimensional real vector space, with basis $1, i$ for example. Similarly, every $n \times n$ complex matrix $Z = [z_{ij}]$ can also be viewed as a $2n \times 2n$ real matrix using one of the following constructions.

We identify each complex number $z = x + yi$ with a 2×2 real matrix by defining a function

$$\rho\colon \mathbb{C} \longrightarrow \mathrm{M}_2(\mathbb{R}); \quad \rho(x + yi) = \begin{bmatrix} x & -y \\ y & x \end{bmatrix}.$$

This can also be expressed as

$$\rho(x + yi) = xI_2 - yJ$$

where $J = \begin{bmatrix} 0 & 1 \\ -1 & 0 \end{bmatrix}$ was introduced in Section 1.5. Then ρ is an injective ring homomorphism, so we can view $\mathbb{C}$ as a subring of $\mathrm{M}_2(\mathbb{R})$ by identifying $\mathbb{C}$ with its image under ρ,

$$\rho\mathbb{C} = \left\{ \begin{bmatrix} a & b \\ c & d \end{bmatrix} \in \mathrm{M}_2(\mathbb{R}) : d = a,\ c = -b \right\}.$$

Notice that complex conjugation corresponds to transposition since

$$\rho(\overline{z}) = \rho(z)^T. \tag{1.8}$$

We will describe two different ways to extend this to a function $\mathrm{M}_n(\mathbb{C}) \longrightarrow \mathrm{M}_{2n}(\mathbb{R})$. For $Z = [z_{rs}] \in \mathrm{M}_n(\mathbb{C})$ with $z_{rs} = x_{rs} + y_{rs}i$, we can write

$$Z = [x_{rs}] + i[y_{rs}]$$

where $X = [x_{rs}]$ and $Y = [y_{rs}]$ are $n \times n$ real matrices.

First we define the function

$$\rho_n \colon \mathrm{M}_n(\mathbb{C}) \longrightarrow \mathrm{M}_{2n}(\mathbb{R})$$

for which $\rho_n(Z)$ is the $n \times n$ matrix of 2×2 real blocks with $\rho(z_{rs})$ as the one in the (r, s) place,

$$\rho_n(Z) = \begin{bmatrix} \rho(z_{11}) & \rho(z_{12}) & \cdots & \rho(z_{1n}) \\ \rho(z_{21}) & \ddots & \ddots & \vdots \\ \vdots & \ddots & \ddots & \vdots \\ \rho(z_{n1}) & \cdots & \cdots & \rho(z_{nn}) \end{bmatrix}$$

Then ρ_n is an injective ring homomorphism which is also continuous. In the notation of Section 1.5, this gives

$$\rho_n(iI_n) = -J_{2n}.$$

More generally, for $n \times n$ real matrices X, Y,

$$\rho_n(X + iY) = \rho_n(X) - J_{2n}\rho_n(Y)J_{2n},$$

where $\rho_n(X)$ and $\rho_n(Y)$ are $2n \times 2n$ real matrices consisting of 2×2 scalar blocks $x_{rs}I_2$ and $y_{rs}I_2$.

Our second function is

$$\rho'_n \colon \mathrm{M}_n(\mathbb{C}) \longrightarrow \mathrm{M}_{2n}(\mathbb{R}); \quad \rho'_n(X + iY) = \begin{bmatrix} X & -Y \\ Y & X \end{bmatrix} \quad (X, Y \in \mathrm{M}_n(\mathbb{R})),$$

which is an injective ring homomorphism. ρ'_n is easily seen to be continuous. Using the matrix J'_{2m} of Section 1.5, we have the following identities for $X, Y \in \mathrm{M}_n(\mathbb{R})$ and $Z \in \mathrm{M}_n(\mathbb{C})$:

$$\begin{aligned} (J'_{2n})^2 &= -I_{2n}, \\ (J'_{2n})^T &= -J'_{2n}, \\ \rho'_n(X + iY) &= \begin{bmatrix} X & O_n \\ O_n & X \end{bmatrix} - \begin{bmatrix} Y & O_n \\ O_n & Y \end{bmatrix} J'_{2n}, \\ \rho'_n(\overline{Z}) &= \rho'_n(Z)^T. \end{aligned}$$

The images of ρ_n and ρ'_n, $\rho_n \mathrm{GL}_n(\mathbb{C}) \leqslant \mathrm{GL}_{2n}(\mathbb{R})$ and $\rho'_n \mathrm{GL}_n(\mathbb{C}) \leqslant \mathrm{GL}_{2n}(\mathbb{R})$, are clearly closed subgroups of $\mathrm{GL}_{2n}(\mathbb{R})$, so any matrix subgroup

$G \leqslant \mathrm{GL}_n(\mathbb{C})$ can be viewed as a matrix subgroup of $\mathrm{GL}_{2n}(\mathbb{R})$ by identifying it with either of its images $\rho_n G$ or $\rho'_n G$. It is sometimes useful to characterise $\rho_n \mathrm{GL}_n(\mathbb{C})$ and $\rho'_n \mathrm{GL}_n(\mathbb{C})$ as in the following result.

Proposition 1.44

We have

$$\rho_n \mathrm{M}_n(\mathbb{C}) = \{A \in \mathrm{M}_{2n}(\mathbb{R}) : AJ_{2n} = J_{2n}A\}, \tag{1.9a}$$

$$\rho_n \mathrm{GL}_n(\mathbb{C}) = \{A \in \mathrm{GL}_{2n}(\mathbb{R}) : AJ_{2n} = J_{2n}A\}, \tag{1.9b}$$

and

$$\rho'_n \mathrm{M}_n(\mathbb{C}) = \{A \in \mathrm{M}_{2n}(\mathbb{R}) : AJ'_{2n} = J'_{2n}A\}, \tag{1.9c}$$

$$\rho'_n \mathrm{GL}_n(\mathbb{C}) = \{A \in \mathrm{GL}_{2n}(\mathbb{R}) : AJ'_{2n} = J'_{2n}A\}. \tag{1.9d}$$

Proof

We give the proof of the second pair of equations, the first pair being similar. Writing

$$A = \begin{bmatrix} A_{11} & A_{12} \\ A_{21} & A_{22} \end{bmatrix}$$

for some $n \times n$ matrices A_{rs}, we see that $AJ'_{2n} = J'_{2n}A$ if and only if

$$\begin{bmatrix} A_{12} & -A_{11} \\ A_{22} & -A_{21} \end{bmatrix} = \begin{bmatrix} -A_{21} & -A_{22} \\ A_{11} & A_{22} \end{bmatrix},$$

from which we obtain

$$A = \begin{bmatrix} A_{11} & -A_{21} \\ A_{21} & A_{11} \end{bmatrix} \in \rho'_n \mathrm{M}_n(\mathbb{C}),$$

giving the result. □

1.7 Continuous Homomorphisms of Matrix Groups

In studying groups, the notion of a homomorphism of groups plays a central rôle. For matrix groups we need to be careful about topological properties as well as the algebraic ones.

Definition 1.45

Let G, H be two matrix groups. A group homomorphism $\varphi\colon G \longrightarrow H$ is a *continuous homomorphism* of matrix groups if it is continuous and its image $\varphi G \leqslant H$ is a closed subspace of H.

Example 1.46

The function

$$\varphi\colon \mathrm{SUT}_2(\mathbb{R}) \longrightarrow \mathrm{U}(1); \quad \varphi\left(\begin{bmatrix} 1 & t \\ 0 & 1 \end{bmatrix}\right) = \begin{bmatrix} e^{2\pi t i} \end{bmatrix}$$

is a continuous surjective group homomorphism, so it is a continuous homomorphism of matrix groups.

To see why the closure condition on the image in the above definition is desirable, consider the following example. Recall that a subset $U \subseteq X$ of a topological space X is said to be *dense* if its closure $\overline{U} = X$.

Example 1.47

Let

$$G = \left\{ \begin{bmatrix} 1 & n \\ 0 & 1 \end{bmatrix} \in \mathrm{SUT}_1(\mathbb{R}) : n \in \mathbb{Z} \right\}.$$

Then G is a closed subgroup of $\mathrm{SUT}_1(\mathbb{R})$, so it is a matrix group.

For any irrational number $r \in \mathbb{R} - \mathbb{Q}$, the function

$$\varphi\colon G \longrightarrow \mathrm{U}(1); \quad \varphi\left(\begin{bmatrix} 1 & n \\ 0 & 1 \end{bmatrix}\right) = \begin{bmatrix} e^{2\pi r n i} \end{bmatrix}$$

is a continuous group homomorphism. But its image is a dense proper subset of $\mathrm{U}(1)$. So φ is not a continuous homomorphism of matrix groups.

The point of this example is that φG has limit points in $\mathrm{U}(1)$ which are not in φG, whereas G is discrete as a subspace of $\mathrm{SUT}_2(\mathbb{R})$.

Whenever we have a homomorphism of matrix groups $\varphi\colon G \longrightarrow H$ which is a homeomorphism (*i.e.*, a bijection with continuous inverse) we say that φ is a *continuous isomorphism of matrix groups* and regard G and H as essentially identical as matrix groups.

Proposition 1.48

Let $\varphi\colon G \longrightarrow H$ be a continuous homomorphism of matrix groups. Then $\ker\varphi \leqslant G$ is a closed subgroup, hence $\ker\varphi$ is a matrix group. The quotient group $G/\ker\varphi$ can be identified with the matrix group φG by the usual quotient isomorphism $\overline{\varphi}\colon G/\ker\varphi \longrightarrow \varphi G$ (which need not be a homomorphism of matrix groups since $G/\ker\varphi$ need not be a matrix group).

Proof

Since φ is continuous, whenever it makes sense in G we have

$$\lim_{n\to\infty} \varphi(A_n) = \varphi(\lim_{n\to\infty} A_n),$$

which implies that a limit of elements of $\ker\varphi$ in G is also in $\ker\varphi$. So $\ker\varphi$ is a closed subset of G. By Proposition 1.31, $\ker\varphi \leqslant G$ is a matrix group. □

Remark 1.49

$G/\ker\varphi$ has a natural *quotient topology* discussed in Section 8.1; this is not obviously a metric topology. Then $\overline{\varphi}$ is always a homeomorphism.

Remark 1.50

Not every closed normal matrix subgroup $N \lhd G$ of a matrix group G gives rise to a matrix group G/N; there are examples for which G/N is a *Lie group* but not a matrix group. This is one of the most important differences between matrix groups and Lie groups and we will see later that every matrix group is a Lie group. One consequence is that certain important matrix groups have quotients which are not matrix groups and therefore have no faithful finite dimensional representations; such groups occur readily in Quantum Physics, where their infinite dimensional representations play an important rôle.

1.8 Matrix Groups for Normed Vector Spaces

Our approach to matrix groups has been in terms of $\mathbb{R}^n$ and $\mathbb{C}^n$, naturally relying on coordinates and the standard notion of distance on these vector spaces. It is often desirable to generalise this to arbitrary finite dimensional real or complex vector spaces and also to allow a more general notion of distance defined in terms of a *norm*. From now on we assume that $\mathbb{k} = \mathbb{R}$ or $\mathbb{C}$.

Definition 1.51

Let V be a finite dimensional $\Bbbk$-vector space with a function $\nu\colon V \longrightarrow \mathbb{R}^+$. Then ν is called a $\Bbbk$-*norm* on V if it satisfies the conditions
i) $\nu(tv) = |t|\,\nu(v)$ for $t \in \Bbbk$, $v \in V$;
ii) $\nu(v_1 + v_2) \leqslant \nu(v_1) + \nu(v_2)$ for $v_1, v_2 \in V$;
iii) for $v \in V$, $\nu(v) = 0$ if and only if $v = 0$.
We denote such a normed vector space by writing (V, ν).

There are many examples of such norms on $\Bbbk^n$ apart from the standard one.

Example 1.52

Consider the function

$$| \ |_1 \colon \Bbbk^n \longrightarrow \mathbb{R}^+; \quad |\mathbf{x}|_1 = \max\{|x_i| : i = 1, 2, \ldots, n\}, \quad \mathbf{x} = \begin{bmatrix} x_1 \\ x_2 \\ \vdots \\ x_n \end{bmatrix}.$$

Then $| \ |_1$ is a $\Bbbk$-norm on $\Bbbk^n$.

Given a finite dimensional normed $\Bbbk$-vector space V, there is a metric ρ_ν on V defined by

$$\rho_\nu(x, y) = \nu(x - y).$$

We can use the associated topology to define continuous functions between V and any other topological space, including other metric spaces. We will regard a normed vector space as a metric space in this way.

Our next result is standard and depends on the fact that $\mathbb{R}$ and $\mathbb{C}$ are complete metric spaces; see [21, 22, 23].

Theorem 1.53

Let (U, μ) and (V, ν) be two finite dimensional normed $\Bbbk$-vector spaces of the same dimension. Then any linear isomorphism $\varphi\colon U \longrightarrow V$ is a homeomorphism, *i.e.*, it is continuous with continuous inverse.

Corollary 1.54

If μ and ν are two $\Bbbk$-norms on a finite dimensional $\Bbbk$-vector space, then they give rise to the same topology.

Proof

The identity function $\mathrm{Id}_V : V \longrightarrow V$ is a linear isomorphism, which must be continuous. This implies that every open set in the topology of ν is open in the topology of μ. Similarly, every open set in the topology of μ is open in the topology of ν. □

Given a finite dimensional normed $\Bbbk$-vector space (V, ν) and a linear transformation $\alpha : V \longrightarrow V$, we define the *operator norm* of α with respect to ν by

$$\|\alpha\|_\nu = \sup\{\nu(\alpha(x)) : x \in V,\ \nu(x) = 1\}.$$

If we denote by $\mathrm{End}_\Bbbk(V)$ the set of all linear transformations $V \longrightarrow V$, this defines a $\Bbbk$-norm on $\mathrm{End}_\Bbbk(V)$ in the sense of Proposition 1.5. This makes $\mathrm{End}_\Bbbk(V)$ into a finite dimensional normed $\Bbbk$-vector space with associated topology.

Proposition 1.55

If μ and ν are two norms on finite dimensional $\Bbbk$-vector space V, then the topologies associated to $(\mathrm{End}_\Bbbk(V), \| \ \|_\mu)$ and $(\mathrm{End}_\Bbbk(V), \| \ \|_\nu)$ are the same.

Proof

Apply Theorem 1.53. □

Inside of $\mathrm{End}_\Bbbk(V)$ we have the group of linear isomorphisms, *i.e.*, invertible linear transformations, $\mathrm{GL}_\Bbbk(V) \subseteq \mathrm{End}_\Bbbk(V)$, the *general linear group* of the $\Bbbk$-vector space V. If ν is a norm on V, $\mathrm{GL}_\Bbbk(V)$ inherits the metric and is an open subset of $\mathrm{End}_\Bbbk(V)$. The following is straightforward to verify.

Proposition 1.56

Let (U, μ) and (V, ν) be two finite dimensional normed $\Bbbk$-vector spaces of the same dimension. If $\varphi : U \longrightarrow V$ is a linear isomorphism, then it induces a continuous group isomorphism with continuous inverse,

$$\varphi_* : \mathrm{GL}_\Bbbk(U) \longrightarrow \mathrm{GL}_\Bbbk(V); \quad \varphi_*(\alpha) = \varphi \circ \alpha \circ \varphi^{-1}.$$

Recall that for $V = \Bbbk^n$ we can identify $\mathrm{End}_\Bbbk(V)$ with $\mathrm{M}_n(\Bbbk)$ and $\mathrm{GL}_\Bbbk(V)$ with $\mathrm{GL}_n(\Bbbk)$. More generally, given a finite basis for V, there is an associated $\Bbbk$-linear isomorphism $V \cong \Bbbk^n$ for some n, a ring isomorphism $\mathrm{End}_\Bbbk(V) \cong \mathrm{M}_n(\Bbbk)$ and a group isomorphism $\mathrm{GL}_\Bbbk(V) \cong \mathrm{GL}_n(\Bbbk)$.

Theorem 1.57

Let (V, ν) be a finite dimensional normed $\Bbbk$-vector space (V, ν) of dimension $\dim_{\Bbbk} V = n$ and let $\varphi \colon V \longrightarrow \Bbbk^n$ be a linear isomorphism. Then $\varphi_* \colon \mathrm{GL}_{\Bbbk}(V) \longrightarrow \mathrm{GL}_n(\Bbbk)$ is a continuous group isomorphism with continuous inverse. Hence $\mathrm{GL}_{\Bbbk}(V)$ is a matrix group.

This result shows that although we could have set up a theory of matrix groups based on finite dimensional normed vector spaces (V, ν), defining matrix groups to be closed subgroups of $\mathrm{GL}_{\Bbbk}(V)$, we would have obtained an essentially equivalent theory to the one we have described. The extra flexibility is sometimes useful and many accounts are based on it.

We end this section with some useful observations about a finite dimensional normed $\Bbbk$-vector space (V, ν) and a $\Bbbk$-vector subspace $W \subseteq V$.

Proposition 1.58

Let $\mathbf{v} \in V$. Then there is a vector $\mathbf{w}_0 \in W$ for which

$$\nu(\mathbf{w}_0 - \mathbf{v}) = \inf\{\nu(\mathbf{w} - \mathbf{v}) : \mathbf{w} \in W\}.$$

Proof

Since $\mathbf{0} \in W$, if we set

$$\lambda_W = \inf\{\nu(\mathbf{w} - \mathbf{v}) : \mathbf{w} \in W\},$$

then $0 \leqslant \lambda_W \leqslant \nu(\mathbf{v})$. Now consider the set

$$S = \{\mathbf{w} \in W : \nu(\mathbf{w}) \leqslant \lambda_W + 2\nu(\mathbf{v})\}.$$

This is a closed and bounded subset of the normed vector space (W, ν) so it is compact. By Corollary 1.23, the continuous function

$$f \colon S \longrightarrow \mathbb{R}; \quad f(\mathbf{w}) = \nu(\mathbf{w} - \mathbf{v})$$

has values bounded below by 0 and attains its infimum. Finally, notice that if $\mathbf{w} \in W$ and $\nu(\mathbf{w} - \mathbf{v}) \leqslant \lambda_W + \nu(\mathbf{v})$ then $\mathbf{w} \in S$, hence

$$\inf\{\nu(\mathbf{w} - \mathbf{v}) : \mathbf{w} \in W\} = \inf\{\nu(\mathbf{w} - \mathbf{v}) : \mathbf{w} \in S\}.$$

So $\nu(\mathbf{w} - \mathbf{v})$ certainly attains its infimum on W. □

Proposition 1.59

Every subspace $W \subseteq V$ is a closed subset.

Proof

If $\mathbf{v} \notin W$, then $\lambda_W > 0$ since otherwise

$$0 = \lambda_W = \nu(\mathbf{w}_0 - \mathbf{v}),$$

which can only happen if $\mathbf{w}_0 = \mathbf{v}$. So the open disc $\mathrm{N}_V(\mathbf{v}; \lambda_W)$ has trivial intersection with W,

$$\mathrm{N}_V(\mathbf{v}; \lambda_W) \cap W = \varnothing.$$

So the complement of W in V is open, hence W is closed. □

1.9 Continuous Group Actions

In ordinary group theory, the notion of a *group action* is fundamental. Suitably formulated, it amounts to the following. An *action* μ of a group G on a set X is a function $\mu\colon G \times X \longrightarrow X$ for which we usually write $\mu(g, x) = gx$ if there is no danger of ambiguity, satisfying the following conditions for all $g, h \in G$ and $x \in X$ and with ι being the identity element of G:

- $(gh)x = g(hx)$, *i.e.*, $\mu(gh, x) = \mu(g, \mu(h, x))$;
- $\iota x = x$.

There are two important notions associated to such an action.

For $x \in X$, the *stabiliser of* x is

$$\mathrm{Stab}_G(x) = \{g \in G : gx = x\} \subseteq G,$$

while the *orbit of* x is

$$\mathrm{Orb}_G(x) = \{gx \in X : g \in G\} \subseteq X.$$

Theorem 1.60

Let G act on X.
i) For $x \in X$, $\mathrm{Stab}_G(x) \leqslant G$, *i.e.*, $\mathrm{Stab}_G(x)$ is a subgroup of G.
ii) For $x, y \in X$, $y \in \mathrm{Orb}_G(x)$ if and only if $\mathrm{Orb}_G(y) = \mathrm{Orb}_G(x)$.
For $x \in X$, there is a bijection

$$\varphi\colon G/\mathrm{Stab}_G(x) \longrightarrow \mathrm{Orb}_G(x); \quad \varphi(g) = gx.$$

Furthermore, this is G-equivariant in the sense that for all $g, h \in G$,

$$\varphi((hg)\,\mathrm{Stab}_G(x)) = h\varphi(g\,\mathrm{Stab}_G(x)).$$

iii) If $y \in \mathrm{Orb}_G(x)$, then for any $t \in G$ with $y = tx$,

$$\mathrm{Stab}_G(y) = t\,\mathrm{Stab}_G(x)t^{-1}.$$

For a topological group there is a notion of *continuous group action* on a topological space.

Definition 1.61

Let G be a topological group and X be a topological space. Then a group action $\mu\colon G \times X \longrightarrow X$ is a *continuous group action* if the function μ is continuous.

In this definition, $G \times X$ has the product topology which is obtained from a suitable metric when G and X are metric spaces.

If X is Hausdorff (in particular if it is a metric space) then any one-element subset $\{x\} \subseteq X$ is closed and the stabiliser of x, $\mathrm{Stab}_G(x) \leqslant G$, is a closed subgroup. This provides a useful source of closed subgroups.

For us, the most important type of action for matrix groups arises when a matrix group G has a continuous homomorphism $\varphi\colon G \longrightarrow \mathrm{GL}_{\Bbbk}(V)$ where (V, ν) is a $\Bbbk$-norm on the finite dimensional $\Bbbk$-vector space V. Then the associated action

$$\mu_\varphi\colon G \times V \longrightarrow V; \quad \mu_\varphi(g, \mathbf{v}) = \varphi(g)(\mathbf{v})$$

is continuous. It is worth remarking that by Proposition 1.59, any vector subspace $W \subseteq V$ is closed, so the stabiliser

$$\mathrm{Stab}_G(W) = \{g \in G : \varphi(g)W = W\} \leqslant G$$

is a closed subgroup, as is

$$\bigcap_{\mathbf{w} \in W} \mathrm{Stab}_G(\mathbf{w}) \leqslant G.$$

See the exercises at the end of this chapter for more on this.

Definition 1.62

If $\varphi\colon G \longrightarrow \mathrm{GL}_{\Bbbk}(V)$ is a continuous group homomorphism, then the associated action μ_φ is called a (*continuous*) *linear action* or *representation* of G on V.

By choosing a basis for V and applying the ideas discussed at the end of Section 1.8 and Theorem 1.57, we may as well assume that $V = \Bbbk^n$. Then a continuous action is essentially the same thing as a continuous group homomorphism $G \longrightarrow \mathrm{GL}_n(\Bbbk)$.

Here is an example that illustrates how ubiquitous this idea is. Recall that for indeterminates $x_1, \ldots, x_k$, a polynomial $f(x_1, \ldots, x_k) \in \mathbb{k}[x_1, \ldots, x_k]$ is *homogeneous* of degree n if $f(x_1, \ldots, x_k)$ is a linear combination of monomials $x_1^{r_1} \cdots x_k^{r_k}$ with degree

$$r_1 + \cdots + r_k = n.$$

We write $\mathbb{k}[x_1, \ldots, x_k]_n$ for the subset of all elements of $\mathbb{k}[x_1, \ldots, x_k]$ homogeneous of degree n. It is easy to see that with addition and scalar multiplication $\mathbb{k}[x_1, \ldots, x_k]$ is a $\mathbb{k}$-vector space and $\mathbb{k}[x_1, \ldots, x_k]_n$ is a finite dimensional subspace with a basis consisting of all monomials of degree n.

Example 1.63

For indeterminates x, y, let $V_n = \mathbb{k}[x, y]_n$ with $\dim_k V_n = (n+1)$. There is a $\mathbb{k}$-linear action $\mathrm{GL}_2(\mathbb{k}) \longrightarrow \mathrm{GL}_{\mathbb{k}}(V_n)$ given by

$$\begin{bmatrix} a & b \\ c & d \end{bmatrix} f(x, y) = f(ax + cy, bx + dy).$$

On a monomial this yields

$$\begin{aligned} \begin{bmatrix} a & b \\ c & d \end{bmatrix} x^r y^{n-r} &= (ax + cy)^r (bx + dy)^{n-r} \\ &= \sum_{i=0}^{r} \sum_{j=0}^{n-r} \binom{r}{i} \binom{n-r}{j} (ax)^i (cy)^{r-i} (bx)^j (dy)^{n-r-j} \\ &= \sum_{k=0}^{n} \left(\sum_{i=0}^{k} \binom{r}{i} \binom{n-r}{k-i} a^i c^{r-i} b^{k-i} d^{n-r-k+i} \right) x^k y^{n-k}, \end{aligned}$$

giving for example,

$$\begin{bmatrix} a & b \\ c & d \end{bmatrix} x^2 y = a^2 b x^3 + (a^2 d + 2abc) x^2 y + (2acd + c^2 b) x y^2 + c^2 d y^3.$$

Taking the basis of monomials of degree n in the order

$$x^n, x^{n-1} y, \ldots, x y^{n-1}, y^n,$$

this can be interpreted as a homomorphism $\varphi_n \colon \mathrm{GL}_2(\mathbb{k}) \longrightarrow \mathrm{GL}_{n+1}(\mathbb{k})$.

Of course, any matrix subgroup $G \leqslant \mathrm{GL}_2(\mathbb{k})$ will also act on V_n by a continuous linear action. When $\mathbb{k} = \mathbb{C}$, the representations of $\mathrm{SU}(2)$ on the V_n are particularly important, see Sternberg [27] for an illuminating discussion of these representations and their applications.

EXERCISES

1.1. Determine the norm $\|A\|$ for each of the following matrices A and real numbers $t, u, v \in \mathbb{R}$.

$$\begin{bmatrix} u & 0 \\ 0 & v \end{bmatrix}, \quad \begin{bmatrix} u & 1 \\ 0 & u \end{bmatrix}, \quad \begin{bmatrix} \cos t & -\sin t \\ \sin t & \cos t \end{bmatrix}, \quad \begin{bmatrix} \cosh t & \sinh t \\ \sinh t & \cosh t \end{bmatrix}.$$

What can be said when $u, v \in \mathbb{C}$?

1.2. Suppose that $A \in \mathrm{M}_n(\mathbb{C})$.
a) If $B \in \mathrm{U}(n)$, show that $\|BAB^{-1}\| = \|A\|$.
b) For a general element $C \in \mathrm{GL}_n(\mathbb{C})$, what can be said about $\|CAC^{-1}\|$?

1.3. [This problem expands on Remark 1.4 and requires knowledge of the diagonalisation of hermitian matrices.] Let $A \in \mathrm{M}_n(\mathbb{C})$.
a) Show that

$$\begin{aligned} \|A\|^2 &= \sup\{\mathbf{x}^* A^* A\mathbf{x} : \mathbf{x} \in \mathbb{C}^n, \ |\mathbf{x}| = 1\} \\ &= \max\{\mathbf{x}^* A^* A\mathbf{x} : \mathbf{x} \in \mathbb{C}^n, \ |\mathbf{x}| = 1\}. \end{aligned}$$

b) Show that the eigenvalues of A^*A are non-negative real numbers. Deduce that if $\lambda \in \mathbb{R}$ is the largest eigenvalue of A^*A then $\|A\| = \sqrt{\lambda}$ and for any unit eigenvector $\mathbf{v} \in \mathbb{C}^n$ of A^*A for the eigenvalue λ, $\|A\| = |A\mathbf{v}|$.

1.4. If $\{A_r\}_{r\geqslant 0}$ is a sequence of matrices $A_r \in \mathrm{M}_n(\mathbb{k})$, prove the following version of the *ratio test*.

a) If $\displaystyle\lim_{r\to\infty} \frac{\|A_{r+1}\|}{\|A_r\|} < 1$, the series $\sum_{r=0}^{\infty} A_r$ converges in $\mathrm{M}_n(\mathbb{k})$.

b) If $\displaystyle\lim_{r\to\infty} \frac{\|A_{r+1}\|}{\|A_r\|} > 1$, the series $\sum_{r=0}^{\infty} A_r$ diverges in $\mathrm{M}_n(\mathbb{k})$.

c) Develop other convergence tests for $\sum_{r=0}^{\infty} A_r$.

1.5. Suppose that $A \in \mathrm{M}_n(\mathbb{k})$ and $\|A\| < 1$.
a) Show that the series

$$\sum_{r=0}^{\infty} A^r = I + A + A^2 + A^3 + \cdots$$

converges in $\mathrm{M}_n(\mathbb{k})$.
b) Show that $(I - A)$ is invertible and find a formula for $(I - A)^{-1}$.
c) If A is nilpotent (*i.e.*, $A^k = O$ for k large), determine $(I - A)^{-1}$ and $\exp(A)$.

1.6. a) Show that the set of all $n \times n$ real orthogonal matrices $\mathrm{O}(n)$ is compact.
b) Show that the set of all $n \times n$ unitary matrices $\mathrm{U}(n)$ is compact.
c) Show that $\mathrm{GL}_n(\Bbbk)$ and $\mathrm{SL}_n(\Bbbk)$ are not compact if $n \geqslant 2$.
d) Investigate which of the other matrix groups of Section 1.5 are compact.

1.7. Recall that in a topological space X, the *closure* $\overline{U}$ of a subset $U \subseteq X$ is the smallest closed subset $V \subseteq X$ for which $U \subseteq V$. Then $U \subseteq X$ is closed in X if and only if $\overline{U} = U$. It is also useful to note that if $u \in U$, then every open set W containing u intersects U.

a) If G is a matrix group and $H \leqslant G$ is a subgroup, show that the closure $\overline{H} \subseteq G$ of H in G is also a subgroup.
b) Generalise (a) to an arbitrary topological group G.
c) Let
$$\Gamma = \{A \in \mathrm{GL}_n(\mathbb{R}) : \det A \in \mathbb{Q}\} \leqslant \mathrm{GL}_n(\mathbb{R}).$$
Show that Γ not a closed subgroup of $\mathrm{GL}_n(\mathbb{R})$ and find its closure $\overline{\Gamma}$ in $\mathrm{GL}_n(\mathbb{R})$.

1.8. Show that a compact subgroup $G \leqslant \mathrm{GL}_n(\mathbb{R})$ is a matrix group.

1.9. Let G be a compact matrix group and suppose that the matrix subgroup $H \leqslant G$ is *discrete* as a subspace of G, *i.e.*, each singleton subset $\{h\} \subseteq H$ is open in H. Show that H is finite.

1.10. Using Example 1.35, verify each of the following for $n \geqslant 1$.
a) $\mathrm{O}(n)$ is a matrix subgroup of $\mathrm{O}(n+1)$;
b) $\mathrm{SO}(n)$ is a matrix subgroup of $\mathrm{SO}(n+1)$;
c) $\mathrm{U}(n)$ is a matrix subgroup of $\mathrm{U}(n+1)$;
d) $\mathrm{SU}(n)$ is a matrix subgroup of $\mathrm{SU}(n+1)$.

1.11. a) If $A \in \mathrm{Symp}_{2m}(\mathbb{R})$, prove that $\det A = \pm 1$.
b) Prove that $\mathrm{Symp}_2(\mathbb{R}) = \mathrm{SL}_2(\mathbb{R})$.

1.12. Recall Section 1.6 and in particular Proposition 1.44. Verify the following sets of equalities:

$$\begin{aligned}
\rho_n \mathrm{U}(n) &= \mathrm{O}(2n) \cap \rho_n \mathrm{GL}_n(\mathbb{C}) \\
&= \mathrm{O}(2n) \cap \mathrm{Symp}_{2n}(\mathbb{R}) \\
&= \rho_n \mathrm{GL}_n(\mathbb{C}) \cap \mathrm{Symp}_{2n}(\mathbb{R}), \\
\rho'_n \mathrm{U}(n) &= \mathrm{O}(2n) \cap \rho'_n \mathrm{GL}_n(\mathbb{C}) \\
&= \mathrm{O}(2n) \cap \mathrm{Symp}'_{2n}(\mathbb{R}) \\
&= \rho'_n \mathrm{GL}_n(\mathbb{C}) \cap \mathrm{Symp}'_{2n}(\mathbb{R}),
\end{aligned}$$

1.13. Let (X_1, ρ_1), (X_2, ρ_2) be two metric spaces. Consider the function

$$\rho\colon (X_1 \times X_2) \times (X_1 \times X_2) \longrightarrow \mathbb{R}^+;$$
$$\rho((x_1, x_2), (y_1, y_2)) = \sqrt{\rho_1(x_1, y_1)^2 + \rho_2(x_2, y_2)^2}.$$

a) Show that $(X_1 \times X_2, \rho)$ is a metric space whose open sets are the same as those of the product topology.
b) Show that a sequence $\{(x_{1,r}, x_{2,r})\}_{r\geqslant 0}$ converges (*i.e.*, has a limit) in $X_1 \times X_2$ if and only if the sequences $\{(x_{1,r})\}_{r\geqslant 0}$, $\{(x_{2,r})\}_{r\geqslant 0}$ converge in X_1 and X_2 respectively.

1.14. Let $\mu\colon G \times X \longrightarrow X$ be a continuous group action as in Definition 1.61 and assume that X is Hausdorff (for example a metric space).
a) If $x \in X$, show that the stabiliser

$$\mathrm{Stab}_G(x) = \{g \in G : gx = x\}$$

is a closed subgroup of G.
b) If $W \subseteq X$ is a closed subset, show that

$$\mathrm{Stab}_G(W) = \{g \in G : gW = W\} \leqslant G, \qquad \bigcap_{w \in W} \mathrm{Stab}_G(w) \leqslant G,$$

are closed subgroups, where $gW = \{gw : w \in W\}$.

1.15. Let $\mathbb{k} = \mathbb{R}$ or $\mathbb{C}$.
a) Making use of a suitable metric ρ on the product space $\mathrm{M}_n(\mathbb{k}) \times \mathbb{k}^n$, show that the product map

$$\varphi\colon \mathrm{M}_n(\mathbb{k}) \times \mathbb{k}^n \longrightarrow \mathbb{k}^n; \quad \varphi(A, \mathbf{x}) = A\mathbf{x},$$

is continuous.
b) Let $G \leqslant \mathrm{GL}_n(\mathbb{k})$ be a matrix subgroup. By restricting the metric ρ and product map φ of (a) to the subset $G \times \mathbb{k}^n$, consider the resulting continuous group action of G on $\mathbb{k}^n$. Show that the stabiliser of $\mathbf{x} \in \mathbb{k}^n$,

$$\mathrm{Stab}_G(\mathbf{x}) = \{A \in G : A\mathbf{x} = \mathbf{x}\},$$

is a matrix subgroup of G. More generally, if $X \subseteq \mathbb{k}^n$ is a closed subset, show that

$$\mathrm{Stab}_G(X) = \{A \in G : AX = X\}$$

is a matrix subgroup of G, where $AX = \{A\mathbf{x} : \mathbf{x} \in X\}$.
c) For the standard basis vector $\mathbf{e}_n \in \mathbb{k}^n$ and

$$X = \{t\mathbf{e}_n : t \in \mathbb{R}\} \subseteq \mathbb{k}^n,$$

determine $\mathrm{Stab}_G(\mathbf{e}_n)$ and $\mathrm{Stab}_G(X)$ for each of the following matrix subgroups $G \leqslant \mathrm{GL}_n(\mathbb{R})$: $\mathrm{GL}_n(\mathbb{R})$, $\mathrm{SL}_n(\mathbb{R})$, $\mathrm{O}(n)$, $\mathrm{SO}(n)$.

1.16. Let $n \geqslant 1$. Consider the $\mathbb{C}$-linear action of $\mathrm{SU}(2)$ on the $(n+1)$-dimensional complex vector space of $V_n = \mathbb{C}[x, y]_n$ introduced in Example 1.63. Let $\varphi_n \colon \mathrm{SU}(2) \longrightarrow \mathrm{GL}_{n+1}(\mathbb{C})$ be the continuous homomorphism obtained by working with matrices relative to the basis of monomials in x and y ordered by increasing degree in x.
a) Determine φ_n explicitly for small values of n.
b) Determine the stabiliser of $x^n \in V_n$.
c) When $n = 2$, determine the stabiliser of $xy \in V_2$.
d) Show that

$$\ker \varphi_n = \begin{cases} \{I\} & \text{if } n \text{ is odd,} \\ \{I, -I\} & \text{if } n \text{ is even.} \end{cases}$$

1.17. For $\mathbb{k} = \mathbb{R}$ or $\mathbb{C}$, let (V, ν) be a finite dimensional normed $\mathbb{k}$-vector space.
a) If (V', ν') is another finite dimensional normed $\mathbb{k}$-vector space, show that any $\mathbb{k}$-linear transformation $\varphi \colon V \longrightarrow V'$ is continuous.
b) If $W \subseteq V$ and $W' \subseteq V$ are $\mathbb{k}$-vector subspaces and $W' \subseteq V$ is a linear complement of W (*i.e.*, $V = W + W'$ and $W \cap W' = \{0\}$), show that the projection maps $\pi \colon V \longrightarrow W$ and $\pi' \colon V \longrightarrow W'$ defined by

$$\pi(w + w') = w, \quad \pi'(w + w') = w' \qquad (w \in W,\ w' \in W'),$$

are continuous. Use this to give another proof of Proposition 1.59.

2
Exponentials, Differential Equations and One-parameter Subgroups

The matrix versions of the familiar real and complex exponential and logarithm functions are fundamental for the study of many aspects of matrix group theory, particularly the *one-parameter subgroups.* Indeed, the *matrix exponential function* provides the link between the *Lie algebra* of a matrix group and the group itself. In the case of a compact connected matrix group, the exponential is even surjective, allowing a parametrisation of such a group by a region in $\mathbb{R}^n$ for some n; see Chapter 10 for details. Just as in the theory of ordinary differential equations, matrix exponential functions also play a central rôle in the theory of certain types of differential equations for matrix-valued functions and these are important in many applications of Lie theory.

2.1 The Matrix Exponential and Logarithm

Throughout this section, we will assume that $\mathbb{k} = \mathbb{R}$ or $\mathbb{C}$. The power series

$$\operatorname{Exp}(X) = \sum_{n\geqslant 0} \frac{1}{n!} X^n, \quad \operatorname{Log}(X) = \sum_{n\geqslant 1} \frac{(-1)^{n-1}}{n} X^n,$$

have radii of convergence (r. o. c.) ∞ and 1 respectively. If $z \in \mathbb{C}$, the series $\operatorname{Exp}(z)$, $\operatorname{Log}(z)$ converge absolutely whenever $|z| <$ r. o. c..

Let $A \in M_n(\Bbbk)$. The matrix-valued series

$$\operatorname{Exp}(A) = \sum_{n\geqslant 0} \frac{1}{n!} A^n = I + A + \frac{1}{2!}A^2 + \frac{1}{3!}A^3 + \cdots,$$

$$\operatorname{Log}(A) = \sum_{n\geqslant 1} \frac{(-1)^{n-1}}{n} A^n = A - \frac{1}{2}A^2 + \frac{1}{3}A^3 - \frac{1}{4}A^4 + \cdots,$$

converge whenever $\|A\| <$ r. o. c.. So $\operatorname{Exp}(A)$ makes sense for every $A \in M_n(\Bbbk)$, while $\operatorname{Log}(A)$ converges if $\|A\| < 1$ but may not do so otherwise.

Proposition 2.1

Let $A \in M_n(\Bbbk)$.
i) For $u, v \in \mathbb{C}$, $\operatorname{Exp}((u+v)A) = \operatorname{Exp}(uA)\operatorname{Exp}(vA)$.
ii) $\operatorname{Exp}(A) \in GL_n(\Bbbk)$ and $\operatorname{Exp}(A)^{-1} = \operatorname{Exp}(-A)$.

Proof

(i) By expanding the first series we obtain

$$\begin{aligned}\operatorname{Exp}((u+v)A) &= \sum_{n\geqslant 0} \frac{1}{n!}(u+v)^n A^n \\ &= \sum_{n\geqslant 0} \frac{(u+v)^n}{n!} A^n.\end{aligned}$$

By a sequence of obvious manipulations that are justified since these series are all absolutely convergent,

$$\begin{aligned}\operatorname{Exp}(uA)\operatorname{Exp}(vA) &= \left(\sum_{r\geqslant 0} \frac{u^r}{r!} A^r\right)\left(\sum_{s\geqslant 0} \frac{v^s}{s!} A^s\right) \\ &= \sum_{\substack{r\geqslant 0 \\ s\geqslant 0}} \frac{u^r v^s}{r!s!} A^{r+s} \\ &= \sum_{n\geqslant 0} \left(\sum_{r=0}^{n} \frac{u^r v^{n-r}}{r!(n-r)!}\right) A^n \\ &= \sum_{n\geqslant 0} \frac{1}{n!} \left(\sum_{r=0}^{n} \binom{n}{r} u^r v^{n-r}\right) A^n \\ &= \sum_{n\geqslant 0} \frac{(u+v)^n}{n!} A^n \\ &= \operatorname{Exp}((u+v)A).\end{aligned}$$

(ii) From part (i),

$$I = \mathrm{Exp}(O) = \mathrm{Exp}((1 + (-1))A) = \mathrm{Exp}(A)\,\mathrm{Exp}(-A),$$

so $\mathrm{Exp}(A)$ has multiplicative inverse $\mathrm{Exp}(-A)$. □

Using these series we can define the matrix version of the *exponential function*

$$\exp\colon \mathrm{M}_n(\Bbbk) \longrightarrow \mathrm{GL}_n(\Bbbk); \quad \exp(A) = \mathrm{Exp}(A).$$

Proposition 2.2

If $A, B \in \mathrm{M}_n(\Bbbk)$ commute then

$$\exp(A + B) = \exp(A)\exp(B).$$

Proof

Again we expand the series and perform a sequence of manipulations, all of which are justified because these series are absolutely convergent. The result is

$$\begin{aligned}
\exp(A)\exp(B) &= \left(\sum_{r\geqslant 0}\frac{1}{r!}A^r\right)\left(\sum_{s\geqslant 0}\frac{1}{s!}B^s\right)\\
&= \sum_{\substack{r\geqslant 0\\ s\geqslant 0}}\frac{1}{r!s!}A^rB^s\\
&= \sum_{n\geqslant 0}\left(\sum_{r=0}^{n}\frac{1}{r!(n-r)!}A^rB^{n-r}\right)\\
&= \sum_{n\geqslant 0}\frac{1}{n!}\left(\sum_{r=0}^{n}\binom{n}{r}A^rB^{n-r}\right)\\
&= \sum_{n\geqslant 0}\frac{1}{n!}(A+B)^n\\
&= \exp(A+B).
\end{aligned}$$

Of course the identity

$$\sum_{r=0}^{n}\binom{n}{r}A^rB^{n-r} = (A+B)^n$$

depends crucially on the fact that A and B commute. □

We define the *logarithm function* by

$$\log\colon \mathrm{N}_{\mathrm{M}_n(\mathbb{k})}(I;1) \longrightarrow \mathrm{M}_n(\mathbb{k}); \quad \log(A) = \mathrm{Log}(A-I).$$

So for $\|A-I\|<1$,

$$\log(A) = \sum_{n\geqslant 1} \frac{(-1)^{n-1}}{n}(A-I)^n.$$

Proposition 2.3

The functions exp and log have the following properties.
i) If $\|A-I\|<1$, then $\exp(\log(A)) = A$.
ii) If $\|\exp(B)-I\|<1$, then $\log(\exp(B)) = B$.

Proof

Both parts follow from two formal identities between power series, namely

$$\sum_{m\geqslant 0} \frac{1}{m!}\left(\sum_{n\geqslant 1} \frac{(-1)^{n-1}}{n}(X-1)^n\right)^m = X,$$

$$\sum_{n\geqslant 1} \frac{(-1)^{n-1}}{n}\left(\sum_{m\geqslant 1} \frac{1}{m!}X^m\right)^n = X.$$

These are proved by comparing coefficients. □

The functions exp and log are continuous and in fact infinitely differentiable on their domains. By continuity of exp at O, there is a $\delta_1 > 0$ such that

$$\mathrm{N}_{\mathrm{M}_n(\mathbb{k})}(O;\delta_1) \subseteq \exp^{-1} \mathrm{N}_{\mathrm{GL}_n(\mathbb{k})}(I;1).$$

In fact we can actually take $\delta_1 = \log 2$ since

$$\exp \mathrm{N}_{\mathrm{M}_n(\mathbb{k})}(O;r) \subseteq \mathrm{N}_{\mathrm{M}_n(\mathbb{k})}(I;e^r-1).$$

Hence we have the following results.

Proposition 2.4

The exponential function exp is injective when restricted to the open subset $\mathrm{N}_{\mathrm{M}_n(\mathbb{k})}(O;\log 2) \subseteq \mathrm{M}_n(\mathbb{k})$, hence it is locally a diffeomorphism at O with local inverse log.

It will sometimes be useful to have a formula for the derivative of exp at an arbitrary $A \in \mathrm{M}_n(\mathbb{k})$. When $B \in \mathrm{M}_n(\mathbb{k})$ commutes with A,

$$\frac{\mathrm{d}}{\mathrm{d}t}_{|t=0} \exp(A+tB) = \lim_{h\to 0} \frac{1}{h}(\exp(A+hB) - \exp(A))$$
$$= \exp(A)B = B\exp(A). \tag{2.1}$$

The general situation is more complicated.

For a variable X consider the series

$$F(X) = \sum_{r\geqslant 0} \frac{1}{(k+1)!} X^k = \frac{\exp(X) - 1}{X}$$

which has infinite radius of convergence. If we have a linear operator Φ on $\mathrm{M}_n(\mathbb{C})$ we can apply the convergent series of operators

$$F(\Phi) = \sum_{r\geqslant 0} \frac{1}{(k+1)!} \Phi^k$$

to elements of $\mathrm{M}_n(\mathbb{C})$. In particular we can consider

$$\Phi(C) = AC - CA = \operatorname{ad} A(C),$$

where

$$\operatorname{ad} A\colon \mathrm{M}_n(\mathbb{C}) \longrightarrow \mathrm{M}_n(\mathbb{C}); \quad \operatorname{ad} A(C) = AC - CA,$$

is viewed as a $\mathbb{C}$-linear operator which acts on $\mathrm{M}_n(\mathbb{C})$ by the *adjoint action* of A. Then

$$F(\operatorname{ad} A)(C) = \sum_{r\geqslant 0} \frac{1}{(k+1)!} (\operatorname{ad} A)^k(C).$$

Proposition 2.5

For $A, B \in \mathrm{M}_n(\mathbb{C})$ we have

$$\frac{\mathrm{d}}{\mathrm{d}t}_{|t=0} \exp(A+tB) = F(\operatorname{ad} A)(B) \exp(A).$$

In particular, if $A = O$ or more generally if $AB = BA$,

$$\frac{\mathrm{d}}{\mathrm{d}t}_{|t=0} \exp(A+tB) = B\exp(A).$$

Proof

The following proof is based on that outlined by Segal in [4], page 80.

We begin by observing that if $D = \dfrac{\mathrm{d}}{\mathrm{d}\,s}$ and $f(s)$ is a smooth function of the real variable s, then

$$F(D)_{|_{s=0}} f(s) = \int_0^1 f(s)\,\mathrm{d}\,s. \tag{2.2}$$

This holds since the Taylor expansion of a smooth function g satisfies

$$\sum_{k\geqslant 1} \frac{1}{k!} D^k g(s) = g(s+1) - g(s),$$

hence taking $g(s) = \int f(s)\,\mathrm{d}\,s$ to be an indefinite integral of f we obtain

$$\sum_{k\geqslant 0} \frac{1}{(k+1)!} D^k f(s) = g(s+1) - g(s).$$

Evaluating at $s = 0$ gives Equation (2.2).

The matrix-valued function

$$\varphi(s) = \exp(sA) B \exp((1-s)A)$$

can be shown (for example, using Theorem 2.12) to satisfy

$$\varphi(s) = \exp(sA) B \exp(-sA) \exp(A) = \exp(s \operatorname{ad} A)(B) \exp(A).$$

So

$$F(D)(\varphi(s)) = \left(\sum_{k\geqslant 0} \frac{((s+1)^{k+1} - s^{k+1})}{(k+1)!} (\operatorname{ad} A)^k \right) (B) \exp(A),$$

giving

$$\begin{aligned} F(D)(\varphi(s))_{|_{s=0}} &= \left(\sum_{k\geqslant 0} \frac{1}{(k+1)!} (\operatorname{ad} A)^k \right) (B) \exp(A) \\ &= F(\operatorname{ad} A)(B) \exp(A). \end{aligned}$$

We also have

$$\frac{\mathrm{d}}{\mathrm{d}\,t}_{|_{t=0}} \exp(A + tB) = \int_0^1 \varphi(s)\,\mathrm{d}\,s,$$

which is obtained by expanding the left hand side as a power series in $A + tB$ and differentiating, then using the identity

$$\int_0^1 \frac{s^m (1-s)^n}{m!\,n!}\,\mathrm{d}\,s = \frac{1}{(m+n+1)!}$$

for $m, n \geqslant 0$ to identify this with the right hand side. The desired formula now follows by combining the last two results. □

2.2 Calculating Exponentials and Jordan Form

Since exponentials of matrices occur throughout this subject matter, it is important to be able to calculate them. It is an easy exercise to show that for $A \in \mathrm{M}_n(\mathbb{C})$ and $B \in \mathrm{GL}_n(\mathbb{C})$,

$$\exp(BAB^{-1}) = B\exp(A)B^{-1}.$$

When A is *diagonalisable*, *i.e.*, $A = C\,\mathrm{diag}(\lambda_1, \ldots, \lambda_n)C^{-1}$ for some $C \in \mathrm{GL}_n(\mathbb{C})$, we have

$$\begin{aligned}\exp(A) &= C\exp(\mathrm{diag}(\lambda_1, \ldots, \lambda_n))C^{-1}\\ &= C\,\mathrm{diag}(e^{\lambda_1}, \ldots, e^{\lambda_n})C^{-1},\end{aligned} \tag{2.3}$$

since

$$\exp(\mathrm{diag}(\lambda_1, \ldots, \lambda_n)) = \mathrm{diag}(e^{\lambda_1}, \ldots, e^{\lambda_n}).$$

This means that the problem of calculating the exponential of a diagonalisable matrix is solved once an explicit diagonalisation is found. Many important types of matrices are indeed diagonalisable, including (skew) symmetric, (skew) hermitian, orthogonal, unitary and normal matrices. However, there are also many non-diagonalisable matrices. The general situation is best discussed in terms of the *Jordan form*, a good reference for which is Strang [28], although many books on linear algebra contain accounts of this material. Similar results hold over any algebraically closed field but we work over the field of complex numbers $\mathbb{C}$.

We start by recalling that for a matrix $A \in \mathrm{M}_n(\mathbb{C})$, the *characteristic polynomial* of A is the monic polynomial $\mathrm{char}_A(X) \in \mathbb{C}[X]$ of degree n given by

$$\mathrm{char}_A(X) = \det(XI_n - A).$$

We will sometimes write

$$\mathrm{char}_A(X) = X^n + c_{n-1}(A)X^{n-1} + \cdots + c_1(A)X + c_0(A),$$

where $c_k(A) \in \mathbb{C}$. We recall the following important result; see [5, 15, 16, 28].

Theorem 2.6 (Cayley–Hamilton Theorem)

A complex matrix $A \in \mathrm{M}_n(\mathbb{C})$ satisfies its own characteristic polynomial, *i.e.*,

$$\mathrm{char}_A(A) = A^n + c_{n-1}(A)A^{n-1} + \cdots + c_1(A)A + c_0(A)I_n = O_n.$$

We also recall the *minimal polynomial* of A, $\min_A(X) \in \mathbb{C}[X]$. This is the non-zero monic polynomial of minimal degree for which

$$\min_A(A) = O_n.$$

This always exists and has the following property.

Proposition 2.7

If $f(X) \in \mathbb{C}[X]$ satisfies $f(A) = O_n$ then $\min_A(X) \mid f(X)$, *i.e.*, there is a $g(X) \in \mathbb{C}[X]$ such that $\min_A(X)g(X) = f(X)$.

Corollary 2.8

$\min_A(X) \mid \mathrm{char}_A(X)$, hence $\deg \min_A(X) \leqslant n$.

The complex roots of $\mathrm{char}_A(X)$ are the eigenvalues of A, so every root of $\min_A(X)$ is an eigenvalue of A. In fact the converse is also true. So if the *distinct* eigenvalues of A are $\lambda_1, \ldots, \lambda_d$ and

$$\mathrm{char}_A(X) = (X - \lambda_1)^{m_1} \cdots (X - \lambda_d)^{m_d}$$

with $m_j \geqslant 1$, then

$$\min_A(X) = (X - \lambda_1)^{m'_1} \cdots (X - \lambda_d)^{m'_d}$$

where $m_j \geqslant m'_j \geqslant 1$.

For $\lambda \in \mathbb{C}$ and $r \geqslant 1$, we have the *Jordan block matrix*

$$J(\lambda, r) = \begin{bmatrix} \lambda & 1 & 0 & \cdots & 0 \\ 0 & \lambda & 1 & \cdots & 0 \\ \ddots & \ddots & \ddots & \ddots & \ddots \\ \vdots & \cdots & 0 & \lambda & 1 \\ 0 & \cdots & & 0 & \lambda \end{bmatrix} \in \mathrm{M}_r(\mathbb{C}).$$

The characteristic polynomial of $J(\lambda, r)$ is

$$\mathrm{char}_{J(\lambda,r)}(X) = \det(XI_r - J(\lambda, r)) = (X - \lambda)^r \tag{2.4a}$$

and by the Cayley–Hamilton Theorem 2.6 (or by direct calculation),

$$(J(\lambda, r) - \lambda I_r)^r = O_r.$$

Notice that

$$(J(\lambda, r) - \lambda I_r)^{r-1} \mathbf{e}_r = \mathbf{e}_1,$$

which implies that

$$(J(\lambda, r) - \lambda I_r)^{r-1} \neq O_r.$$

Hence we also have

$$\min_{J(\lambda,r)}(X) = (X - \lambda)^r. \tag{2.4b}$$

The main result on Jordan form is the following.

Theorem 2.9

Let $A \in \mathrm{M}_n(\mathbb{C})$. Then there is an invertible matrix $P \in \mathrm{GL}_n(\mathbb{C})$ for which $A = PJ(A)P^{-1}$ and $J(A) \in \mathrm{M}_n(\mathbb{C})$ is the matrix having block form

$$J(A) = \begin{bmatrix} J(\lambda_1, r_1) & O & O & \cdots & O \\ O & J(\lambda_2, r_2) & O & \cdots & O \\ \ddots & \ddots & \ddots & \ddots & \ddots \\ O & \cdots & \cdots & O & J(\lambda_m, r_m) \end{bmatrix}.$$

This form is unique except for the order in which the Jordan blocks $J(\lambda_j, r_j)$ occur.

Corollary 2.10

A is diagonalisable if and only if $A = PJ(A)P^{-1}$ where $J(A)$ has $r_1 = \cdots = r_m = 1$.

In this *Jordan form* for A, the λ_j are eigenvalues of A and in fact the characteristic polynomial of A is

$$\mathrm{char}_A(X) = (X - \lambda_1)^{r_1} \cdots (X - \lambda_m)^{r_m}. \tag{2.5}$$

However the minimal polynomial is more complicated to specify. First notice that in general the λ_j are not all distinct. For example, the matrix

$$A = \begin{bmatrix} 7 & 0 & 0 & 0 \\ 0 & 7 & 1 & 0 \\ 0 & 0 & 7 & 0 \\ 0 & 0 & 0 & 5 \end{bmatrix} = \begin{bmatrix} J(7,1) & O & 0 \\ O & J(7,2) & O \\ 0 & O & J(5,1) \end{bmatrix}$$

has two Jordan blocks for the eigenvalue 7. In this case, the characteristic polynomial is

$$\mathrm{char}_A(X) = (X-7)^1(X-7)^2(X-5) = (X-7)^3(X-5),$$

while the minimal polynomial is

$$\min_A(X) = (X-7)^2(X-5).$$

Let us calculate the exponential of this matrix. We have

$$\exp(A) = \begin{bmatrix} \exp(J(7,1)) & O & 0 \\ O & \exp(J(7,2)) & O \\ 0 & O & \exp(J(5,1)) \end{bmatrix},$$

so it suffices to determine $\exp(J(7,2))$. Since for $k \geqslant 1$

$$J(7,2)^k = \begin{bmatrix} 7^k & k7^{k-1} \\ 0 & 7^k \end{bmatrix},$$

we find

$$\begin{aligned}
\exp(J(7,2)) &= \sum_{k=0}^{\infty} \frac{1}{k!} J(7,2)^k \\
&= I_2 + \sum_{k=1}^{\infty} \frac{1}{k!} \begin{bmatrix} 7^k & k7^{k-1} \\ 0 & 7^k \end{bmatrix} \\
&= \operatorname{diag}(e^7, e^7) + \begin{bmatrix} 0 & \sum_{k=1}^{\infty} 7^{k-1}/(k-1)! \\ 0 & 0 \end{bmatrix} \\
&= e^7 I_2 + e^7 \begin{bmatrix} 0 & 1 \\ 0 & 0 \end{bmatrix} \\
&= e^7 \begin{bmatrix} 1 & 1 \\ 0 & 1 \end{bmatrix}.
\end{aligned}$$

So we have

$$\exp(A) = \begin{bmatrix} e^7 & 0 & 0 & 0 \\ 0 & e^7 & e^7 & 0 \\ 0 & 0 & e^7 & 0 \\ 0 & 0 & 0 & e^5 \end{bmatrix}.$$

In the general case, for each eigenvalue λ_j of A, the factor $(X - \lambda_j)^{m'_j}$ of $\min_A(X)$ is determined by taking

$$m'_j = \max\{r_i : \lambda_i = \lambda_j\}. \tag{2.6}$$

2.3 Differential Equations in Matrices

Definition 2.11

A *differentiable curve* in $\mathrm{M}_n(\Bbbk)$ is a function

$$\alpha\colon (a,b) \longrightarrow \mathrm{M}_n(\Bbbk)$$

for which the derivative $\alpha'(t)$ exists for each $t \in (a,b)$. Here $\alpha'(t)$ is defined as an element of $\mathrm{M}_n(\Bbbk)$ by

$$\alpha'(t) = \lim_{s\to t} \frac{1}{(s-t)}(\alpha(s)-\alpha(t)),$$

provided this limit exists.

Let $A \in \mathrm{M}_n(\mathbb{R})$. Let $(a,b) \subseteq \mathbb{R}$ be the open interval with endpoints a, b and $a < b$; we will usually assume that $a < 0 < b$. We will use the notation

$$\alpha'(t) = \frac{\mathrm{d}}{\mathrm{d}t}\alpha(t)$$

for the derivative of α.

Consider the first order differential equation

$$\alpha'(t) = \alpha(t)A, \tag{2.7}$$

in which $\alpha\colon (a,b) \longrightarrow \mathrm{M}_n(\mathbb{R})$ is assumed to be a differentiable curve.

If $n = 1$ then taking $A = a$ to be a non-zero real number we know that the general solution is $\alpha(t) = ce^{at}$ where $\alpha(0) = c$. Hence there is a unique solution subject to this boundary condition, namely the function of t given by the power series

$$\alpha(t) = \sum_{k\geqslant 0} \frac{ca^k}{k!} t^k.$$

This is indicative of the general case $n \geqslant 1$.

Theorem 2.12

For $A, C \in \mathrm{M}_n(\mathbb{R})$ with A non-zero, and $a < 0 < b$, the differential equation of (2.7) has a unique solution $\alpha\colon (a,b) \longrightarrow \mathrm{M}_n(\mathbb{R})$ for which $\alpha(0) = C$. Furthermore, if C is invertible then so is $\alpha(t)$ for $t \in (a,b)$, hence $\alpha\colon (a,b) \longrightarrow \mathrm{GL}_n(\mathbb{R})$.

Proof

First we will solve the equation subject to the boundary condition $\alpha(0) = I$. By Section 2.1, for each $t \in (a,b)$, the series

$$\sum_{k\geqslant 0} \frac{t^k}{k!} A^k = \sum_{k\geqslant 0} \frac{1}{k!}(tA)^k = \exp(tA)$$

converges, so the function

$$\alpha\colon (a,b) \longrightarrow \mathrm{M}_n(\mathbb{R}); \quad \alpha(t) = \exp(tA),$$

is defined and differentiable with

$$\alpha'(t) = \sum_{k\geqslant 1} \frac{t^{k-1}}{(k-1)!} A^k = \exp(tA)A = A\exp(tA).$$

Hence α satisfies the above differential equation with boundary condition $\alpha(0) = I$. Notice also that whenever $s, t, (s+t) \in (a,b)$,

$$\alpha(s+t) = \alpha(s)\alpha(t).$$

In particular, this shows that $\alpha(t)$ is always invertible with $\alpha(t)^{-1} = \alpha(-t)$.

One solution subject to $\alpha(0) = C$ is easily seen to be $\alpha(t) = C\exp(tA)$. If β is a second such solution then $\gamma(t) = \beta(t)\exp(-tA)$ satisfies

$$\begin{aligned}\gamma'(t) &= \beta'(t)\exp(-tA) + \beta(t)\frac{\mathrm{d}}{\mathrm{d}t}\exp(-tA)\\ &= \beta'(t)\exp(-tA) - \beta(t)\exp(-tA)A\\ &= \beta(t)A\exp(-tA) - \beta(t)\exp(-tA)A\\ &= O.\end{aligned}$$

Hence $\gamma(t)$ is a constant function with $\gamma(t) = \gamma(0) = C$. Thus $\beta(t) = C\exp(tA)$, and this is the unique solution subject to $\beta(0) = C$. If C is invertible then so is $C\exp(tA)$ for all t. □

2.4 One-parameter Subgroups in Matrix Groups

Let $G \leqslant \mathrm{GL}_n(\Bbbk)$ be a matrix group. Since $G \subseteq \mathrm{M}_n(\Bbbk)$, the next definition is an obvious modification of Definition 2.11, in particular the derivative is defined by

$$\gamma'(t) = \lim_{s\to t} \frac{1}{(s-t)}(\gamma(s) - \gamma(t)) \in \mathrm{M}_n(\Bbbk),$$

provided this limit exists.

Definition 2.13

A *differentiable curve* in G is a function $\gamma\colon (a,b) \longrightarrow G$ for which the derivative $\gamma'(t)$ exists at each $t \in (a,b)$.

Since $G \subseteq \mathrm{M}_n(\Bbbk)$ such a curve is also a curve in $\mathrm{M}_n(\Bbbk)$. We will usually assume that $a < 0 < b$ when considering such curves.

Now suppose that $\varepsilon > 0$ or $\varepsilon = \infty$.

Definition 2.14

A *one-parameter semigroup in G* is a continuous function $\gamma\colon (-\varepsilon,\varepsilon) \longrightarrow G$ which is differentiable at 0 and also satisfies

$$\gamma(s+t) = \gamma(s)\gamma(t)$$

whenever $s, t, (s+t) \in (-\varepsilon, \varepsilon)$. We will refer to the last condition as the *homomorphism property.*

If $\varepsilon = \infty$ then $\gamma\colon \mathbb{R} \longrightarrow G$ is called a *one-parameter group in G* or *one-parameter subgroup of G.*

Notice that for a one-parameter semigroup in G, $\gamma(0) = I$.

Proposition 2.15

Let $\gamma\colon (-\varepsilon,\varepsilon) \longrightarrow G$ be a one-parameter semigroup in G. Then for every $t \in (-\varepsilon,\varepsilon)$, γ is differentiable at t and

$$\gamma'(t) = \gamma'(0)\gamma(t) = \gamma(t)\gamma'(0).$$

Proof

For small $h \in \mathbb{R}$,

$$\gamma(h)\gamma(t) = \gamma(h+t) = \gamma(t+h) = \gamma(t)\gamma(h).$$

Hence

$$\begin{aligned}\gamma'(t) &= \lim_{h\to 0} \frac{1}{h}(\gamma(t+h) - \gamma(t)) \\ &= \lim_{h\to 0} \frac{1}{h}(\gamma(h) - I)\gamma(t) \\ &= \gamma'(0)\gamma(t),\end{aligned}$$

and similarly $\gamma'(t) = \gamma(t)\gamma'(0)$. □

Of course, part of the importance of this result is the implication that γ is a differentiable curve in G even though we only assumed it was continuous.

Proposition 2.16

Let $\gamma\colon (-\varepsilon, \varepsilon) \longrightarrow G$ be a one-parameter semigroup in G. Then there is a unique extension to a one-parameter group $\widetilde{\gamma}\colon \mathbb{R} \longrightarrow G$ in G, *i.e.*, a function $\widetilde{\gamma}$ for which $\widetilde{\gamma}(t) = \gamma(t)$ for all $t \in (-\varepsilon, \varepsilon)$.

Proof

Let $t \in \mathbb{R}$. Then for a large enough natural number m, $t/m \in (-\varepsilon, \varepsilon)$, hence $\gamma(t/m), \gamma(t/m)^m \in G$. Similarly, for a second such natural number n, $\gamma(t/n), \gamma(t/n)^n \in G$. Since $mn \geqslant m$ and $mn \geqslant n$, we also have $t/mn \in (-\varepsilon, \varepsilon)$ and therefore

$$\begin{aligned} \gamma(t/n)^n &= \gamma(mt/mn)^n \\ &= \gamma(t/mn)^{mn} \\ &= \gamma(nt/mn)^m \\ &= \gamma(t/m)^m. \end{aligned}$$

Thus $\gamma(t/n)^n = \gamma(t/m)^m$, which shows that we obtain a well-defined element of G for every real number t. This defines a function

$$\widetilde{\gamma}\colon \mathbb{R} \longrightarrow G; \quad \widetilde{\gamma}(t) = \gamma(t/n)^n \quad \text{for large } n.$$

It is easy to see that $\widetilde{\gamma}$ is a one-parameter group in G. □

We can now determine the form of all one-parameter groups in G.

Theorem 2.17

Let $\gamma\colon \mathbb{R} \longrightarrow G$ be a one-parameter group in G. Then it has the form

$$\gamma(t) = \exp(tA)$$

for some $A \in \mathrm{M}_n(\Bbbk)$.

Proof

Let $A = \gamma'(0)$. By Proposition 2.15, γ satisfies the differential equation

$$\gamma'(t) = \gamma(t)A, \quad \gamma(0) = I.$$

By Theorem 2.12, this equation has the unique solution $\gamma(t) = \exp(tA)$. □

Remark 2.18

We cannot yet reverse this process and decide for which $A \in \mathrm{M}_n(\Bbbk)$ the one-parameter group

$$\gamma\colon \mathbb{R} \longrightarrow \mathrm{GL}_n(\Bbbk); \quad \gamma(t) = \exp(tA)$$

actually takes values in G. The answer involves the *Lie algebra* of G which will be defined in Chapter 3. Notice that we also have the curious phenomenon that although the definition of a one-parameter group only involves first order differentiability, the general form $\exp(tA)$ is always infinitely differentiable and indeed *analytic* as a function of t. This is an important characteristic of much of Lie theory, namely that conditions of first order differentiability (and sometimes merely continuity) often lead to much stronger conclusions.

2.5 One-parameter Subgroups and Differential Equations

In this section we show how ideas about one-parameter subgroups can be applied to solving differential equations. A good source for applications of linear algebra is the book of Strang [28].

Consider the following differential equation for a function $\mathbf{v}\colon \mathbb{R} \longrightarrow \mathbb{C}^n$ which is assumed to be differentiable:

$$\mathbf{v}'(t) = A\mathbf{v}(t), \quad \mathbf{v}(0) = \mathbf{v}_0. \tag{2.8}$$

Here $A \in \mathrm{M}_n(\mathbb{C})$. In examples we will often find that solutions take values in $\mathbb{R}^n$, however it is more convenient to work over $\mathbb{C}$ since eigenvalues and eigenvectors are often involved in calculations.

If $\mathbf{v}_0 \neq \mathbf{0}$, then we can look for a solution of the form

$$\mathbf{v}(t) = \alpha(t)\mathbf{v}_0, \tag{2.9}$$

where $\alpha\colon \mathbb{R} \longrightarrow \mathrm{GL}_n(\Bbbk)$ is a one-parameter subgroup as discussed in Section 2.4 and satisfying

$$\alpha'(t) = \alpha(t)A.$$

By Theorem 2.17, for $t \in \mathbb{R}$ we have

$$\alpha(t) = \exp(tA),$$

hence a solution of Equation (2.8) is

$$\mathbf{v}(t) = \exp(tA)\mathbf{v}_0. \tag{2.10}$$

In fact, this is the unique solution.

Example 2.19

For a skew symmetric matrix $S \in \mathrm{M}_n(\mathbb{R})$ and non-zero $\mathbf{v}_0 \in \mathbb{R}^n$, the differential equation

$$\mathbf{v}'(t) = S\mathbf{v}(t), \quad \mathbf{v}(0) = \mathbf{v}_0.$$

has a solution for which

$$|\mathbf{v}(t)| = |\mathbf{v}_0| \quad (t \in \mathbb{R}).$$

Proof

By Equation (3.13), for all $t \in \mathbb{R}$ we have $\exp(tS) \in \mathrm{SO}(n)$, hence

$$|\mathbf{v}(t)| = |\exp(tS)\mathbf{v}(0)| = |\mathbf{v}(0)|.$$

Notice that since $|\mathbf{v}(t)|^2 = |\mathbf{v}(0)|^2$ is a constant,

$$\begin{aligned} 2\mathbf{v}(t)\cdot\mathbf{v}'(t) &= \mathbf{v}(t)\cdot\mathbf{v}'(t) + \mathbf{v}'(t)\cdot\mathbf{v}(t) \\ &= \frac{\mathrm{d}}{\mathrm{d}t}(\mathbf{v}(t)\cdot\mathbf{v}(t)) \\ &= \frac{\mathrm{d}}{\mathrm{d}t}|\mathbf{v}(t)|^2 \\ &= 0. \end{aligned}$$

This shows that $\mathbf{v}'(t)$ is tangent to the sphere centred at the origin and of radius $|\mathbf{v}(0)|$ at the point $\mathbf{v}(t)$. □

Here is an explicit example of this type.

Example 2.20

When $n = 3$ and $S = \begin{bmatrix} 0 & 1 & 0 \\ -1 & 0 & 2 \\ 0 & -2 & 0 \end{bmatrix}$, the differential equation

$$\mathbf{v}'(t) = S\mathbf{v}(t), \quad \mathbf{v}(0) = \begin{bmatrix} 0 \\ 0 \\ 1 \end{bmatrix},$$

has the solution

$$\mathbf{v}(t) = \frac{1}{5}\begin{bmatrix} 2 - 2\cos\sqrt{5}t \\ 2\sqrt{5}\sin\sqrt{5}t \\ 1 + 4\cos\sqrt{5}t \end{bmatrix}.$$

Proof

One approach to computing $\exp(tS)$ involves determining the eigenvalues of S and then diagonalising it over $\mathbb{C}$; the details are left as an exercise for the reader. We obtain the special orthogonal matrix

$$\exp(tS) = \frac{1}{5}\begin{bmatrix} 4 + \cos\sqrt{5}t & \sqrt{5}\sin\sqrt{5}t & 2 - 2\cos\sqrt{5}t \\ -\sqrt{5}\sin\sqrt{5}t & 5\cos\sqrt{5}t & 2\sqrt{5}\sin\sqrt{5}t \\ 2 - 2\cos\sqrt{5}t & -2\sqrt{5}\sin\sqrt{5}t & 1 + 4\cos\sqrt{5}t \end{bmatrix},$$

giving for the required solution

$$\mathbf{v}(t) = \exp(tS)\mathbf{v}(0) = \frac{1}{5}\begin{bmatrix} 2 - 2\cos\sqrt{5}t \\ 2\sqrt{5}\sin\sqrt{5}t \\ 1 + 4\cos\sqrt{5}t \end{bmatrix}. \qquad \square$$

When the matrix A in Equation (2.8) is diagonalisable, this approach works well, provided an explicit diagonalisation can actually be found. In particular, symmetric, skew symmetric, hermitian and skew hermitian matrices are always diagonalisable over $\mathbb{C}$, as are matrices without multiple eigenvalues. For more general matrices with multiple eigenvalues it may be necessary to use Jordan forms as discussed in Section 2.2. This is illustrated in the next example.

Example 2.21

If we take

$$A = \begin{bmatrix} 2 & 1 & 0 \\ 0 & 2 & 0 \\ 0 & 0 & 2 \end{bmatrix}, \quad \mathbf{v}_0 \neq \mathbf{0},$$

then the differential equation

$$\mathbf{v}'(t) = A\mathbf{v}(t), \quad \mathbf{v}(0) = \mathbf{v}_0$$

has solution

$$\mathbf{v}(t) = \begin{bmatrix} e^{2t} & te^{2t} & 0 \\ 0 & e^{2t} & 0 \\ 0 & 0 & e^{2t} \end{bmatrix}\mathbf{v}_0.$$

Proof

This follows from Equation (3.6). Here we take

$$D = \operatorname{diag}(2t, 2t, 2t), \quad N = \begin{bmatrix} 0 & t & 0 \\ 0 & 0 & 0 \\ 0 & 0 & 0 \end{bmatrix},$$

and find that $N^2 = O$. This gives

$$\begin{aligned} \exp(tA) &= e^{2t} I_3 + \sum_{k \geqslant 0} \frac{1}{(k+1)!} (k+1) N \operatorname{diag}(2t, 2t, 2t)^k \\ &= e^{2t} I_3 + N \sum_{k \geqslant 0} \frac{1}{k!} \operatorname{diag}(2t, 2t, 2t)^k \\ &= \begin{bmatrix} e^{2t} & te^{2t} & 0 \\ 0 & e^{2t} & 0 \\ 0 & 0 & e^{2t} \end{bmatrix}, \end{aligned}$$

from which the stated solution can be deduced. □

Now suppose we have an equation of the form Equation (2.8). Write

$$\mathbf{v}(t) = \begin{bmatrix} v_1(t) \\ \vdots \\ v_n(t) \end{bmatrix},$$

where $v_k \colon \mathbb{R} \longrightarrow \mathbb{C}$ $(k = 1, \ldots, n)$. Since

$$\mathbf{v}^{(r)}(t) = \frac{\mathrm{d}^r}{\mathrm{d}\,t^r} \mathbf{v}(t) = A^r \mathbf{v}(t),$$

using the Cayley–Hamilton Theorem 2.6 we obtain

$$\mathbf{v}^{(n)}(t) + c_{n-1}(A)\mathbf{v}^{(n-1)}(t) + \cdots + c_1(A)\mathbf{v}^{(1)}(t) + c_0(A)\mathbf{v}(t) = \mathbf{0}, \qquad (2.11\text{a})$$

hence each of the coordinate functions v_k satisfies the ordinary differential equation

$$v_k^{(n)}(t) + c_{n-1}(A)v_k^{(n-1)}(t) + \cdots + c_1(A)v_k^{(1)}(t) + c_0(A)v_k(t) = 0. \qquad (2.11\text{b})$$

Of course, there are also boundary conditions coming from the initial condition $\mathbf{v}(0) = \mathbf{v}_0$.

Example 2.22

In Example 2.21, if $\mathbf{v}_0 = \begin{bmatrix} a \\ b \\ c \end{bmatrix}$, then the coordinate functions of the solution are

$$v_1(t) = ae^{2t} + bte^{2t}, \quad v_2(t) = be^{2t}, \quad v_3(t) = ce^{2t},$$

all of which satisfy the differential equations

$$v_k^{(3)}(t) - 6v_k^{(2)}(t) + 12v_k^{(1)}(t) - 8v_k(t) = 0 \quad (k = 1, 2, 3).$$

Of course the solution of a ordinary differential equation of form (2.11b) subject to suitable boundary conditions should be familiar, the novel point being the solution of an equivalent coupled system of first order differential equations for the v_k as a single vector differential equation solved using a one-parameter subgroup in $\mathrm{GL}_n(\mathbb{C})$.

The geometry underlying this can sometimes be indicated in a diagram, at least in the case where everything is happening in $\mathbb{R}^2$. The *phase portrait* shows at each point with coordinates (x, y) an arrow corresponding to the vector $A \begin{bmatrix} x \\ y \end{bmatrix}$, while the solution through a point can also be plotted as a *flow line*.

Example 2.23

The differential equation

$$\begin{bmatrix} x'(t) \\ y'(t) \end{bmatrix} = \begin{bmatrix} 4x(t) + 2y(t) \\ 4y(t) \end{bmatrix}$$

has solutions of the form

$$\begin{bmatrix} x(t) \\ y(t) \end{bmatrix} = \begin{bmatrix} e^{4t} & 2te^{4t} \\ 0 & e^{4t} \end{bmatrix} \begin{bmatrix} x_0 \\ y_0 \end{bmatrix} = \begin{bmatrix} e^{4t}x_0 + 2te^{4t}y_0 \\ e^{4t}y_0 \end{bmatrix},$$

for arbitrary initial vectors $\begin{bmatrix} x_0 \\ y_0 \end{bmatrix}$.

This follows from the fact that

$$\exp\left(\begin{bmatrix} 4t & 2t \\ 0 & 4t \end{bmatrix}\right) = e^{4t}I_2 + e^{4t}I_2\left(2t\begin{bmatrix} 0 & 1 \\ 0 & 0 \end{bmatrix}\right) = \begin{bmatrix} e^{4t} & 2te^{4t} \\ 0 & e^{4t} \end{bmatrix}.$$

It is well known that the general solution of an ordinary differential equation of the form given in (2.11) is built up as a sum of polynomial functions multiplied by exponential functions.

EXERCISES

2.1. For $t \in \mathbb{R}$, determine each of the matrices

$$\exp\left(\begin{bmatrix} 0 & t \\ -t & 0 \end{bmatrix}\right), \quad \exp\left(\begin{bmatrix} 0 & t \\ t & 0 \end{bmatrix}\right), \quad \exp\left(\begin{bmatrix} t & 0 \\ -2 & t \end{bmatrix}\right).$$

2.2. Let $\Bbbk = \mathbb{R}$ or $\mathbb{C}$ and $A \in \mathrm{M}_n(\Bbbk)$.
a) Show that for $B \in \mathrm{GL}_n(\Bbbk)$,

$$\exp(BAB^{-1}) = B\exp(A)B^{-1}.$$

b) If D is diagonalisable with $D = C\operatorname{diag}(\lambda_1, \ldots, \lambda_n)C^{-1}$ for some $C \in \mathrm{GL}_n(\Bbbk)$, show that

$$\exp(D) = C\operatorname{diag}(e^{\lambda_1}, \ldots, e^{\lambda_n})C^{-1}.$$

c) Use (b) to find the matrices

$$\exp\left(\begin{bmatrix} 0 & t \\ -t & 0 \end{bmatrix}\right), \quad \exp\left(\begin{bmatrix} 0 & t \\ t & 0 \end{bmatrix}\right).$$

2.3. For $\Bbbk = \mathbb{R}, \mathbb{C}$ and $n \geqslant 1$, let $N \in \mathrm{M}_n(\Bbbk)$.
a) If N is strictly upper triangular, show that $\exp(N)$ is unipotent.
b) Determine $\exp(N)$ when N is an upper triangular matrix with a fixed number t in every entry on its main diagonal.

2.4. a) If $S \in \mathrm{M}_n(\mathbb{R})$ is skew symmetric, show that $\exp(S)$ is orthogonal, *i.e.*, $\exp(S)^T = \exp(S)^{-1}$.
b) If $S \in \mathrm{M}_n(\mathbb{C})$ is skew hermitian, show that $\exp(S)$ is unitary, *i.e.*, $\exp(S)^* = \exp(S)^{-1}$.

2.5. a) Solve the differential equation

$$\begin{bmatrix} x'(t) \\ y'(t) \end{bmatrix} = \begin{bmatrix} -1 & -2 \\ 0 & 1 \end{bmatrix} \begin{bmatrix} x(t) \\ y(t) \end{bmatrix}, \quad \begin{bmatrix} x(0) \\ y(0) \end{bmatrix} = \begin{bmatrix} 1 \\ 2 \end{bmatrix}.$$

by finding a solution of the form

$$\begin{bmatrix} x(t) \\ y(t) \end{bmatrix} = \alpha(t) \begin{bmatrix} 1 \\ 2 \end{bmatrix}$$

with $\alpha\colon \mathbb{R} \longrightarrow \mathrm{GL}_2(\mathbb{R})$. Sketch the trajectory of this solution as a curve in the xy-plane. Investigate what happens for other initial

values $x(0), y(0)$.
b) Repeat this with the equations

$$\begin{bmatrix} x'(t) \\ y'(t) \end{bmatrix} = \begin{bmatrix} 0 & -2 \\ 1 & -2 \end{bmatrix} \begin{bmatrix} x(t) \\ y(t) \end{bmatrix}, \quad \begin{bmatrix} x(0) \\ y(0) \end{bmatrix} = \begin{bmatrix} 0 \\ 1 \end{bmatrix};$$

$$\begin{bmatrix} x'(t) \\ y'(t) \end{bmatrix} = \begin{bmatrix} 0 & -1 \\ -1 & 0 \end{bmatrix} \begin{bmatrix} x(t) \\ y(t) \end{bmatrix}, \quad \begin{bmatrix} x(0) \\ y(0) \end{bmatrix} = \begin{bmatrix} 1 \\ 1 \end{bmatrix}.$$

2.6. If $A \in \mathrm{M}_n(\mathbb{R})$ is skew symmetric, show that for a solution $\mathbf{x}\colon \mathbb{R} \longrightarrow \mathbb{R}^n$ of the differential equation

$$\mathbf{x}'(t) = A\mathbf{x}(t),$$

$|\mathbf{x}(t)|$ is constant. In particular, if $|\mathbf{x}(0)| = 1$ then for all $t \in \mathbb{R}$,

$$\mathbf{x}'(t) \in \mathrm{T}_{\mathbf{x}(t)}\, \mathbb{S}^{n-1},$$

the tangent space to the unit sphere at $\mathbf{x}(t)$.

2.7. Let $\Bbbk = \mathbb{R}$ or $\mathbb{C}$.
a) Let $J(\lambda, r) \in \mathrm{M}_r(\Bbbk)$ be a Jordan block matrix. Show that there is a sequence of diagonalisable matrices $\{A_n\}_{n\geqslant 1}$ with $A_n \in \mathrm{M}_r(\Bbbk)$ and $A_n \to J(\lambda, r)$ as $n \to \infty$. If $\lambda \neq 0$, show that we can assume that $A_n \in \mathrm{GL}_r(\Bbbk)$ for all n.
b) Deduce that every matrix $A \in \mathrm{M}_r(\Bbbk)$ is a limit of diagonalisable matrices in $\mathrm{M}_r(\Bbbk)$ and if A is invertible, show that it is a limit of invertible diagonalisable matrices.

3
Tangent Spaces and Lie Algebras

In this chapter we show how to 'linearise' a matrix group G by considering its tangent space at the identity, which has the algebraic structure of a *Lie algebra*; the definition and basic properties of Lie algebra are introduced in Section 3.1. Amazingly, the Lie algebra of G captures enough of the properties of G to act as a more manageable substitute for many purposes, at least when G is *simply connected.* The geometric aspects of this will be studied in Chapter 7 when we investigate G as a *Lie group.*

3.1 Lie Algebras

In this section we collect together some basic concepts and results of the theory of Lie algebras which will be required at various times. We make no attempt at completeness, merely giving brief indications of how the algebraic theory develops.

Definition 3.1

A $\Bbbk$-*Lie algebra* or *Lie algebra over* $\Bbbk$ consists of a vector space $\mathfrak{a}$ over a field $\Bbbk$, together with a $\Bbbk$-bilinear map $[\ ,\ ]\colon \mathfrak{a} \times \mathfrak{a} \longrightarrow \mathfrak{a}$ called the *Lie bracket*, such that for $x, y, z \in \mathfrak{a}$,

$$[x, y] = -[y, x], \qquad \text{(Skew symmetry)}$$
$$[x, [y, z]] + [y, [x, z]] + [z, [x, y]] = 0. \qquad \text{(Jacobi identity)}$$

Here $\Bbbk$-bilinear means that for $x_1, x_2, x, y_1, y_2, y \in \mathfrak{a}$ and $r_1, r_2, r, s_1, s_2, s \in \Bbbk$,

$$[r_1x_1 + r_2x_2, y] = r_1[x_1, y] + r_2[x_2, y],$$
$$[x, s_1y_1 + s_2y_2] = s_1[x, y_1] + s_2[x, y_2].$$

Example 3.2

Let $\Bbbk = \mathbb{R}$ and $\mathfrak{a} = \mathbb{R}^3$ and set

$$[\mathbf{x}, \mathbf{y}] = \mathbf{x} \times \mathbf{y},$$

the *vector product* or *cross product* of $\mathbf{x}$ and $\mathbf{y}$. For the standard basis vectors $\mathbf{e}_1, \mathbf{e}_2, \mathbf{e}_3$ we have the formulæ

$$\begin{cases} [\mathbf{e}_1, \mathbf{e}_2] = -[\mathbf{e}_2, \mathbf{e}_1] = \mathbf{e}_3, \\ [\mathbf{e}_2, \mathbf{e}_3] = -[\mathbf{e}_3, \mathbf{e}_2] = \mathbf{e}_1, \\ [\mathbf{e}_3, \mathbf{e}_1] = -[\mathbf{e}_1, \mathbf{e}_3] = \mathbf{e}_2. \end{cases} \tag{3.1}$$

Then $\mathbb{R}^3$ equipped with this bracket operation is an $\mathbb{R}$-Lie algebra. As we will see later, this is the Lie algebra of $\mathrm{SO}(3)$ and also of $\mathrm{SU}(2)$ in disguise.

Given two matrices $A, B \in \mathrm{M}_n(\Bbbk)$, their *commutator* or *Lie bracket* is the matrix

$$[A, B] = AB - BA.$$

This defines a $\Bbbk$-bilinear function

$$[\ ,\ ]\colon \mathrm{M}_n(\Bbbk) \times \mathrm{M}_n(\Bbbk) \longrightarrow \mathrm{M}_n(\Bbbk); \quad (A, B) \mapsto [A, B]$$

which is easily seen to satisfy the conditions of Definition 3.1.

Example 3.3

The $\Bbbk$-vector space $\mathrm{M}_n(\Bbbk)$ with the commutator bracket $[\ ,\ ]$ is a $\Bbbk$-Lie algebra.

Remark 3.4

Recall that A, B *commute* if $AB = BA$. Then $[A, B] = O_n$ if and only if A, B commute. So the Lie algebra structure on $\mathrm{M}_n(\Bbbk)$ gives a measure of how pairs of matrices fail to commute.

For later use, in the remainder of this section we develop some of the basic algebraic notions of Lie algebra theory. This material is analogous to the basic theory of groups or rings in that it introduces Lie subalgebras, Lie ideals, homomorphisms, etc.

Definition 3.5

If $\mathfrak{a}$ is a $\Bbbk$-Lie algebra with bracket $[\ ,\ ]$, then a $\Bbbk$-subspace $\mathfrak{b} \subseteq \mathfrak{a}$ is a $\Bbbk$-*Lie subalgebra* of $\mathfrak{a}$ if it is closed under taking commutators of pairs of elements in $\mathfrak{b}$, *i.e.*, if whenever $x, y \in \mathfrak{b}$ then $[x, y] \in \mathfrak{b}$; we write $\mathfrak{b} \leqslant \mathfrak{a}$.

Of course, $\mathfrak{a}$ is a $\Bbbk$-Lie subalgebra of itself, and $\{0\} \subseteq \mathfrak{a}$ is also a Lie subalgebra.

A Lie algebra in which all brackets are trivial is called *abelian*.

Suppose that $\mathfrak{a} \subseteq \mathrm{M}_n(\Bbbk)$ is a $\Bbbk$-vector subspace. Recalling Definition 3.5, we see that $\mathfrak{a}$ is a $\Bbbk$-Lie subalgebra of $\mathrm{M}_n(\Bbbk)$ if it is closed under taking commutators of pairs of elements in $\mathfrak{a}$, *i.e.*, whenever $A, B \in \mathfrak{a}$, then $[A, B] \in \mathfrak{a}$.

Definition 3.6

If $\mathfrak{a}$ is a $\Bbbk$-Lie algebra with bracket $[\ ,\ ]$, then a $\Bbbk$-vector subspace $\mathfrak{n} \subseteq \mathfrak{a}$ is a *Lie ideal* of $\mathfrak{a}$ if $[z, x] \in \mathfrak{n}$ for all $z \in \mathfrak{n}$ and $x \in \mathfrak{a}$; we then write $\mathfrak{n} \lhd \mathfrak{a}$. A Lie ideal $\mathfrak{n} \lhd \mathfrak{a}$ is *proper* if $\mathfrak{n} \neq \mathfrak{a}$ and it is *non-trivial* if $\mathfrak{n} \neq \{0\}$; the ideal $\{0\} \lhd \mathfrak{a}$ is the *trivial ideal*.

Definition 3.7

Let $\mathfrak{a}$ be a $\Bbbk$-Lie algebra with bracket $[\ ,\ ]$.

- The *centre* of $\mathfrak{a}$ is

$$\mathfrak{z}(\mathfrak{a}) = \{z \in \mathfrak{a} : \forall x \in \mathfrak{a},\ [z, x] = 0\}.$$

- The *commutator* or *derived* subalgebra $\mathfrak{a}' \leqslant \mathfrak{a}$ is the vector subspace spanned by all the brackets $[x, y]$ $(x, y \in \mathfrak{a})$.

Proposition 3.8

$\mathfrak{z}(\mathfrak{a})$ and $\mathfrak{a}'$ are Lie ideals of $\mathfrak{a}$, *i.e.*, $\mathfrak{z}(\mathfrak{a}) \lhd \mathfrak{a}$ and $\mathfrak{a}' \lhd \mathfrak{a}$.

Proof

It is easy to see that $\mathfrak{z}(\mathfrak{a})$ is a vector subspace. For $z \in \mathfrak{z}(\mathfrak{a})$ and $x \in \mathfrak{a}$, $[x, z] = 0$ so $\mathfrak{z}(\mathfrak{a}) \lhd \mathfrak{a}$.

For $\mathfrak{a}'$, if $x, y, z \in \mathfrak{a}$ then $[x, [y, z]] \in \mathfrak{a}'$, so $\mathfrak{a}' \lhd \mathfrak{a}$. □

Here is a useful result about the derived subalgebra which is left as an exercise.

Proposition 3.9

Let $\Phi \colon \mathfrak{a} \longrightarrow \mathfrak{b}$ be a surjective homomorphism of $\Bbbk$-Lie algebras. Then $\mathfrak{b}$ is abelian if and only if $\mathfrak{a}' \subseteq \ker \Phi$.

Definition 3.10

A $\Bbbk$-Lie algebra $\mathfrak{a}$ is *simple* if every proper ideal $\mathfrak{b} \lhd \mathfrak{a}$ is trivial, *i.e.*, $\mathfrak{b} = \{0\}$.

Definition 3.11

A $\Bbbk$-Lie algebra $\mathfrak{a}$ has *trivial centre* or is *centreless* if $\mathfrak{z}(\mathfrak{g}) = \{0\}$.

Definition 3.12

Let $\mathfrak{g}, \mathfrak{h}$ be $\Bbbk$-Lie algebras. A $\Bbbk$-linear transformation $\Phi \colon \mathfrak{g} \longrightarrow \mathfrak{h}$ is a *homomorphism of Lie algebras* if for all $x, y \in \mathfrak{g}$,

$$\Phi([x, y]) = [\Phi(x), \Phi(y)].$$

Such a homomorphism is an *isomorphism of Lie algebras* if it is also a $\Bbbk$-linear isomorphism.

For an ideal $\mathfrak{n} \lhd \mathfrak{a}$, the quotient vector space $\mathfrak{a}/\mathfrak{n}$ inherits a bracket operation defined by

$$[x + \mathfrak{n}, y + \mathfrak{n}] = [x, y] + \mathfrak{n}$$

for cosets $x + \mathfrak{n}, y + \mathfrak{n} \in \mathfrak{a}/\mathfrak{n}$. This is easily seen to make $\mathfrak{a}/\mathfrak{n}$ into a Lie algebra so that the quotient linear transformation $\mathfrak{a} \longrightarrow \mathfrak{a}/\mathfrak{n}$ is a homomorphism of Lie algebras. $\mathfrak{a}/\mathfrak{n}$ is referred to as the *quotient Lie algebra* of $\mathfrak{a}$ with respect to the ideal $\mathfrak{n}$. The following result is easily verified.

Proposition 3.13

Let $\Phi\colon \mathfrak{a} \longrightarrow \mathfrak{b}$ be a homomorphism of $\Bbbk$-Lie algebras. Then $\ker\Phi \lhd \mathfrak{a}$ and there is an isomorphism of Lie algebras $\operatorname{im}\Phi \cong \mathfrak{a}/\ker\Phi$.

The notion of *semi-direct product* of Lie algebras in the next definition is analogous to the group theoretic concept of Definition 1.36.

Definition 3.14

Let $\mathfrak{a}$ be $\Bbbk$-Lie algebra, $\mathfrak{b} \leqslant \mathfrak{a}$ and $\mathfrak{n} \lhd \mathfrak{a}$. Then $\mathfrak{a}$ is known as the *semi-direct product* of $\mathfrak{b}$ and $\mathfrak{n}$ if $\mathfrak{b} \cap \mathfrak{n} = \{0\}$ and $\mathfrak{a} = \mathfrak{b} + \mathfrak{n}$ as $\Bbbk$-vector spaces; we then write $\mathfrak{a} = \mathfrak{b} \ltimes \mathfrak{n}$.

As in group theory, there is a notion of *direct product* of two Lie algebras $\mathfrak{a}$ and $\mathfrak{b}$. This is the vector space

$$\mathfrak{a} \times \mathfrak{b} = \{(x, y) : x \in \mathfrak{a},\ y \in \mathfrak{b}\}$$

endowed with the bracket

$$[(x_1, y_1), (x_2, y_2)] = ([x_1, x_2], [y_1, y_2]).$$

This is a special case of the semi-direct product.

Finally, the following is a useful notation. For subspaces $U, V \leqslant \mathfrak{g}$ of a $\Bbbk$-Lie algebra $\mathfrak{g}$, set

$$[U, V] = \{\sum_j c_j[x_j, y_j] : c_j \in \Bbbk,\ x_j \in U,\ y_j \in V\} \subseteq \mathfrak{g}.$$

3.2 Curves, Tangent Spaces and Lie Algebras

Throughout this section, let $G \leqslant \mathrm{GL}_n(\Bbbk)$ be a matrix group.

Definition 3.15

The *tangent space* to G at $U \in G$ is

$$\mathrm{T}_U\, G = \{\gamma'(0) \in \mathrm{M}_n(\Bbbk) : \gamma \text{ a differentiable curve in } G \text{ with } \gamma(0) = U\}.$$

Proposition 3.16

$\mathrm{T}_U\, G$ is a real vector subspace of $\mathrm{M}_n(\Bbbk)$.

Proof

Suppose that α, β are differentiable curves in G for which $\alpha(0) = \beta(0) = U$. Then

$$\gamma\colon \operatorname{dom}\alpha \cap \operatorname{dom}\beta \longrightarrow G; \quad \gamma(t) = \alpha(t)U^{-1}\beta(t),$$

is also a differentiable curve in G with $\gamma(0) = U$. The Product Rule gives

$$\gamma'(t) = \alpha'(t)U^{-1}\beta(t) + \alpha(t)U^{-1}\beta'(t),$$

hence

$$\gamma'(0) = \alpha'(0)U^{-1}\beta(0) + \alpha(0)U^{-1}\beta'(0) = \alpha'(0) + \beta'(0),$$

which shows that T_U is closed under addition.

Similarly, if $r \in \mathbb{R}$ and α is a differentiable curve in G with $\alpha(0) = U$, then $\eta(t) = \alpha(rt)$ defines another such curve. Since $\eta'(0) = r\alpha'(0)$, we see that $\mathrm{T}_U G$ is closed under real scalar multiplication. □

Definition 3.17

The *dimension* of the real matrix group G is $\dim G = \dim_{\mathbb{R}} \mathrm{T}_I G$.

We will use the notation $\mathfrak{g} = \mathrm{T}_I G$ for this real vector subspace of $\mathrm{M}_n(\mathbb{k})$. In fact, $\mathfrak{g}$ has the algebraic structure of a real Lie algebra as we now show.

Theorem 3.18

i) If $G \leqslant \mathrm{GL}_n(\mathbb{k})$ is a matrix subgroup, then $\mathfrak{g}$ is an $\mathbb{R}$-Lie subalgebra of $\mathrm{M}_n(\mathbb{k})$.
ii) If $G \leqslant \mathrm{GL}_m(\mathbb{C})$ is a matrix subgroup and $\mathfrak{g}$ is a $\mathbb{C}$-subspace of $\mathrm{M}_m(\mathbb{C})$, then $\mathfrak{g}$ is a $\mathbb{C}$-Lie subalgebra.

Proof

(i) We will show that for two differentiable curves α and β in G which satisfy $\alpha(0) = \beta(0) = I_n$, there is another such curve γ with $\gamma'(0) = [\alpha'(0), \beta'(0)]$.

Consider the function

$$F\colon \operatorname{dom}\alpha \times \operatorname{dom}\beta \longrightarrow G; \quad F(s,t) = \alpha(s)\beta(t)\alpha(s)^{-1}.$$

This is clearly continuous and differentiable with respect to each of the variables s, t. For each $s \in \operatorname{dom}\alpha$, the function $F(s, \ )\colon \operatorname{dom}\beta \longrightarrow G$ is a differentiable curve in G with $F(s, 0) = I_n$. On differentiating we find

$$\frac{\mathrm{d}\,F(s,t)}{\mathrm{d}\,t}\bigg|_{t=0} = \alpha(s)\beta'(0)\alpha(s)^{-1},$$

and so

$$\alpha(s)\beta'(0)\alpha(s)^{-1} \in \mathfrak{g}.$$

Since $\mathfrak{g}$ is a closed subspace of $\mathrm{M}_n(\mathbb{k})$, whenever this limit exists we also have

$$\lim_{s\to 0} \frac{1}{s}\left(\alpha(s)\beta'(0)\alpha(s)^{-1} - \beta'(0)\right) \in \mathfrak{g}.$$

We will use the following easily verified matrix version of the usual rule for differentiating an inverse:

$$\frac{\mathrm{d}}{\mathrm{d}t}\left(\alpha(t)^{-1}\right) = -\alpha(t)^{-1}\alpha'(t)\alpha(t)^{-1}. \tag{3.2}$$

We have

$$\begin{aligned}
\lim_{s\to 0} \frac{1}{s}&\left(\alpha(s)\beta'(0)\alpha(s)^{-1} - \beta'(0)\right) \\
&= \frac{\mathrm{d}}{\mathrm{d}s}\Big|_{s=0} \alpha(s)\beta'(0)\alpha(s)^{-1} \\
&= \alpha'(0)\beta'(0)\alpha(0)^{-1} \\
&\qquad - \alpha(0)\beta'(0)\alpha(0)^{-1}\alpha'(0)\alpha(0)^{-1} \\
&\qquad\qquad \text{[by Equation (3.2)]} \\
&= \alpha'(0)\beta'(0)\alpha(0)^{-1} - \alpha(0)\beta'(0)\alpha'(0) \\
&= \alpha'(0)\beta'(0) - \beta'(0)\alpha'(0) \\
&= [\alpha'(0), \beta'(0)].
\end{aligned}$$

This shows that $[\alpha'(0), \beta'(0)] \in \mathfrak{g}$, hence it must be of the form $\gamma'(0)$ for some differentiable curve.

Part (ii) is left as an exercise. □

So for each matrix group G there is a Lie algebra $\mathfrak{g} = \mathrm{T}_I G$. A suitable type of homomorphism $G \longrightarrow H$ between matrix groups gives rise to a linear transformation $\mathfrak{g} \longrightarrow \mathfrak{h}$ which is a homomorphism of Lie algebras in the sense of Definition 3.12.

Definition 3.19

Let $G \leqslant \mathrm{GL}_n(\mathbb{k})$, $H \leqslant \mathrm{GL}_m(\mathbb{k})$ be matrix groups and $\varphi\colon G \longrightarrow H$ be a continuous map. Then φ is said to be a *differentiable map* if it satisfies the following two conditions:

- for every differentiable curve $\gamma\colon (a,b) \longrightarrow G$, the curve $\varphi\circ\gamma\colon (a,b) \longrightarrow H$ is differentiable and has derivative

$$(\varphi\circ\gamma)'(t) = \frac{\mathrm{d}}{\mathrm{d}t}\varphi(\gamma(t));$$

- if two differentiable curves $\alpha, \beta\colon (a,b) \longrightarrow G$ satisfy
$$\alpha(0) = \beta(0), \quad \alpha'(0) = \beta'(0),$$
then
$$(\varphi \circ \alpha)'(0) = (\varphi \circ \beta)'(0).$$

A differentiable map which is also a group homomorphism is called a *differentiable homomorphism.* A continuous homomorphism of matrix groups that is also a differentiable map is called a *Lie homomorphism.*

Later we will see that the technical restriction in this definition is unnecessary. For now we note that if $\varphi\colon G \longrightarrow H$ is the restriction of a differentiable map $\Phi\colon \mathrm{GL}_n(\mathbf{k}) \longrightarrow \mathrm{GL}_m(\mathbf{k})$ then φ is also a differentiable map.

Proposition 3.20

Let G, H, K be matrix groups with $\varphi\colon G \longrightarrow H$ and $\theta\colon H \longrightarrow K$ differentiable homomorphisms. Then the following are true.
i) For each $A \in G$ there is an $\mathbb{R}$-linear transformation $\mathrm{d}\,\varphi\colon \mathrm{T}_A\, G \longrightarrow \mathrm{T}_{\varphi(A)}\, H$ given by
$$\mathrm{d}\,\varphi_A(\gamma'(0)) = (\varphi \circ \gamma)'(0),$$
for every differentiable curve $\gamma\colon (a,b) \longrightarrow G$ with $\gamma(0) = A$.
ii) We have
$$\mathrm{d}\,\theta_{\varphi(A)} \circ \mathrm{d}\,\varphi_A = \mathrm{d}(\theta \circ \varphi)_A.$$
iii) For the identity map $\mathrm{Id}_G\colon G \longrightarrow G$ and $A \in G$,
$$\mathrm{d}\,\mathrm{Id}_G = \mathrm{Id}_{\mathrm{T}_A\, G}.$$

Proof

(i) The definition of $\mathrm{d}\,\varphi_A$ makes sense since by the definition of differentiability, given $X \in \mathrm{T}_A\, G$, for any curve γ with
$$\gamma(0) = A, \quad \gamma'(0) = X,$$
the derivative $(\varphi \circ \gamma)'(0)$ depends only on X and not on γ. Linearity is established using similar ideas to the proof of Proposition 3.16.

The identities of (ii) and (iii) are straightforward to verify. □

If $\varphi\colon G \longrightarrow H$ is a differentiable homomorphism then since $\varphi(I) = I$, $\mathrm{d}\,\varphi_I\colon \mathrm{T}_I\, G \longrightarrow \mathrm{T}_I\, H$ is a linear transformation called the *derivative* of φ which is usually denoted $\mathrm{d}\,\varphi\colon \mathfrak{g} \longrightarrow \mathfrak{h}$.

Theorem 3.21

Let G, H be matrix groups and $\varphi\colon G \longrightarrow H$ be a differentiable homomorphism. Then the derivative $\mathrm{d}\varphi\colon \mathfrak{g} \longrightarrow \mathfrak{h}$ is a homomorphism of Lie algebras.

Proof

Following ideas and notation in the proof of Theorem 3.18, for differentiable curves α, β in G with $\alpha(0) = \beta(0) = I$, we can use the composite function $\varphi \circ F$ given by

$$\varphi \circ F(s,t) = \varphi(F(s,t)) = \varphi(\alpha(s))\varphi(\beta(t))\varphi(\alpha(s))^{-1},$$

to deduce that

$$\mathrm{d}\varphi([\alpha'(0), \beta'(0)]) = [\mathrm{d}\varphi(\alpha'(0)), \mathrm{d}\varphi(\beta'(0))]. \qquad \square$$

As a special case, for each $A \in G$ the conjugation map

$$\chi_A\colon G \longrightarrow G; \quad \chi_A(U) = AUA^{-1}$$

has as its derivative at I the *adjoint action* of A on $\mathfrak{g}$, namely the linear transformation

$$\mathrm{Ad}_A = \mathrm{d}\chi_A\colon \mathfrak{g} \longrightarrow \mathfrak{g}; \quad \mathrm{Ad}_A(X) = AXA^{-1}.$$

This satisfies

$$\mathrm{Ad}_{AB} = \mathrm{Ad}_A \circ \mathrm{Ad}_B \quad (A, B \in G),$$

and on choosing a basis for $\mathfrak{g}$ defines a continuous (and in fact differentiable) group homomorphism

$$\mathrm{Ad}\colon G \longrightarrow \mathrm{GL}_{\dim G}(\Bbbk); \quad A \mapsto \mathrm{Ad}_A\,.$$

Given a differentiable curve $\alpha\colon (-\varepsilon, \varepsilon) \longrightarrow G$ with $\alpha(0) = I$, there is another curve

$$\tilde{\alpha}\colon (-\varepsilon, \varepsilon) \longrightarrow \mathrm{GL}_{\dim G}(\Bbbk); \quad \tilde{\alpha}(t) = \mathrm{Ad}_{\alpha(t)},$$

whose derivative at 0 is given by

$$\tilde{\alpha}'(0)(X) = [\alpha'(0), X].$$

This defines the adjoint action of $U \in \mathfrak{g}$ on $\mathfrak{g}$,

$$\mathrm{ad}_U\colon \mathfrak{g} \longrightarrow \mathfrak{g}; \quad \mathrm{ad}_U(X) = [U, X].$$

It is easy to verify that

$$\mathrm{ad}_{[U,V]}(X) = \mathrm{ad}_U \circ \mathrm{ad}_V(X) - \mathrm{ad}_V \circ \mathrm{ad}_U(X) \quad (U, V, X \in \mathfrak{g}).$$

So once a basis of $\mathfrak{g}$ is chosen, the adjoint action gives rise to a homomorphism of Lie algebras (with $n = \dim \mathfrak{g}$)

$$\mathrm{ad}\colon \mathfrak{g} \longrightarrow \mathrm{M}_n(\Bbbk); \quad U \mapsto \mathrm{ad}_U\,.$$

Ad and ad are commonly referred to as the *adjoint representations* of G and $\mathfrak{g}$.

3.3 The Lie Algebras of Some Matrix Groups

In this section we determine the Lie algebras of some important families of real and complex matrix groups. As usual $\Bbbk = \mathbb{R}$ or $\mathbb{C}$.

General and Special Linear Groups

We will start with the real matrix group $\mathrm{GL}_n(\mathbb{R}) \subseteq \mathrm{M}_n(\mathbb{R})$. For $A \in \mathrm{M}_n(\mathbb{R})$ and $\varepsilon > 0$ there is a differentiable curve

$$\alpha\colon (-\varepsilon, \varepsilon) \longrightarrow \mathrm{M}_n(\mathbb{R}); \quad \alpha(t) = I + tA.$$

For $t \neq 0$, the roots of the equation $\det(t^{-1}I + A) = 0$ are of the form $t = -1/\lambda$ where λ is a non-zero eigenvalue of A. Hence if

$$\varepsilon < \min\left\{\frac{1}{|\lambda|} : \lambda \text{ a non-zero eigenvalue of } A\right\},$$

then $\operatorname{im}\alpha \subseteq \mathrm{GL}_n(\mathbb{R})$, so we will view α as a function $\alpha\colon (-\varepsilon, \varepsilon) \longrightarrow \mathrm{GL}_n(\mathbb{R})$. Calculating the derivative we find that $\alpha'(t) = A$, hence $\alpha'(0) = A$. This shows that $A \in \mathrm{T}_I \mathrm{GL}_n(\mathbb{R})$. Since $A \in \mathrm{M}_n(\mathbb{R})$ was arbitrary, we have

$$\begin{cases} \mathfrak{gl}_n(\mathbb{R}) = \mathrm{T}_I \mathrm{GL}_n(\mathbb{R}) = \mathrm{M}_n(\mathbb{R}), \\ \dim \mathrm{GL}_n(\mathbb{R}) = n^2. \end{cases} \tag{3.3}$$

Similarly,

$$\begin{cases} \mathfrak{gl}_n(\mathbb{C}) = \mathrm{T}_I \mathrm{GL}_n(\mathbb{C}) = \mathrm{M}_n(\mathbb{C}), \\ \dim_{\mathbb{C}} \mathrm{GL}_n(\mathbb{C}) = n^2, \\ \dim \mathrm{GL}_n(\mathbb{C}) = 2n^2. \end{cases} \tag{3.4}$$

For $\mathrm{SL}_n(\mathbb{R}) \leqslant \mathrm{GL}_n(\mathbb{R})$, suppose that $\alpha\colon (a, b) \longrightarrow \mathrm{SL}_n(\mathbb{R})$ is a curve lying in $\mathrm{SL}_n(\mathbb{R})$ and satisfying $\alpha(0) = I$. For $t \in (a, b)$ we have $\det \alpha(t) = 1$, so

$$\frac{\mathrm{d}(\det \alpha(t))}{\mathrm{d}t} = 0.$$

Lemma 3.22

We have

$$\frac{\mathrm{d}(\det \alpha(t))}{\mathrm{d}t}\Big|_{t=0} = \operatorname{tr} \alpha'(0).$$

Proof

Recall that for $A \in \mathrm{M}_n(\Bbbk)$,

$$\operatorname{tr} A = \sum_{i=1}^{n} A_{ii}.$$

It is easy to verify that the operation $\partial = \dfrac{\mathrm{d}}{\mathrm{d}\,t}_{|t=0}$ on functions has the *derivation property*

$$\partial(\gamma_1\gamma_2) = (\partial\gamma_1)\gamma_2(0) + \gamma_1(0)\partial\gamma_2. \tag{3.5}$$

Put $a_{ij} = \alpha(t)_{ij}$ and notice that when $t = 0$, $a_{ij} = \delta_{ij}$. Write C_{ij} for the cofactor matrix obtained from $\alpha(t)$ by deleting the ith row and jth column. By expanding along the nth row we obtain

$$\det \alpha(t) = \sum_{j=1}^{n} (-1)^{n+j} a_{nj} \det C_{nj}.$$

Then

$$\begin{aligned}
\partial \det \alpha(t) &= \sum_{j=1}^{n} (-1)^{n+j} \left((\partial a_{nj}) \det C_{nj} + a_{nj}(\partial \det C_{nj})\right) \\
&= \sum_{j=1}^{n} (-1)^{n+j} (\partial a_{nj}) \det C_{nj} + \partial \det C_{nn}.
\end{aligned}$$

For $t = 0$, $\det C_{nj} = \delta_{jn}$ since $\alpha(0) = I$, hence

$$\partial \det(\alpha(t)) = \partial a_{nn} + \partial \det C_{nn}.$$

We can repeat this calculation with the $(n-1) \times (n-1)$ matrix C_{nn} and so on. This yields

$$\begin{aligned}
\partial \det(\alpha(t)) &= \partial a_{nn} + \partial a_{(n-1)(n-1)} + \partial \det C_{(n-1)(n-1)} \\
&\;\;\vdots \\
&= \partial a_{nn} + \partial a_{(n-1)(n-1)} + \cdots + \partial a_{11} \\
&= \operatorname{tr} \alpha'(0),
\end{aligned}$$

giving the result. □

So we have $\operatorname{tr} \alpha'(0) = 0$ and hence

$$\mathfrak{sl}_n(\mathbb{R}) = \mathrm{T}_I \operatorname{SL}_n(\mathbb{R}) \subseteq \ker \operatorname{tr} \subseteq \mathrm{M}_n(\mathbb{R}).$$

If $A \in \ker \operatorname{tr} \subseteq \mathrm{M}_n(\mathbb{R})$, the function

$$\alpha\colon (-\varepsilon, \varepsilon) \longrightarrow \mathrm{M}_n(\mathbb{R}); \quad \alpha(t) = \exp(tA) = \sum_{k\geqslant 0} \frac{t^k}{k!} A^k,$$

is defined for every $\varepsilon > 0$ and satisfies the boundary conditions

$$\alpha(0) = I, \quad \alpha'(0) = A.$$

We will make use of the following lemma.

Lemma 3.23

For $A \in \mathrm{M}_n(\mathbb{C})$ we have

$$\det \exp(A) = e^{\operatorname{tr} A}.$$

We will give two proofs here, while another appears in Section 7.5.

Proof

Approach using differential equations: Consider the curve

$$\gamma\colon \mathbb{R} \longrightarrow \mathrm{GL}_1(\mathbb{C}) = \mathbb{C}^{\times}; \quad \gamma(t) = \det \exp(tA).$$

By applying Lemma 3.22 to the curve $t \mapsto \det \exp(tA)$ we obtain

$$\begin{aligned}\gamma'(t) &= \lim_{h\to 0} \frac{1}{h} (\det \exp((t+h)A) - \det \exp(tA)) \\ &= \det \exp(tA) \lim_{h\to 0} \frac{1}{h} (\det \exp(hA) - 1) \\ &= \det \exp(tA) \operatorname{tr} A \\ &= \gamma \operatorname{tr} A.\end{aligned}$$

So α satisfies the same differential equation and initial condition as the curve $t \mapsto e^{t \operatorname{tr} A}$. Hence

$$\alpha(t) = \det \exp(tA) = e^{t \operatorname{tr} A}$$

by the uniqueness part of Theorem 2.12. □

Proof

Approach using Jordan form: If $S \in \mathrm{GL}_n(\mathbb{C})$,

$$\begin{aligned}\det \exp(SAS^{-1}) &= \det \left(S \exp(A) S^{-1}\right) \\ &= \det S \det \exp(A) \det S^{-1} \\ &= \det \exp A,\end{aligned}$$

and

$$e^{\operatorname{tr} SAS^{-1}} = e^{\operatorname{tr} A}.$$

So it suffices to prove the identity for SAS^{-1} for a suitably chosen invertible matrix S. Using the theory of *Jordan forms*, described in Section 2.2, there is a suitable choice of such an S for which

$$B = SAS^{-1} = D + N,$$

where D is diagonal, N is strictly upper triangular (*i.e.*, $N_{ij} = 0$ whenever $i \geqslant j$) and with D, N commuting. Then N is nilpotent, *i.e.*, $N^k = O_n$ for k large.

We have

$$\begin{aligned}
\exp(B) &= \sum_{k\geqslant 0} \frac{1}{k!}(D+N)^k \\
&= \sum_{k\geqslant 0} \frac{1}{k!} D^k + \sum_{k\geqslant 0} \frac{1}{(k+1)!} \left(\sum_{j=0}^{k} \binom{k+1}{j} D^j N^{k+1-j} \right) \\
&= \exp(D) + \sum_{k\geqslant 0} \frac{1}{(k+1)!} N \left(\sum_{j=0}^{k} \binom{k+1}{j} D^j N^{k-j} \right).
\end{aligned} \tag{3.6}$$

For each $k \geqslant 0$, the matrix

$$N \sum_{j=0}^{k} \binom{k+1}{j} D^j N^{k-j}$$

is strictly upper triangular, hence

$$\exp(B) = \exp(D) + N',$$

where N' is strictly upper triangular. If $D = \operatorname{diag}(\lambda_1, \ldots, \lambda_n)$, by calculating the determinant we find that

$$\begin{aligned}
\det \exp(A) &= \det \exp(B) \\
&= \det \exp(D) \\
&= \det \operatorname{diag}(e^{\lambda_1}, \ldots, e^{\lambda_n}) \\
&= e^{\lambda_1} \cdots e^{\lambda_n} \\
&= e^{\lambda_1 + \cdots + \lambda_n}.
\end{aligned}$$

Since $\operatorname{tr} D = \lambda_1 + \cdots + \lambda_n$, this implies that

$$\det \exp(A) = e^{\operatorname{tr} D}.$$

□

Using this lemma and the function α, we obtain

$$\begin{cases} \mathfrak{sl}_n(\mathbb{R}) = \mathrm{T}_I \, \mathrm{SL}_n(\mathbb{R}) = \ker \mathrm{tr} \subseteq \mathrm{M}_n(\mathbb{R}), \\ \dim \mathrm{SL}_n(\mathbb{R}) = n^2 - 1. \end{cases} \tag{3.7}$$

Working over $\mathbb{C}$ we also have

$$\begin{cases} \mathfrak{sl}_n(\mathbb{C}) = \mathrm{T}_I \, \mathrm{SL}_n(\mathbb{C}) = \ker \mathrm{tr} \subseteq \mathrm{M}_n(\mathbb{C}), \\ \dim_{\mathbb{C}} \mathrm{SL}_n(\mathbb{C}) = n^2 - 1, \\ \dim \mathrm{SL}_n(\mathbb{C}) = 2n^2 - 2. \end{cases} \tag{3.8}$$

Affine Groups

Recall the affine group $\mathrm{Aff}_n(\mathbb{k}) \leqslant \mathrm{GL}_n(\mathbb{k})$ which by Proposition 1.37 is the semi-direct product $\mathrm{GL}_n(\mathbb{k}) \ltimes \mathrm{Trans}_n(\mathbb{k})$. Using similar methods to those we used for $\mathrm{GL}_n(\mathbb{k})$ we find that

$$\mathfrak{aff}_n(\mathbb{k}) = \left\{ \begin{bmatrix} A & \mathbf{t} \\ \mathbf{0} & 0 \end{bmatrix} : A \in \mathfrak{gl}_n(\mathbb{k}), \ \mathbf{t} \in \mathbb{k}^n \right\} = \left\{ \begin{bmatrix} A & \mathbf{t} \\ \mathbf{0} & 0 \end{bmatrix} : A \in \mathrm{M}_n(\mathbb{k}), \ \mathbf{t} \in \mathbb{k}^n \right\}.$$

Notice that $\mathfrak{aff}_n(\mathbb{k})$ is a $\mathbb{k}$-vector space and indeed it is a $\mathbb{k}$-Lie algebra. The Lie algebra structure here is interesting since although the translation subgroup $\mathrm{Trans}_n(\mathbb{k})$ is abelian and hence its Lie algebra $\mathfrak{trans}_n(\mathbb{k})$ has trivial bracket, we find that

$$\begin{aligned} \begin{bmatrix} A_1 & \mathbf{t}_1 \\ \mathbf{0} & 0 \end{bmatrix} \begin{bmatrix} A_2 & \mathbf{t}_2 \\ \mathbf{0} & 0 \end{bmatrix} - \begin{bmatrix} A_2 & \mathbf{t}_2 \\ \mathbf{0} & 0 \end{bmatrix} \begin{bmatrix} A_1 & \mathbf{t}_1 \\ \mathbf{0} & 0 \end{bmatrix} &= \begin{bmatrix} A_1A_2 & A_1\mathbf{t}_2 \\ \mathbf{0} & 0 \end{bmatrix} - \begin{bmatrix} A_2A_1 & A_2\mathbf{t}_1 \\ \mathbf{0} & 0 \end{bmatrix} \\ &= \begin{bmatrix} [A_1, A_2] & A_1\mathbf{t}_2 - A_2\mathbf{t}_1 \\ \mathbf{0} & 0 \end{bmatrix}. \end{aligned}$$

This shows in particular that $\mathfrak{trans}_n(\mathbb{k})$ is a $\mathbb{k}$-Lie ideal in $\mathfrak{aff}_n(\mathbb{k})$ since it satisfies

$$[a, t] \in \mathfrak{trans}_n(\mathbb{k}) \quad (a \in \mathfrak{aff}_n(\mathbb{k}), \ t \in \mathfrak{trans}_n(\mathbb{k})).$$

Recalling Definition 3.14, we see that $\mathfrak{aff}_n(\mathbb{k})$ is the semi-direct product

$$\mathfrak{aff}_n(\mathbb{k}) = \mathfrak{gl}_n(\mathbb{k}) \ltimes \mathfrak{trans}_n(\mathbb{k}).$$

To summarise, for $\mathbb{k} = \mathbb{R}$ or $\mathbb{C}$ we have

$$\begin{cases} \mathfrak{aff}_n(\mathbb{k}) = \mathrm{T}_I \, \mathrm{Aff}_n(\mathbb{k}) = \mathfrak{gl}_n(\mathbb{k}) \ltimes \mathfrak{trans}_n(\mathbb{k}), \\ \dim_{\mathbb{k}} \mathrm{Aff}_n(\mathbb{k}) = n^2 + n, \\ \dim \mathrm{Aff}_n(\mathbb{k}) = (n^2 + n) \dim_{\mathbb{R}} \mathbb{k}. \end{cases} \tag{3.9}$$

Upper Triangular and Unipotent Groups

For $n \geqslant 1$, recall the upper triangular and unipotent subgroups $\mathrm{UT}_n(\mathbb{k})$ and $\mathrm{SUT}_n(\mathbb{k})$ of $\mathrm{GL}_n(\mathbb{k})$. Let

$$\alpha\colon (-\varepsilon, \varepsilon) \longrightarrow \mathrm{UT}_n(\mathbb{R})$$

be a differentiable curve with $\alpha(0) = I$. Then $\alpha'(t)$ is upper triangular. Moreover, using the argument for $\mathrm{GL}_n(\mathbb{k})$ we see that given any upper triangular matrix $A \in \mathrm{M}_n(\mathbb{k})$, there is a curve

$$\alpha\colon (-\varepsilon, \varepsilon) \longrightarrow \mathrm{UT}_n(\mathbb{k}); \quad \alpha(t) = I + tA,$$

where $\varepsilon > 0$ has to be chosen small and $\alpha'(0) = A$. Then we have

$$\begin{cases} \mathfrak{ut}_n(\mathbb{k}) = \text{set of upper triangular matrices in } \mathrm{M}_n(\mathbb{k}), \\ \dim \mathfrak{ut}_n(\mathbb{k}) = \dbinom{n+1}{2} \dim_{\mathbb{R}} \mathbb{k}. \end{cases} \tag{3.10}$$

An upper triangular matrix $A \in \mathrm{M}_n(\mathbb{k})$ is *strictly upper triangular* if all its diagonal entries are 0, *i.e.*, $a_{ii} = 0$. Then

$$\begin{cases} \mathfrak{sut}_n(\mathbb{k}) = \text{set of strictly upper triangular matrices in } \mathrm{M}_n(\mathbb{k}), \\ \dim \mathfrak{sut}_n(\mathbb{k}) = \dbinom{n}{2} \dim_{\mathbb{R}} \mathbb{k}. \end{cases} \tag{3.11}$$

Orthogonal and special orthogonal groups

Let $\mathrm{O}(n)$ be the $n \times n$ orthogonal group, *i.e.*,

$$\mathrm{O}(n) = \{A \in \mathrm{GL}_n(\mathbb{R}) : A^T A = I\} \leqslant \mathrm{GL}_n(\mathbb{R}).$$

Given a curve $\alpha\colon (a, b) \longrightarrow \mathrm{O}(n)$ satisfying $\alpha(0) = I$ we have

$$\frac{\mathrm{d}}{\mathrm{d}t}\alpha(t)^T \alpha(t) = O,$$

and so

$$\alpha'(t)^T \alpha(t) + \alpha(t)^T \alpha'(t) = O,$$

implying

$$\alpha'(0)^T + \alpha'(0) = O.$$

Thus we must have $\alpha'(0)^T = -\alpha'(0)$, *i.e.*, $\alpha'(0)$ is skew symmetric. Thus

$$\mathfrak{o}(n) = \mathrm{T}_I\,\mathrm{O}(n) \subseteq \text{Sk-Sym}_n(\mathbb{R}),$$

the set of $n \times n$ real skew symmetric matrices.

On the other hand, if $A \in \text{Sk-Sym}_n(\mathbb{R})$, for $\varepsilon > 0$ we can consider the curve

$$\alpha\colon (-\varepsilon, \varepsilon) \longrightarrow \mathrm{GL}_n(\mathbb{R}); \quad \alpha(t) = \exp(tA).$$

Then

$$\begin{aligned}\alpha(t)^T\alpha(t) = \exp(tA)^T\exp(tA) &= \exp(tA^T)\exp(tA)\\ &= \exp(-tA)\exp(tA)\\ &= I.\end{aligned}$$

Hence we can view α as a curve $\alpha\colon (-\varepsilon, \varepsilon) \longrightarrow \mathrm{O}(n)$. Since $\alpha'(0) = A$, this shows that

$$\text{Sk-Sym}_n(\mathbb{R}) \subseteq \mathfrak{o}(n) = \mathrm{T}_I\,\mathrm{O}(n)$$

and so

$$\mathfrak{o}(n) = \mathrm{T}_I\,\mathrm{O}(n) = \text{Sk-Sym}_n(\mathbb{R}).$$

Notice that if $A \in \text{Sk-Sym}_n(\mathbb{R})$ then

$$\operatorname{tr} A = \operatorname{tr} A^T = \operatorname{tr}(-A) = -\operatorname{tr} A,$$

hence $\operatorname{tr} A = 0$. By Lemma 3.23, $\det\exp(tA) = 1$, so $\alpha\colon (-\varepsilon, \varepsilon) \longrightarrow \mathrm{SO}(n)$ where $\mathrm{SO}(n)$ is the $n \times n$ special orthogonal group. It is also easy to show that

$$\dim \text{Sk-Sym}_n(\mathbb{R}) = \binom{n}{2}.$$

So we have actually shown

$$\begin{cases}\mathfrak{so}(n) = \mathrm{T}_I\,\mathrm{SO}(n) = \mathfrak{o}(n) = \mathrm{T}_I\,\mathrm{O}(n) = \text{Sk-Sym}_n(\mathbb{R}),\\ \dim \mathrm{SO}(n) = \binom{n}{2}.\end{cases} \tag{3.12}$$

We have also shown that if $A \in \text{Sk-Sym}_n(\mathbb{R})$, then

$$\exp(tA) \in \mathrm{SO}(n) \quad (t \in \mathbb{R}). \tag{3.13}$$

Unitary and special unitary groups

Now consider the $n \times n$ unitary group

$$\mathrm{U}(n) = \{A \in \mathrm{GL}_n(\mathbb{C}) : A^*A = I\}.$$

For a curve α in $\mathrm{U}(n)$ satisfying $\alpha(0) = I$, we obtain

$$\alpha'(0)^* + \alpha'(0) = 0$$

and so $\alpha'(0)^* = -\alpha'(0)$, *i.e.*, $\alpha(0)$ is *skew hermitian*. So

$$\mathfrak{u}(n) = \mathrm{T}_I \, \mathrm{U}(n) \subseteq \text{Sk-Herm}_n(\mathbb{C}),$$

the set of all $n \times n$ skew hermitian matrices.

If $H \in \text{Sk-Herm}_n(\mathbb{C})$ then the curve

$$\eta\colon (-\varepsilon, \varepsilon) \longrightarrow \mathrm{GL}_n(\mathbb{C}); \quad \eta(t) = \exp(tH)$$

satisfies

$$\begin{aligned}\eta(t)^*\eta(t) = \exp(tH)^* \exp(tH) &= \exp(tH^*) \exp(tH) \\ &= \exp(-tH) \exp(tH) \\ &= I.\end{aligned}$$

Hence we can view η as a curve $\eta\colon (-\varepsilon, \varepsilon) \longrightarrow \mathrm{U}(n)$. Since $\eta'(0) = H$, this shows that

$$\text{Sk-Herm}_n(\mathbb{C}) \subseteq \mathfrak{u}(n) = \mathrm{T}_I \, \mathrm{U}(n).$$

Hence since

$$\dim \text{Sk-Herm}_n(\mathbb{C}) = n + 2\binom{n}{2} = n^2,$$

we have

$$\begin{cases} \mathfrak{u}(n) = \mathrm{T}_I \, \mathrm{U}(n) = \text{Sk-Herm}_n(\mathbb{C}), \\ \dim \mathrm{U}(n) = n^2. \end{cases} \tag{3.14}$$

The special unitary group $\mathrm{SU}(n)$ can be handled in a similar way. Again we have

$$\mathfrak{su}(n) = \mathrm{T}_I \, \mathrm{SU}(n) \subseteq \text{Sk-Herm}_n(\mathbb{C}).$$

But also if $\eta\colon (a, b) \longrightarrow \mathrm{SU}(n)$ is a curve with $\eta(0) = I$, then as in the analysis for $\mathrm{SL}_n(\mathbb{R})$, we have $\operatorname{tr} \eta'(0) = 0$. Writing

$$\text{Sk-Herm}^0_n(\mathbb{C}) = \{H \in \text{Sk-Herm}_n(\mathbb{C}) : \operatorname{tr} H = 0\},$$

we see that

$$\dim \text{Sk-Herm}^0_n(\mathbb{C}) = \dim \text{Sk-Herm}_n(\mathbb{C}) - 1 = n^2 - 1,$$

and $\mathfrak{su}(n) \subseteq \text{Sk-Herm}_n^0(\mathbb{C})$. On the other hand, if $H \in \text{Sk-Herm}_n^0(\mathbb{C})$ then the curve

$$\eta\colon (-\varepsilon, \varepsilon) \longrightarrow \mathrm{U}(n); \quad \eta(t) = \exp(tH),$$

takes values in $\mathrm{SU}(n)$ by Lemma 3.23 and satisfies $\eta'(0) = H$. Hence

$$\begin{cases} \mathfrak{su}(n) = \mathrm{T}_I\,\mathrm{SU}(n) \subseteq \text{Sk-Herm}_n^0(\mathbb{C}), \\ \dim \mathrm{SU}(n) = n^2 - 1. \end{cases} \tag{3.15}$$

3.4 Some Observations on the Exponential Function of a Matrix Group

Later we will see that for a matrix group $G \leqslant \mathrm{GL}_n(\mathbb{R})$, the following statements are always true and are often helpful in determining Lie algebras of matrix groups using the approaches seen in this section.

- The function

$$\exp_G\colon \mathfrak{g} \longrightarrow \mathrm{GL}_n(\mathbb{R}); \quad \exp_G(X) = \exp(X),$$

  has image contained in G, $\exp_G \mathfrak{g} \subseteq G$; so we will usually write $\exp_G\colon \mathfrak{g} \longrightarrow G$ for the exponential of G and sometimes even just exp.
- If G is compact and connected then $\exp_G \mathfrak{g} = G$.
- There is an open disc $\mathrm{N}_{\mathfrak{g}}(O; r) \subseteq \mathfrak{g}$ on which exp is injective and gives a homeomorphism

$$\exp\colon \mathrm{N}_{\mathfrak{g}}(O; r) \longrightarrow \exp \mathrm{N}_{\mathfrak{g}}(O; r),$$

  where $\exp \mathrm{N}_{\mathfrak{g}}(O; r) \subseteq G$ is in fact an open subset.

Here is an example which shows that the exponential map need not be surjective for a connected matrix group.

Example 3.24

Consider the exponential $\exp_{\mathrm{SL}_2(\mathbb{R})}\colon \mathfrak{sl}_2(\mathbb{R}) \longrightarrow \mathrm{SL}_2(\mathbb{R})$. Then for $\delta \geqslant 0$, the matrix

$$\begin{bmatrix} -(2+\delta) & 0 \\ 0 & \dfrac{-1}{(2+\delta)} \end{bmatrix} \in \mathrm{SL}_2(\mathbb{R})$$

is not in the image of $\exp_{\mathrm{SL}_2(\mathbb{R})}$.

Proof

Let

$$A \in \mathfrak{sl}_2(\mathbb{R}) = \{A \in \mathrm{M}_2(\mathbb{R}) : \operatorname{tr} A = 0\}.$$

By the Cayley–Hamilton Theorem 2.6, we have

$$A^2 + (\det A)I_2 = 0.$$

If $\det A = 0$ then the only eigenvalue of A is 0 and for any $t \in \mathbb{R}$ we have

$$\exp(tA) = I_2 + tA,$$

from which we obtain

$$\operatorname{tr}\exp(tA) = 2.$$

If $\det A \neq 0$, then A has two distinct non-zero eigenvalues, namely

$$\begin{cases} \pm\sqrt{\det A}\, i & \text{if } \det A > 0, \\ \pm\sqrt{-\det A} & \text{if } \det A < 0. \end{cases}$$

In either case there would be an invertible matrix $P \in \mathrm{GL}_2(\mathbb{C})$ for which

$$P^{-1}AP = \begin{cases} \operatorname{diag}(\sqrt{\det A}\, i, -\sqrt{\det A}\, i) & \text{if } \det A > 0, \\ \operatorname{diag}(\sqrt{-\det A}, -\sqrt{-\det A}) & \text{if } \det A < 0. \end{cases}$$

So for $t \in \mathbb{R}$ we have

$$P^{-1}\exp(tA)P = \begin{cases} \operatorname{diag}\left(e^{t\sqrt{\det A}\, i}, e^{-t\sqrt{\det A}\, i}\right) & \text{if } \det A > 0, \\ \operatorname{diag}\left(e^{t\sqrt{-\det A}}, e^{-t\sqrt{-\det A}}\right) & \text{if } \det A < 0, \end{cases}$$

which yields

$$\exp(tA) = \begin{cases} \cos(t\sqrt{\det A})I_2 + \dfrac{\sin(t\sqrt{\det A})}{\sqrt{\det A}}A & \text{if } \det A > 0, \\ \cosh(t\sqrt{-\det A})I_2 + \dfrac{\sinh(t\sqrt{-\det A})}{\sqrt{-\det A}}A & \text{if } \det A < 0. \end{cases}$$

This in turn gives

$$\operatorname{tr}\exp(tA) = \begin{cases} 2\cos(t\sqrt{\det A})I_2 & \text{if } \det A > 0, \\ 2\cosh(t\sqrt{-\det A})I_2 & \text{if } \det A < 0. \end{cases}$$

Hence, whenever $\operatorname{tr} A = 0$ and $t \in \mathbb{R}$, $\operatorname{tr}\exp(tA) \geqslant -2$. Since

$$\operatorname{tr}\begin{bmatrix} -(2+\delta) & 0 \\ 0 & \dfrac{-1}{(2+\delta)} \end{bmatrix} < -2,$$

the matrix $\begin{bmatrix} -(2+\delta) & 0 \\ 0 & -1/(2+\delta) \end{bmatrix}$ cannot be of the form $\exp(A)$ for any real traceless matrix A. □

3.5 SO(3) and SU(2)

In this section we will discuss the groups SO(3) and SU(2) and their Lie algebras in detail. It is more usual to do this using the identification of SU(2) with the unit *quaternions* which we define in Chapter 4, but here we will develop the ideas without that interpretation. The Lie algebras $\mathfrak{so}(3)$ and $\mathfrak{su}(2)$ are both 3-dimensional real vector spaces, for example having the following bases:

$$\mathfrak{so}(3): \quad P = \begin{bmatrix} 0 & -1 & 0 \\ 1 & 0 & 0 \\ 0 & 0 & 0 \end{bmatrix}, \quad Q = \begin{bmatrix} 0 & 0 & -1 \\ 0 & 0 & 0 \\ 1 & 0 & 0 \end{bmatrix}, \quad R = \begin{bmatrix} 0 & 0 & 0 \\ 0 & 0 & -1 \\ 0 & 1 & 0 \end{bmatrix},$$

$$\mathfrak{su}(2): \quad H = \frac{1}{2}\begin{bmatrix} i & 0 \\ 0 & -i \end{bmatrix}, \quad E = \frac{1}{2}\begin{bmatrix} 0 & 1 \\ -1 & 0 \end{bmatrix}, \quad F = \frac{1}{2}\begin{bmatrix} 0 & i \\ i & 0 \end{bmatrix}.$$

The non-trivial Lie brackets amongst these are

$$[P,Q] = R, \quad [Q,R] = P, \quad [R,P] = Q, \tag{3.16a}$$

$$[H,E] = F, \quad [E,F] = H, \quad [F,H] = E. \tag{3.16b}$$

This implies that the $\mathbb{R}$-linear isomorphism

$$\varphi\colon \mathfrak{su}(2) \longrightarrow \mathfrak{so}(3); \quad \varphi(xH+yE+zF) = xP+yQ+zR \quad (x,y,z \in \mathbb{R}), \tag{3.17}$$

satisfies

$$\varphi([U,V]) = [\varphi(U), \varphi(V)],$$

and so is an *isomorphism of* $\mathbb{R}$*-Lie algebras*. Thus these Lie algebras look the same algebraically. This suggests that there might be a close relationship between the groups themselves. Before describing this, notice also that for the Lie algebra of Example 3.2, the $\mathbb{R}$-linear transformation

$$\theta_0\colon \mathbb{R}^3 \longrightarrow \mathfrak{so}(3); \quad \theta_0(xe_1 + ye_2 + ze_3) = xP + yQ + zR,$$

is an isomorphism of $\mathbb{R}$-Lie algebras by the equations of (3.1).

Now we will construct a Lie homomorphism $SU(2) \longrightarrow SO(3)$ whose derivative is φ. Recall the *adjoint action* of Ad of SU(2) on $\mathfrak{su}(2)$ by

$$\mathrm{Ad}_A(U) = AUA^{-1} = AUA^* \quad (A \in SU(2),\ U \in \mathfrak{su}(2)).$$

Then each Ad_A is an $\mathbb{R}$-linear isomorphism $\mathfrak{su}(2) \longrightarrow \mathfrak{su}(2)$.

We can define a *real inner product* $(\ \mid\ )$ on $\mathfrak{su}(2)$ by

$$(X \mid Y) = -\operatorname{tr}(XY) \quad (X, Y \in \mathfrak{su}(2)).$$

The elements

$$\hat{H} = \sqrt{2}H = \frac{1}{\sqrt{2}}\begin{bmatrix} i & 0 \\ 0 & -i \end{bmatrix},$$

$$\hat{E} = \sqrt{2}E = \frac{1}{\sqrt{2}}\begin{bmatrix} 0 & 1 \\ -1 & 0 \end{bmatrix},$$

$$\hat{F} = \sqrt{2}F = \frac{1}{\sqrt{2}}\begin{bmatrix} 0 & i \\ i & 0 \end{bmatrix},$$

form an orthonormal basis $\{\hat{H}, \hat{E}, \hat{F}\}$ of $\mathfrak{su}(2)$ with respect to the inner product $(\ |\ )$, *i.e.*,

$$(\hat{H} \mid \hat{H}) = (\hat{E} \mid \hat{E}) = (\hat{F} \mid \hat{F}) = 1, \tag{3.18a}$$

$$(\hat{H} \mid \hat{E}) = (\hat{H} \mid \hat{F}) = (\hat{E} \mid \hat{F}) = 0. \tag{3.18b}$$

We can define an $\mathbb{R}$-linear isomorphism

$$\theta\colon \mathbb{R}^3 \longrightarrow \mathfrak{su}(2); \quad \theta(x\mathbf{e}_1 + y\mathbf{e}_2 + z\mathbf{e}_3) = x\hat{H} + y\hat{E} + z\hat{F}, \tag{3.19}$$

which is also an isometry, *i.e.*,

$$(\theta(\mathbf{x}) \mid \theta(\mathbf{y})) = \mathbf{x} \cdot \mathbf{y} \quad (\mathbf{x}, \mathbf{y} \in \mathbb{R}^3).$$

Remark 3.25

It would perhaps be more natural to rescale the inner product $(\ |\ )$ so that H, E, F were all unit vectors. This would certainly make many of the formulæ that follow neater as well as making the Lie bracket in $\mathrm{SU}(2)$ correspond exactly with the vector product in $\mathbb{R}^3$. However, our choice of $(\ |\ )$ agrees with the conventional one for $\mathrm{SU}(n)$ defined in Chapter 11.

Proposition 3.26

$(\ |\ )$ is a *real symmetric bilinear form* on $\mathfrak{su}(2)$ which is also *positive definite.* It is *invariant* in the sense that

$$([Z, X] \mid Y) + (X \mid [Z, Y]) = 0 \quad (X, Y, Z \in \mathfrak{su}(2)).$$

Proof

The $\mathbb{R}$-bilinearity is clear, as is the symmetry. For positive definiteness, notice that for $x, x', y, y', z, z' \in \mathbb{R}$,

$$(x\hat{H} + y\hat{E} + z\hat{F} \mid x'\hat{H} + y'\hat{E} + z'\hat{F}) = xx' + yy' + zz'$$

and in particular,

$$(x\hat{H} + y\hat{E} + z\hat{F} \mid x\hat{H} + y\hat{E} + z\hat{F}) = x^2 + y^2 + z^2 \geqslant 0,$$

with equality precisely when $x = y = z = 0$.

The invariance property is checked with a straightforward calculation. □

Also, for $A \in \mathrm{SU}(2)$ and $X, Y \in \mathfrak{su}(2)$,

$$\begin{aligned}(AXA^* \mid AYA^*) = -\operatorname{tr}(AXA^*AYA^*) &= -\operatorname{tr}(AXYA^*)\\ &= -\operatorname{tr}(AXYA^{-1})\\ &= -\operatorname{tr}(XY)\\ &= (X \mid Y),\end{aligned}$$

hence Ad_A is actually an orthogonal linear transformation with respect to this inner product. Using the orthonormal basis $\hat{H}, \hat{E}, \hat{F}$, we can identify $\mathfrak{su}(2)$ with $\mathbb{R}^3$ and $(\ \mid\ )$ with the usual inner product $\cdot$, then each Ad_A corresponds to an element of $\mathrm{O}(3)$ which we will still write as Ad_A. It is then easy to see that the function

$$\overline{\mathrm{Ad}}\colon \mathrm{SU}(2) \longrightarrow \mathrm{O}(3); \quad \overline{\mathrm{Ad}}(A) = \mathrm{Ad}_A \in \mathrm{O}(3),$$

is a continuous homomorphism of groups. In Examples 9.14 and 9.15 we will show that $\mathrm{SU}(2)$ and $\mathrm{SO}(3)$ are path connected. Since $\overline{\mathrm{Ad}}(I) = I$, this implies that $\overline{\mathrm{Ad}}\,\mathrm{SU}(2) \subseteq \mathrm{SO}(3)$. Because of this, it is convenient to redefine $\overline{\mathrm{Ad}}$ by setting

$$\overline{\mathrm{Ad}}\colon \mathrm{SU}(2) \longrightarrow \mathrm{SO}(3); \quad \overline{\mathrm{Ad}}(A) = \mathrm{Ad}_A\,.$$

Proposition 3.27

The continuous homomorphism of matrix groups

$$\overline{\mathrm{Ad}}\colon \mathrm{SU}(2) \longrightarrow \mathrm{SO}(3); \quad \overline{\mathrm{Ad}}(A) = \mathrm{Ad}_A,$$

is smooth, has $\ker \overline{\mathrm{Ad}} = \{\pm I\}$ and is surjective.

Proof

The identification of the kernel is an easy exercise. The remaining statements can be proved using ideas from Section 9.1. We will give a direct proof that $\overline{\mathrm{Ad}}$ is surjective to illustrate some important special geometric aspects of this example.

We can view an element of $\mathfrak{su}(2)$ as a vector in $\mathbb{R}^3$ by identifying the orthonormal basis vectors $\hat{H}$, $\hat{E}$, $\hat{F}$ with $\mathbf{e}_1$, $\mathbf{e}_2$, $\mathbf{e}_3$. From Equations (3.16), the non-trivial brackets of these basis elements are as follows:

$$[\hat{H}, \hat{E}] = \sqrt{2}\hat{F}, \quad [\hat{E}, \hat{F}] = \sqrt{2}\hat{H}, \quad [\hat{F}, \hat{H}] = \sqrt{2}\hat{E}. \tag{3.20}$$

So apart from the factors of $\sqrt{2}$, this behaves exactly like the vector product on $\mathbb{R}^3$.

Lemma 3.28

For $U_1 = x_1\hat{H} + y_1\hat{E} + z_1\hat{F}$, $U_2 = x_2\hat{H} + y_2\hat{E} + z_2\hat{F} \in \mathfrak{su}(2)$,

$$[U_1, U_2] = \sqrt{2}\left(\begin{vmatrix} y_1 & z_1 \\ y_2 & z_2 \end{vmatrix} \hat{H} - \begin{vmatrix} x_1 & z_1 \\ x_2 & z_2 \end{vmatrix} \hat{E} + \begin{vmatrix} x_1 & y_1 \\ x_2 & y_2 \end{vmatrix} \hat{F}\right).$$

Proof

The result is implied by the calculation

$$\begin{aligned} x\mathbf{e}_1 + y\mathbf{e}_2 + z\mathbf{e}_3 &= (x_1\mathbf{e}_1 + y_1\mathbf{e}_2 + z_1\mathbf{e}_3) \times (x_2\mathbf{e}_1 + y_2\mathbf{e}_2 + z_2\mathbf{e}_3) \\ &= \begin{vmatrix} y_1 & z_1 \\ y_2 & z_2 \end{vmatrix} \mathbf{e}_1 - \begin{vmatrix} x_1 & z_1 \\ x_2 & z_2 \end{vmatrix} \mathbf{e}_2 + \begin{vmatrix} x_1 & y_1 \\ x_2 & y_2 \end{vmatrix} \mathbf{e}_3. \end{aligned}$$ □

In a similar fashion, we can express a product of elements of $\mathfrak{su}(2)$ in terms of the dot and cross products. Note however, that if $U_1, U_2 \in \mathfrak{su}(2)$ then in general $U_1U_2 \notin \mathfrak{su}(2)$.

Lemma 3.29

If $U_1, U_2 \in \mathfrak{su}(2)$ with $U_1 = x_1\hat{H} + y_1\hat{E} + z_1\hat{F}$ and $U_2 = x_2\hat{H} + y_2\hat{E} + z_2\hat{F}$, then

$$\begin{aligned} U_1U_2 = &-\frac{(x_1x_2 + y_1y_2 + z_1z_2)}{2} I \\ &+ \frac{1}{\sqrt{2}}\left(\begin{vmatrix} y_1 & z_1 \\ y_2 & z_2 \end{vmatrix} \hat{H} - \begin{vmatrix} x_1 & z_1 \\ x_2 & z_2 \end{vmatrix} \hat{E} + \begin{vmatrix} x_1 & y_1 \\ x_2 & y_2 \end{vmatrix} \hat{F}\right) \\ = &-\frac{(U_1 \mid U_2)}{2} I + \frac{1}{2}[U_1, U_2]. \end{aligned}$$

Proof

Calculation! □

Notice that when $U_1, U_2 \in \mathfrak{su}(2)$ are *orthogonal* with respect to $(\ |\ )$ in the sense that $(U_1 \mid U_2) = 0$, then

$$U_1U_2 = \frac{1}{2}[U_1, U_2] \in \mathfrak{su}(2). \tag{3.21}$$

Next we will examine the effect of $A \in \mathrm{SU}(2)$ acting as an $\mathbb{R}$-linear transformation on $\mathfrak{su}(2)$ which we will identify with $\mathbb{R}^3$. Note that A can be uniquely written as

$$A = \begin{bmatrix} u & v \\ -\overline{v} & \overline{u} \end{bmatrix} \tag{3.22}$$

for $u, v \in \mathbb{C}$ and $|u|^2 + |v|^2 = 1$. This allows us to express A in the form

$$A = \cos\theta I + S,$$

where S is skew hermitian and $\mathrm{Re}\, u = \cos\theta$ for $\theta \in [0, \pi]$, so $\sin\theta \geqslant 0$. A calculation gives

$$S^2 = -((\mathrm{Im}\, u)^2 + |v|^2)I = -\sin^2\theta I, \tag{3.23a}$$
$$(S \mid S) = 2\sin^2\theta. \tag{3.23b}$$

Since $A \in \mathrm{SU}(2)$, we have

$$A^{-1} = A^* = \cos\theta I - S.$$

Notice that for any $t \in \mathbb{R}$,

$$\mathrm{Ad}_A(tS) = A(tS)A^{-1} = tS.$$

On the other hand, if $U \in \mathfrak{su}(2)$ with $(S \mid U) = 0$, then by the above results,

$$\begin{aligned}
\mathrm{Ad}_A(U) &= (\cos\theta I + S)U(\cos\theta I - S) \\
&= (\cos\theta U + SU)(\cos\theta I - S) \\
&= \cos^2\theta U + \cos\theta SU - \cos\theta US - SUS \\
&= \cos^2\theta U + \cos\theta[S, U] - SUS.
\end{aligned}$$

A further calculation using properties of the vector product shows that

$$SUS = \frac{(S \mid S)}{2}U.$$

By Equation (3.23b), whenever $(S \mid U) = 0$ we have

$$\begin{aligned}
\mathrm{Ad}_A(U) &= (\cos^2\theta - \sin^2\theta)U + \cos\theta[S, U] \\
&= (\cos 2\theta)U + \cos\theta[S, U] \\
&= (\cos 2\theta)U + \sqrt{2}\cos\theta\sin\theta[\hat{S}, U] \\
&= \cos 2\theta\, U + \sin 2\theta \hat{S} \times U,
\end{aligned}$$

where $\hat{S} = \dfrac{1}{\sqrt{2}\sin\theta}S$ is of unit length. Noting that U and $\hat{S} \times U$ are orthogonal to S, we see that the effect of Ad_A on U is to rotate it in the plane orthogonal to S (and spanned by U and $\hat{S} \times U$) through the angle θ.

We can now see that every element $R \in \mathrm{SO}(3)$ has the form Ad_A for some $A \in \mathrm{SU}(2)$. This follows from the facts that the eigenvalues of R have modulus 1 and $\det R = 1$. Together these show that at least one of the eigenvalues of R must be 1 with corresponding eigenvector $\mathbf{v}$ say, while the other two have the form $e^{\pm\varphi i} = \cos\varphi \pm i\sin\varphi$ for some φ. Now we can take $A = \cos(\varphi/2)I + S$, where $S \in \mathfrak{su}(2)$ is chosen to correspond to a multiple of $\mathbf{v}$ and $(S \mid S) = 2\sin^2(\varphi/2)$. If we choose $-\varphi$ in place of φ we obtain $-A$ in place of A.

This completes the proof of Proposition 3.27. □

Let $B \in \mathfrak{su}(2)$. Then starting with the curve

$$\beta\colon \mathbb{R} \longrightarrow \mathrm{SU}(2); \quad \beta(t) = \exp(tB),$$

we can construct another curve

$$\overline{\beta}\colon \mathbb{R} \longrightarrow \mathrm{SO}(3); \quad \overline{\beta}(t) = \overline{\mathrm{Ad}}_{\beta(t)}.$$

We can differentiate $\overline{\beta}$ at $t = 0$ to obtain an element of $\mathfrak{so}(3)$ given by the formula:

$$\begin{aligned}\overline{\beta}'(0)(X) &= \frac{\mathrm{d}}{\mathrm{d}t}\exp(tB)X\exp(-tB)_{|t=0} \\ &= BX - XB = [B, X].\end{aligned}$$

For example, when $B = H$ we have

$$[H, H] = 0, \quad [H, E] = F, \quad [H, F] = -E,$$

hence the matrix of H acting on $\mathfrak{su}(2)$ relative to the basis H, E, F is

$$R = \begin{bmatrix} 0 & 0 & 0 \\ 0 & 0 & -1 \\ 0 & 1 & 0 \end{bmatrix}.$$

Similarly,

$$[E, H] = -F, \quad [E, E] = 0, \quad [E, F] = H,$$

giving the matrix

$$Q = \begin{bmatrix} 0 & 0 & 1 \\ 0 & 0 & 0 \\ -1 & 0 & 0 \end{bmatrix},$$

and

$$[F, H] = E, \quad [F, E] = -H, \quad [F, F] = 0,$$

giving

$$P = \begin{bmatrix} 0 & -1 & 0 \\ 1 & 0 & 0 \\ 0 & 0 & 0 \end{bmatrix}.$$

The corresponding derivative map is then

$$\mathrm{d}\,\overline{\mathrm{Ad}}\colon \mathfrak{su}(2) \longrightarrow \mathfrak{so}(3); \quad \mathrm{d}\,\overline{\mathrm{Ad}}(xH + yE + zF) = xR + yQ + zP.$$

Apart from the change in order, this is the 'obvious' isomorphism between these two Lie algebras.

To summarise, we have proved the following important result.

Theorem 3.30

$\overline{\mathrm{Ad}}\colon \mathrm{SU}(2) \longrightarrow \mathrm{SO}(3)$ is a surjective Lie homomorphism with $\ker \overline{\mathrm{Ad}} = \{\pm I\}$. Furthermore, the derivative $\mathrm{d}\,\overline{\mathrm{Ad}}\colon \mathfrak{su}(2) \longrightarrow \mathfrak{so}(3)$ is an isomorphism of $\mathbb{R}$-Lie algebras.

In Chapter 5 we will meet the *spinor groups* and some generalisations of this double covering. A related double covering involving $\mathrm{SL}_2(\mathbb{C})$ and the Lorentz group Lor will be discussed in Section 6.3.

3.6 The Complexification of a Real Lie Algebra

The ideas of this section are especially important in the representation theory of Lie groups and Lie algebras (see Serre [26] for example) and for completeness we provide a brief discussion.

Definition 3.31

Given a finite dimensional $\mathbb{R}$-Lie algebra $\mathfrak{g}$, a $\mathbb{C}$-Lie algebra $\mathfrak{g}'$ which contains $\mathfrak{g}$ as an $\mathbb{R}$-Lie subalgebra and for which $\dim_{\mathbb{C}} \mathfrak{g}' = \dim_{\mathbb{R}} \mathfrak{g}$ is called a *complexification* of $\mathfrak{g}$.

Theorem 3.32

i) Every finite dimensional $\mathbb{R}$-Lie algebra $\mathfrak{g}$ has a complexification.
ii) If $\mathfrak{g}'$ and $\mathfrak{g}''$ are two complexifications of $\mathfrak{g}$ then there is an isomorphism of $\mathbb{C}$-Lie algebras $\mathfrak{g}' \longrightarrow \mathfrak{g}''$ which extends the identity function on $\mathfrak{g} \subseteq \mathfrak{g}'$.

Because of the uniqueness guaranteed by part (ii) this result, we can write $\mathfrak{g}_{\mathbb{C}}$ for such a complexification of $\mathfrak{g}$ since this is well defined up to an isomorphism of $\mathbb{C}$-Lie algebras. Our next result gives a useful criterion for finding complexifications.

Proposition 3.33

Let $\mathfrak{h}$ be a finite dimensional $\mathbb{C}$-Lie algebra and $\mathfrak{g} \subseteq \mathfrak{h}$ be an $\mathbb{R}$-Lie subalgebra. If $\mathfrak{g}' \subseteq \mathfrak{h}$ is the smallest $\mathbb{C}$-vector subspace containing $\mathfrak{g}$, then $\mathfrak{g}'$ is a complexification of $\mathfrak{g}$. In particular, if $\{u_1, \ldots, u_d\}$ is an $\mathbb{R}$-basis for $\mathfrak{g}$ which is $\mathbb{C}$-linearly independent in $\mathfrak{h}$, then $\{u_1, \ldots, u_d\}$ is a basis for $\mathfrak{g}'$.

Recall that $\dim \mathrm{U}(n) = \dim \mathfrak{u}(n) = n^2$. Then $\mathfrak{u}(n)$ is an $\mathbb{R}$-Lie subalgebra of $\mathfrak{gl}_n(\mathbb{C}) = \mathrm{M}_n(\mathbb{C})$ where $\dim_{\mathbb{R}} \mathfrak{gl}_n(\mathbb{C}) = 2n^2$. Of course, $\mathfrak{gl}_n(\mathbb{C})$ is also a $\mathbb{C}$-Lie algebra of dimension $\dim_{\mathbb{C}} \mathfrak{gl}_n(\mathbb{C}) = n^2$.

Proposition 3.34

The following are complexifications:

$$\mathfrak{gl}_n(\mathbb{C}) = \mathfrak{gl}_n(\mathbb{R})_{\mathbb{C}} = \mathfrak{u}(n)_{\mathbb{C}}, \quad \mathfrak{sl}_n(\mathbb{C}) = \mathfrak{sl}_n(\mathbb{R})_{\mathbb{C}} = \mathfrak{su}(n)_{\mathbb{C}}.$$

Proof

The following statements can be shown to be true by exhibiting explicit $\mathbb{R}$-bases of $\mathfrak{u}(n)$ and $\mathfrak{su}(n)$, then verifying that they are $\mathbb{C}$-bases for $\mathfrak{gl}_n(\mathbb{C})$ and $\mathfrak{sl}_n(\mathbb{C})$.

Let E^{rs} be the $n \times n$ matrix for which

$$E^{rs}{}_{ij} = \delta_{ir}\delta_{js} = \begin{cases} 1 & \text{if } i = r \text{ and } j = s, \\ 0 & \text{otherwise.} \end{cases} \tag{3.24}$$

Then $\mathfrak{u}(n)$ has an $\mathbb{R}$-basis consisting of the elements

$$\begin{cases} D_k = iE^{k\,k} & (1 \leqslant k \leqslant n), \\ P_{k\,\ell} = E^{k\,\ell} - E^{\ell\,k} & (1 \leqslant \ell < k \leqslant n), \\ Q_{k\,\ell} = iE^{k\,\ell} + iE^{\ell\,k} & (1 \leqslant \ell < k \leqslant n), \end{cases} \tag{3.25}$$

while $\mathfrak{su}(n)$ has an $\mathbb{R}$-basis consisting of the elements

$$\begin{cases} H_k = iE^{k\,k} - iE^{(k-1)\,(k-1)} & (1 \leqslant k \leqslant n-1), \\ P_{k\,\ell} = E^{k\,\ell} - E^{\ell\,k} & (1 \leqslant \ell < k \leqslant n), \\ Q_{k\,\ell} = iE^{k\,\ell} + iE^{\ell\,k} & (1 \leqslant \ell < k \leqslant n). \end{cases} \tag{3.26}$$

This suffices to establish the Proposition. □

The Lie algebras $\mathfrak{gl}_n(\mathbb{C})$ and $\mathfrak{sl}_n(\mathbb{C})$ have the following $\mathbb{C}$-bases.

$$\mathfrak{gl}_n(\mathbb{C}) \colon \begin{cases} E^{k\,k} & (1 \leqslant k \leqslant n), \\ E^{k\,\ell} & (1 \leqslant \ell, k \leqslant n,\ k \neq \ell). \end{cases} \tag{3.27}$$

$$\mathfrak{sl}_n(\mathbb{C}) \colon \begin{cases} E^{k\,k} - E^{(k-1)\,(k-1)} & (2 \leqslant k \leqslant n), \\ E^{k\,\ell} & (1 \leqslant \ell, k \leqslant n,\ k \neq \ell). \end{cases} \tag{3.28}$$

Here is a general result on the existence of complexifications.

Proposition 3.35

Every $\mathbb{R}$-Lie subalgebra $\mathfrak{g} \leqslant \mathfrak{u}(n)$ has a complexification $\mathfrak{g}_{\mathbb{C}}$ which is a $\mathbb{C}$-Lie subalgebra of $\mathfrak{gl}_n(\mathbb{C})$. If $G \leqslant \mathrm{SU}(n)$ is a Lie subgroup, then its Lie algebra $\mathfrak{g} \leqslant \mathfrak{su}(n)$ has a complexification $\mathfrak{g}_{\mathbb{C}} \leqslant \mathfrak{su}(n)_{\mathbb{C}} = \mathfrak{sl}_n(\mathbb{C})$.

We know several 3-dimensional real Lie algebras, namely $\mathfrak{so}(3)$, $\mathfrak{su}(2)$, $\mathfrak{sl}_2(\mathbb{R})$ and (up to isomorphism) a complex one

$$\mathfrak{sl}_2(\mathbb{C}) \cong \mathfrak{sl}_2(\mathbb{R})_{\mathbb{C}} \cong \mathfrak{su}(2)_{\mathbb{C}} \cong \mathfrak{so}(3)_{\mathbb{C}}.$$

We know that $\mathfrak{so}(3) \cong \mathfrak{su}(2)$ as real Lie algebras, but have not yet considered whether $\mathfrak{so}(3) \cong \mathfrak{sl}_2(\mathbb{R})$. To do this, we need to investigate these Lie algebras further.

Since the Lie algebra $\mathfrak{so}(3)$ is essentially the vector space $\mathbb{R}^3$ with the vector cross product as its bracket, it is easy to see that given a pair of non-zero elements $X, Y \in \mathfrak{so}(3)$ which are linearly independent, their bracket $[X, Y]$ is non-zero and the vectors $X, Y, [X, Y]$ are linearly independent, and hence form a basis for $\mathfrak{so}(3)$. We will show that this is false for the Lie algebra $\mathfrak{sl}_2(\mathbb{R})$, so these Lie algebras cannot be isomorphic.

The pair of matrices

$$A = \begin{bmatrix} 1 & 0 \\ 0 & -1 \end{bmatrix}, \quad B = \begin{bmatrix} 0 & 1 \\ 0 & 0 \end{bmatrix}$$

in $\mathfrak{sl}_2(\mathbb{R})$ has bracket

$$[A, B] = \begin{bmatrix} 0 & 2 \\ 0 & 0 \end{bmatrix},$$

and the matrices $A, B, [A, B]$ are linearly dependent. As a consequence of this, we can deduce some information about the corresponding matrix groups.

Proposition 3.36

The real Lie algebras $\mathfrak{so}(3)$ and $\mathfrak{sl}_2(\mathbb{R})$ are not isomorphic. Hence there can be no Lie homomorphisms of any of the forms

$$\mathrm{SL}_2(\mathbb{R}) \longrightarrow \mathrm{SO}(3),\ \mathrm{SL}_2(\mathbb{R}) \longrightarrow \mathrm{SU}(2),\ \mathrm{SO}(3) \longrightarrow \mathrm{SL}_2(\mathbb{R}),\ \mathrm{SU}(2) \longrightarrow \mathrm{SL}_2(\mathbb{R})$$

with injective derivatives.

EXERCISES

3.1. Let $\Bbbk = \mathbb{R}$ or $\mathbb{C}$.

a) Consider the 2-dimensional $\Bbbk$-Lie algebras

$$\mathfrak{a} = \Bbbk^2, \quad \mathfrak{b} = \left\{ \begin{bmatrix} u & v \\ 0 & 0 \end{bmatrix} : u, v \in \Bbbk \right\},$$

with the obvious brackets which make $\mathfrak{a}$ abelian and $\mathfrak{b} \leqslant \mathrm{M}_2(\Bbbk)$. Show that any 2-dimensional $\Bbbk$-Lie algebra $\mathfrak{g}$ is isomorphic to $\mathfrak{a}$ if it is abelian and $\mathfrak{b}$ otherwise.

b) Find a matrix group $G \leqslant \mathrm{GL}_2(\Bbbk)$ whose Lie algebra is $\mathfrak{b}$.

3.2. Let G be a matrix group and $U \in G$.

a) Show that each of the functions

$$\begin{aligned} \mathrm{L}_U\colon G \longrightarrow G;\quad & \mathrm{L}_U(A) = UA, \\ \mathrm{R}_U\colon G \longrightarrow G;\quad & \mathrm{R}_U(A) = AU, \\ \mathrm{C}_U\colon G \longrightarrow G;\quad & \mathrm{C}_U(A) = UAU^{-1}, \end{aligned}$$

is a differentiable map and determine its derivative at I.

b) Using (a), show that there are $\mathbb{R}$-linear isomorphisms

$$\lambda_U\colon \mathrm{T}_I G \longrightarrow \mathrm{T}_U G, \quad \rho_U\colon \mathrm{T}_I G \longrightarrow \mathrm{T}_U G, \quad \chi_U\colon \mathrm{T}_I G \longrightarrow \mathrm{T}_I G,$$

such that for all $U, V \in G$,

$$\lambda_{UV} = \lambda_U \circ \lambda_V, \quad \rho_{UV} = \rho_V \circ \rho_U, \quad \chi_{UV} = \chi_U \circ \chi_V.$$

3.3. For each of the following matrix groups G, find the Lie algebra $\mathfrak{g}$.

$$G_1 = \{A \in \mathrm{GL}_2(\mathbb{R}) : A^T Q_1 A = Q_1\}, \qquad Q_1 = \begin{bmatrix} 1 & 0 \\ 0 & 0 \end{bmatrix};$$

$$G_2 = \{A \in \mathrm{GL}_2(\mathbb{R}) : A^T Q_2 A = Q_2\}, \qquad Q_2 = \begin{bmatrix} 1 & 0 \\ 0 & -1 \end{bmatrix};$$

$$G_3 = \{A \in \mathrm{GL}_3(\mathbb{R}) : A^T Q_3 A = Q_3\}, \qquad Q_3 = \begin{bmatrix} 1 & 0 & 0 \\ 0 & 0 & 0 \\ 0 & 0 & -1 \end{bmatrix};$$

$$G_4 = \mathrm{Aff}_n(\Bbbk) \quad (n = 1, 2, \ldots);$$
$$G_5 = \mathrm{Symp}_{2m}(\mathbb{R}) \quad (m = 1, 2, \ldots).$$

3.4. Prove Proposition 3.9.

3.5. Let G be a matrix group with Lie algebra $\mathfrak{g}$ and let $X, Y \in \mathfrak{g}$. Show that $[X, Y] = 0$ if and only if $\exp(sX)\exp(tY) = \exp(tY)\exp(sX)$ for all $s, t \in \mathbb{R}$.

3.6. Consider the set of all $n \times n$ real special orthogonal matrices $\mathrm{SO}(n)$ and its subset

$$U = \{A \in \mathrm{SO}(n) : \det(I + A) \neq 0\} \subseteq \mathrm{SO}(n).$$

Define the function

$$\Phi\colon U \longrightarrow \mathrm{M}_n(\mathbb{R}); \quad \Phi(A) = (I - A)(I + A)^{-1}.$$

[Φ is known as the *real Cayley transform*.]
a) Show that $\operatorname{im}\Phi = \text{Sk-Sym}_n(\mathbb{R})$, the set of all $n \times n$ real skew symmetric matrices. Hence we might as well write $\Phi\colon U \longrightarrow \text{Sk-Sym}_n(\mathbb{R})$.
b) Find the inverse map $\Phi^{-1}\colon \text{Sk-Sym}_n(\mathbb{R}) \longrightarrow U$.
c) Use (b) to determine the dimension of $\mathrm{SO}(n)$.

3.7. Consider the set of all $n \times n$ unitary matrices $\mathrm{U}(n)$ and its subset

$$V = \{A \in \mathrm{U}(n) : \det(I + A) \neq 0\} \subseteq \mathrm{U}(n).$$

Define the function

$$\Theta\colon V \longrightarrow \mathrm{M}_n(\mathbb{C}); \quad \Theta(A) = (I - A)(I + A)^{-1}.$$

[Θ is known as the *complex Cayley transform*.]
a) Show that $\operatorname{im}\Theta = \text{Sk-Herm}_n(\mathbb{C})$, the set of all $n \times n$ skew hermitian matrices. Hence we might as well write $\Theta\colon V \longrightarrow \text{Sk-Herm}_n(\mathbb{C})$.

b) Find the inverse map $\Theta^{-1}\colon \text{Sk-Herm}_n(\mathbb{C}) \longrightarrow V$.
c) Use (b) to determine the dimension of $\mathrm{U}(n)$.
d) In the case $n = 2$, show that $\Theta(V \cap \mathrm{SU}(2)) \subseteq \text{Sk-Herm}_2^0(\mathbb{C})$ and $\Theta^{-1}\,\text{Sk-Herm}_2^0(\mathbb{C}) \subseteq \mathrm{SU}(2)$. Is this true for $n > 2$?

3.8. For $n \geqslant 1$, prove the following:
a) $\mathrm{O}(n)$ is the semi-direct product $\{1, -1\} \ltimes \mathrm{SO}(n)$;
b) $\mathrm{U}(n)$ is the semi-direct product $\mathbb{T} \ltimes \mathrm{SU}(n)$, where

$$\mathbb{T} = \{z \in \mathbb{C} : |z| = 1\}$$

is the unit circle;
c) $\mathrm{GL}_n(\mathbb{R})$ is the semi-direct product $\mathbb{R}^\times \ltimes \mathrm{SL}_n(\mathbb{R})$;
d) $\mathrm{GL}_n(\mathbb{C})$ is the semi-direct product $\mathbb{C}^\times \ltimes \mathrm{SL}_n(\mathbb{C})$.

3.9. Verify the formula of Lemma 3.29.

4
Algebras, Quaternions and Quaternionic Symplectic Groups

In this chapter we begin by studying algebras over a field, with their groups of units providing many interesting groups. In particular, we study division algebras and their linear algebra. Then we introduce the *quaternions* which form the only non-commutative example of a real division algebra. There is an associated family of compact connected matrix groups defined using the quaternions, the *quaternionic symplectic groups* which provide another infinite family of compact simply connected matrix groups.

4.1 Algebras

In this section $\Bbbk$ will denote any field, although our main interest will be in the cases $\Bbbk = \mathbb{R}$ and $\Bbbk = \mathbb{C}$.

Definition 4.1

A *finite dimensional* (*associative* and *unital*) *algebra* A is a finite dimensional $\Bbbk$-vector space which is also an associative and unital ring such that for all $r, s \in \Bbbk$ and $a, b \in A$,

$$(ra)(sb) = (rs)(ab).$$

Here ra and sb are scalar products in the vector space structure, while $(rs)(ab)$ is the scalar product of $rs \in \Bbbk$ with the ring product $ab \in A$.

In such a $\Bbbk$-algebra A, if $1 \in \Bbbk$ is the unit of A, then for $t \in \Bbbk$, the element $t1 \in A$ satisfies

$$(t1)a = ta = t(a1) = a(t1).$$

If $\dim_{\Bbbk} A > 0$, then $1 \neq 0$, and the function

$$\eta\colon \Bbbk \longrightarrow A; \quad \eta(t) = t1$$

is an injective ring homomorphism. We usually write t for $\eta(t) = t1$.

If A is a commutative ring then A is a *commutative* $\Bbbk$*-algebra.*

If every non-zero element $u \in A$ is a unit, *i.e.*, is invertible, then A is a $\Bbbk$*-division algebra* or *division algebra over* $\Bbbk$. A commutative division algebra is a *field* while a non-commutative division algebra is called a *skew field.* In French *corps* ($\sim$ field) is often used to refer to a possibly non-commutative division algebra.

Example 4.2

For $n \geqslant 1$, $\mathrm{M}_n(\Bbbk)$ is a $\Bbbk$-algebra. Here we have $\eta(t) = tI_n$. For $n > 1$, $\mathrm{M}_n(\Bbbk)$ is non-commutative.

Example 4.3

The ring of complex numbers $\mathbb{C}$ is an $\mathbb{R}$-algebra. Here we have $\eta(t) = t$. Notice that $\mathbb{C}$ is a commutative division algebra.

Example 4.4

Let G be a finite group. Then for $\Bbbk = \mathbb{R}$ or $\mathbb{C}$, the *group algebra* $\Bbbk[G]$ has a basis consisting of the elements g of G, while the addition and multiplication are

$$(\sum_{g \in G} x_g g) + (\sum_{g \in G} y_g g) = \sum_{g \in G} (x_g + y_g) g,$$

$$(\sum_{g \in G} x_g g)(\sum_{h \in G} y_h h) = \sum_{g \in G} \left(\sum_{h \in G} x_{gh^{-1}} y_h \right) g.$$

The unit is $1 = 1_G$, the identity element of G.

In any $\mathbb{k}$-algebra A, the set of units of A forms a group $A^\times$ under multiplication, and this contains $\mathbb{k}^\times$ as a central subgroup, $\mathbb{k}^\times \leqslant A^\times$. If A is non-commutative, we might hope to find interesting subgroups of $A^\times$. Later we will see that the Clifford algebras provide a good illustration of this idea.

Example 4.5

For $A = \mathrm{M}_n(\mathbb{k})$, $\mathrm{M}_n(\mathbb{k})^\times = \mathrm{GL}_n(\mathbb{k})$, while $\mathbb{k}^\times$ is identified with the subgroup of invertible scalar matrices tI_n $(t \in \mathbb{k}^\times)$.

Example 4.6

Let $\mathbb{k}[G]$ be the group algebra of a finite group G. Then the basis elements $g \in G$ form a finite subgroup $G \leqslant \mathbb{k}[G]^\times$.

Definition 4.7

Let A and B be two $\mathbb{k}$-algebras. A $\mathbb{k}$-linear transformation $\varphi\colon A \longrightarrow B$ that is also a ring homomorphism is called a *$\mathbb{k}$-algebra homomorphism* or *homomorphism of $\mathbb{k}$-algebras.* A homomorphism of $\mathbb{k}$-algebras which is also an isomorphism of rings or equivalently of $\mathbb{k}$-vector spaces is called an *isomorphism of $\mathbb{k}$-algebras.* An isomorphism $\alpha\colon A \longrightarrow A$ is called an *automorphism of $\mathbb{k}$-algebras.*

Notice that the unit $\eta\colon \mathbb{k} \longrightarrow A$ is always a homomorphism of $\mathbb{k}$-algebras. There are obvious notions of kernel and image for such homomorphisms, and of subalgebra.

Definition 4.8

Given two $\mathbb{k}$-algebras A, B, their *direct product* has underlying set $A \times B$ with sum and product

$$(a_1, b_1) + (a_2, b_2) = (a_1 + a_2, b_1 + b_2), \quad (a_1, b_1)(a_2, b_2) = (a_1a_2, b_1b_2).$$

The zero is $(0,0)$ while the unit is $(1,1)$. It is easy to see that there is an isomorphism of $\mathbb{k}$-algebras $A \times B \cong B \times A$.

Given a $\mathbb{k}$-algebra A, it is also possible to consider the ring $\mathrm{M}_n(A)$ consisting of $m \times m$ matrices with entries in A; this is also a $\mathbb{k}$-algebra of dimension

$$\dim_{\mathbb{k}} \mathrm{M}_m(A) = m^2 \dim_{\mathbb{k}} A.$$

It is often the case that a $\Bbbk$-algebra A contains a subalgebra $\Bbbk_1 \subseteq A$ which is also a field. In that case A can be viewed as a vector space over $\Bbbk_1$ in two different ways, corresponding to left and right multiplication by elements of $\Bbbk_1$. Then for $t \in \Bbbk_1$, $a \in A$,

$$t \cdot a = ta; \qquad \text{(Left scalar multiplication)}$$

$$a \cdot t = at. \qquad \text{(Right scalar multiplication)}$$

These give different $\Bbbk_1$-vector space structures unless all elements of $\Bbbk_1$ commute with all elements of A, in which case $\Bbbk_1$ is said to be a *central subfield* of A. We sometimes write ${}_{\Bbbk_1}A$ and $A_{\Bbbk_1}$ to indicate which structure is being considered. $\Bbbk_1$ is itself a finite dimensional commutative $\Bbbk$-algebra of some dimension $\dim_{\Bbbk} \Bbbk_1$.

Proposition 4.9

Each of the $\Bbbk_1$-vector spaces ${}_{\Bbbk_1}A$ and $A_{\Bbbk_1}$ is finite dimensional and in fact

$$\dim_{\Bbbk} A = \dim_{\Bbbk_1}({}_{\Bbbk_1}A) \dim_{\Bbbk} \Bbbk_1 = \dim_{\Bbbk_1}(A_{\Bbbk_1}) \dim_{\Bbbk} \Bbbk_1.$$

Example 4.10

Let $\Bbbk = \mathbb{R}$ and $A = \mathrm{M}_2(\mathbb{R})$ where $\dim_{\mathbb{R}} A = 4$. If

$$\Bbbk_1 = \left\{ \begin{bmatrix} x & y \\ -y & x \end{bmatrix} : x, y \in \mathbb{R} \right\} \subseteq \mathrm{M}_2(\mathbb{R}),$$

then $\Bbbk_1 \cong \mathbb{C}$ so is a subfield of $\mathrm{M}_2(\mathbb{R})$, but it is not a central subfield. We also have $\dim_{\Bbbk_1} A = 2$.

Example 4.11

Let $\Bbbk = \mathbb{R}$ and $A = \mathrm{M}_2(\mathbb{C})$, so $\dim_{\mathbb{R}} A = 8$. Let

$$\Bbbk_1 = \left\{ \begin{bmatrix} x & y \\ -y & x \end{bmatrix} : x, y \in \mathbb{R} \right\} \subseteq \mathrm{M}_2(\mathbb{C}).$$

Then $\Bbbk_1 \cong \mathbb{C}$ so is subfield of $\mathrm{M}_2(\mathbb{C})$, but it is not a central subfield. Here $\dim_{\Bbbk_1} A = 4$.

Given a $\Bbbk$-algebra A and a subfield $\Bbbk_1 \subseteq A$ containing $\Bbbk$ (possibly equal to $\Bbbk$), an element $a \in A$ acts on A by left multiplication:

$$a \cdot u = au \quad (u \in A).$$

This is always a $\Bbbk$-linear transformation of A, and if we view A as the $\Bbbk_1$-vector space $A_{\Bbbk_1}$, it is always a $\Bbbk_1$-linear transformation. Given a $\Bbbk_1$-basis $\{v_1, \ldots, v_m\}$ for $A_{\Bbbk_1}$, there is an $m \times m$ matrix $\lambda(a)$ with entries in $\Bbbk_1$ defined by

$$av_j = \sum_{r=1}^{m} \lambda(a)_{rj} v_r.$$

It is easy to check that

$$\lambda\colon A \longrightarrow \mathrm{M}_m(\Bbbk_1); \quad a \longmapsto \lambda(a)$$

is a homomorphism of $\Bbbk$-algebras, called the *left regular representation* of A over $\Bbbk_1$ with respect to the basis $\{v_1, \ldots, v_m\}$.

Lemma 4.12

$\lambda\colon A \longrightarrow \mathrm{M}_m(\Bbbk_1)$ has trivial kernel $\ker\lambda = \{0\}$, hence it is an injection.

Proof

If $a \in \ker\lambda$ then $a1 = 0$, so $a = 0$. □

On restricting the left regular representation to the group of units $A^\times$, we obtain an injective group homomorphism

$$\lambda^\times\colon A^\times \longrightarrow \mathrm{GL}_m(\Bbbk_1); \quad \lambda^\times(a)(u) = au,$$

where $\Bbbk_1 \subseteq A$ is a subfield containing $\Bbbk$ and we have chosen a $\Bbbk_1$-basis of $A_{\Bbbk_1}$. Because

$$A^\times \cong \operatorname{im}\lambda^\times \leqslant \mathrm{GL}_m(\Bbbk_1),$$

$A^\times$ and its subgroups give groups of matrices.

Given a $\Bbbk$-basis of A, we obtain a group homomorphism

$$\rho^\times\colon A^\times \longrightarrow \mathrm{GL}_n(\Bbbk); \quad \rho^\times(a)(u) = ua^{-1}.$$

We can combine $\lambda^\times$ and $\rho^\times$ to obtain two further group homomorphisms

$$\lambda^\times \times \rho^\times\colon A^\times \times A^\times \longrightarrow \mathrm{GL}_n(\Bbbk); \quad \lambda^\times \times \rho^\times(a,b)(u) = aub^{-1},$$
$$\Delta\colon A^\times \longrightarrow \mathrm{GL}_n(\Bbbk); \quad \Delta(a)(u) = aua^{-1}.$$

Notice that these homomorphisms have non-trivial kernels since for $0 \neq t \in \mathbb{R}$,

$$(t,t) \in \ker\lambda^\times \times \rho^\times, \quad t \in \ker\Delta.$$

Definition 4.13

A $\mathbb{k}$-algebra A is *simple* if it has only one proper two-sided ideal, namely (0), hence every non-trivial $\mathbb{k}$-algebra homomorphism $\theta\colon A \longrightarrow B$ is an injection.

Example 4.14

Let $\mathbb{k}$ be a field and $p(x) \in \mathbb{k}[x]$ be an irreducible polynomial of degree $d > 0$. Then the quotient ring $\mathbb{k}[x]/(p(x))$ is a field and so is a simple $\mathbb{k}$-algebra of dimension $\dim_{\mathbb{k}} \mathbb{k}[x]/(p(x)) = d$.

Proposition 4.15

Let $\mathbb{k}$ be a field.
i) For a division algebra $\mathbb{D}$ over $\mathbb{k}$, $\mathbb{D}$ is simple.
ii) For a simple $\mathbb{k}$-algebra A, $\mathrm{M}_n(A)$ is simple. In particular, $\mathrm{M}_n(\mathbb{k})$ is a simple $\mathbb{k}$-algebra.

In fact, there is a classification of all such simple $\mathbb{k}$-algebras, for details see [15].

Theorem 4.16 (Wedderburn's Theorem)

Let A be a finite dimensional simple algebra over $\mathbb{k}$.
i) A is isomorphic to some $\mathrm{M}_n(\mathbb{D})$ where $\mathbb{D}$ is a finite dimensional division algebra over $\mathbb{k}$, hence its dimension is $\dim_{\mathbb{k}} A = n^2 \dim_{\mathbb{k}} \mathbb{D}$.
ii) If $\mathbb{k}$ is algebraically closed then A is isomorphic to some $\mathrm{M}_n(\mathbb{k})$, hence it has dimension $\dim_{\mathbb{k}} A = n^2$.
iii) Every finite dimensional division algebra $\mathbb{D}$ over $\mathbb{k}$ has dimension of the form $\dim_{\mathbb{k}} \mathbb{D} = d^2$ for some natural number d.

When $\mathbb{k} = \mathbb{R}$, it turns out that there is only one non-commutative division algebra, namely the division algebra of quaternions, $\mathbb{H}$, described in Section 4.4.

Definition 4.17

Let A be a $\mathbb{k}$-algebra.

- An element $e \in A$ is an *idempotent* if $e^2 = e$.
- An idempotent $e \in A$ is *central* if $ae = ea$ for every $a \in A$.
- A collection of idempotents $e_1, \ldots, e_\ell$ is *orthogonal* if $e_i e_j = 0$ when $i \neq j$.

- An idempotent $e \in A$ is *indecomposable* if whenever $e = e_1 + e_2$ for idempotents e_1, e_2, then $e_1 = 0$ or $e_2 = 0$.

Proposition 4.18

A finite dimensional $\Bbbk$-algebra A is a product of ℓ algebras if and only if there are orthogonal central idempotents $e_1, \ldots, e_\ell$ such that

$$e_1 + \cdots + e_\ell = 1.$$

Proof

If $A = A_1 \times \cdots \times A_\ell$ then

$$e_k = (0, \ldots, 0, 1, 0, \ldots, 0)$$

(with a single 1 in the kth place) is a central idempotent and $e_1 + \cdots + e_\ell = 1$; furthermore, $e_i e_j = 0$ whenever $i \neq j$.

Conversely, if such central orthogonal idempotents exist, we can set

$$A_k = Ae_k = \{ae_k : a \in A\}$$

and this is easily seen to be an algebra. Then the correspondence

$$a \longleftrightarrow (ae_1, \ldots, ae_\ell)$$

shows that A is the product of the algebras A_k. Notice that

$$1 \longleftrightarrow (e_1^2, \ldots, e_\ell^2) = (e_1, \ldots, e_\ell)$$

under this identification. □

Lemma 4.19

Let A be a $\Bbbk$-algebra and $e \in A$ be a central idempotent. Then $(1 - e)$ is an idempotent and the idempotents $e, (1 - e)$ are orthogonal. Moreover, Ae is an ideal in A which is trivial if and only if $e = 0$.

Proof

We have

$$(1 - e)^2 = 1 + e^2 - 2e = 1 + e - 2e = 1 - e,$$

and

$$e(1-e) = (1-e)e = e - e^2 = e - e = 0.$$

If $a, b \in A$, then we have

$$b(ae) = (ba)e \in Ae, \quad (ae)b = aeb = abe = (ab)e \in Ae,$$

showing that Ae is an ideal. □

Definition 4.20

A $\mathbb{k}$-algebra A is *semi-simple* if it is a product of simple algebras A_k,

$$A = A_1 \times \cdots \times A_\ell.$$

Proposition 4.21

A finite dimensional $\mathbb{k}$-algebra A is semi-simple if and only if there are indecomposable orthogonal central idempotents $e_1, \ldots, e_\ell$ such that

$$e_1 + \cdots + e_\ell = 1.$$

Proof

If $A = A_1 \times \cdots \times A_\ell$ with each A_k simple, then each of the corresponding central orthogonal idempotents e_i must be indecomposable since otherwise A_i would have a proper ideal. □

Example 4.22

Consider the commutative 2-dimensional $\mathbb{C}$-algebra

$$A = \{(u, v) : u, v \in \mathbb{C}\}$$

equipped with the obvious addition but multiplication

$$(u_1, v_1)(u_2, v_2) = (u_1u_2 - v_1v_2, u_1v_2 + v_1u_2).$$

The unit is $1 = (1, 0)$. Notice that if we restrict attention to

$$A_0 = \{(u, v) : u, v \in \mathbb{R}\},$$

this is isomorphic to the $\mathbb{R}$-algebra of complex numbers $\mathbb{C}$, using the correspondence

$$(u, v) \longleftrightarrow u + vi.$$

However, A contains the orthogonal idempotents

$$e = (1/2, -i/2), \quad 1 - e = (1/2, i/2).$$

Hence,

$$\begin{aligned} A &= Ae \times A(1-e) \\ &= \left\{\left(\frac{u+vi}{2}, \frac{-ui+v}{2}\right) : u, v \in \mathbb{C}\right\} \times \left\{\left(\frac{u-vi}{2}, \frac{ui+v}{2}\right) : u, v \in \mathbb{C}\right\}. \end{aligned}$$

Each of the factors here is isomorphic to $\mathbb{C}$ as an $\mathbb{R}$-algebra.

Example 4.23

Let $\mathbb{k}$ be a field and $p(x), q(x) \in \mathbb{k}[x]$ be coprime polynomials of positive degree. Consider the $\mathbb{k}$-algebra $A = \mathbb{k}[x]/(p(x)q(x))$ of dimension

$$\dim_{\mathbb{k}} A = \deg p(x) + \deg q(x).$$

Then there are polynomials $u(x), v(x) \in \mathbb{k}[x]$ for which

$$u(x)p(x) + v(x)q(x) = 1.$$

The polynomials $v(x)q(x)$ and $u(x)p(x)$ satisfy

$$\begin{aligned} (v(x)q(x))^2 &= v(x)q(x)(1 - u(x)p(x)) \\ &\equiv v(x)q(x) \bmod (p(x)q(x)) \end{aligned}$$

and similarly

$$\begin{aligned} (u(x)p(x))^2 &= u(x)p(x)(1 - v(x)q(x)) \\ &\equiv u(x)p(x) \bmod (p(x)q(x)), \end{aligned}$$

so their residue classes

$$e_1 = v(x)q(x) + (p(x)q(x)), \quad e_2 = u(x)p(x) + (p(x)q(x))$$

are idempotents in A for which $e_1 + e_2 = 1$. Thus A can be expressed as a product

$$\mathbb{k}[x]/(p(x)q(x)) = \mathbb{k}[x]/(p(x)) \times \mathbb{k}[x]/(q(x)).$$

By Example 4.14, if $p(x)$ or $q(x)$ is irreducible then the corresponding factor $\mathbb{k}[x]/(p(x))$ or $\mathbb{k}[x]/(q(x))$ is simple. If both polynomials are irreducible then A is semi-simple.

Example 4.24

Let $\mathbb{k}[G]$ be the group algebra of a finite group G. Then for $\mathbb{k} = \mathbb{R}$ or $\mathbb{C}$, $\mathbb{k}[G]$ is semi-simple. When $\mathbb{k} = \mathbb{C}$, indecomposable orthogonal idempotents can be written down in terms of the *irreducible complex characters* of G. These are important in the study of finite dimensional representations of G. For details see [15] or any book on the representation theory of finite groups.

Example 4.25

Consider $\mathbb{C}[\mathrm{S}_3]$, the group algebra of S_3, the symmetric group on three objects. There are three central orthogonal idempotents here and

$$\mathbb{C}[\mathrm{S}_3] = A_1 \times A_2 \times A_3,$$

where $A_1 \cong A_2 \cong \mathbb{C}$ and $A_3 \cong \mathrm{M}_2(\mathbb{C})$. These idempotents can be found using character theory and turn out to be

$$\begin{aligned} e_1 &= \frac{1}{6}[\,1 + (1\,2\,3) + (1\,3\,2) + (1\,2) + (1\,3) + (2\,3)\,], \\ e_2 &= \frac{1}{6}[\,1 + (1\,2\,3) + (1\,3\,2) - (1\,2) - (1\,3) - (2\,3)\,], \\ e_3 &= \frac{1}{3}[\,2 - (1\,2\,3) - (1\,3\,2)\,]. \end{aligned}$$

In fact this also gives a decomposition of $\mathbb{R}[\mathrm{S}_3]$ since each of these idempotents is an element of this $\mathbb{R}$-algebra. We obtain

$$\mathbb{R}[\mathrm{S}_3] = A'_1 \times A'_2 \times A'_3,$$

where $A'_1 \cong A'_2 \cong \mathbb{R}$ but this time A'_3 has to be one of the $\mathbb{R}$-algebras $\mathrm{M}_2(\mathbb{R})$ or $\mathbb{H}$ (it is in fact $\mathrm{M}_2(\mathbb{R})$).

Sometimes it is useful to have methods of deciding whether a $\mathbb{k}$-algebra can be a division algebra. Here is one useful criterion that can be applied to show that the factor A'_3 in Example 4.25 cannot be $\mathbb{H}$.

Lemma 4.26

Let $\mathbb{D}$ be a $\mathbb{k}$-division algebra. Suppose that $\alpha \in \mathbb{D}$ is a root of a polynomial $f(X) \in \mathbb{k}[X]$ which factorises completely into linear factors over $\mathbb{k}$. Then $\alpha \in \mathbb{k}$.

Proof

Let $r_1, \ldots, r_d \in \mathbb{k}$ be the roots of $f(X)$ in $\mathbb{k}$ (with multiplicity), so that

$$f(X) = (X - r_1) \cdots (X - r_d).$$

Then since $f(\alpha) = 0$ and α commutes with elements of $\mathbb{k}$,

$$(\alpha - r_1) \cdots (\alpha - r_d) = 0.$$

For each $j = 1, \ldots, d$, either $\alpha = r_j$ or we can multiply by $(\alpha - r_j)^{-1}$. If $\alpha \neq r_j$ for all j, then we obtain $1 = 0$, hence for at least one value of j we must have $\alpha = r_j$. □

Remark 4.27

If a $\mathbb{k}$-algebra A is a product $A = A_1 \times \cdots \times A_\ell$, then its group of units is also a product of groups,

$$A^\times = A_1^\times \times \cdots \times A_\ell^\times.$$

So if looking for interesting groups of units in algebras, we only need to consider simple algebras. It turns out that the real Clifford algebras introduced in Section 5.1 occur as factors of finite group algebras, for example the spinor groups could be found in this way.

Not all algebras are semi-simple as the following examples show. To understand them, notice that if $A = A_1 \times \cdots \times A_\ell$ is a product of simple algebras A_j, then for each $k = 1, \ldots, \ell$, there is a two-sided ideal $J_k \lhd A$ with quotient algebra $A/J_k \cong A_k$. This provides a useful criterion for determining when an algebra is *not* semi-simple.

Lemma 4.28

Let A be a finite dimensional $\mathbb{k}$-algebra and suppose that A has only one proper ideal $J \lhd A$ for which the quotient A/J is simple. If $J \neq (0)$ then A is not semi-simple.

Proof

If A is semi-simple then it must be simple, hence its only proper ideal is (0). □

A very general class of examples is furnished by the next result.

Proposition 4.29

Let $p(X) \in \mathbb{k}[x]$ be an irreducible polynomial of positive degree. Then the commutative $\mathbb{k}$-algebra $A = \mathbb{k}[x]/(p(x)^m)$ is

- simple if $m = 1$;

- not semi-simple if $m > 1$.

Proof

To see this, first notice that if $m = 1$, then A is actually a field, hence is a simple $\Bbbk$-algebra. If $m > 1$, then the ideal $\overline{J} = \overline{(p(x))} \lhd \Bbbk[x]/(p(x)^m)$ is the unique maximal ideal in $A = \Bbbk[x]/(p(x)^m)$, where $\overline{f(x)} = f(x) + (p(x)^m)$ is the residue class of $f(x)$ modulo $(p(x)^m)$. As $A/\overline{(p(x))} \cong \Bbbk[x]/(p(x))$ is a field it is simple, allowing us to apply Lemma 4.28.

When $m > 1$, the group of units can be shown to be

$$A^\times = \Bbbk[x]/(p(x)^m)^\times \cong (\Bbbk[x]/(p(x)))^\times \times (1 + \overline{(p(x))}),$$

where

$$1 + \overline{(p(x))} = \{\overline{1 + f(x)p(x)} \in \Bbbk[x]/(p(x)^m) : f(x) \in \Bbbk[x]\}$$

is a subgroup of $(\Bbbk[x]/(p(x)^m))^\times$. There is an isomorphism of groups

$$(1 + \overline{(p(x))}) \cong \Bbbk[x]/(p(x)^{m-1}); \quad \overline{1 + f(x)p(x)} \longleftrightarrow f(x) + (p(x)^{m-1}),$$

giving rise to an isomorphism of groups

$$\Bbbk[x]/(p(x)^m)^\times \cong (\Bbbk[x]/(p(x)))^\times \times \Bbbk[x]/(p(x)^{m-1}),$$

where $\Bbbk[x]/(p(x)^{m-1})$ is a group under addition. □

Here is an explicit example of Proposition 4.29.

Example 4.30

Let $\Bbbk = \mathbb{R}$ and $A = \mathbb{R}[x]/((x^2+1)^2)$. Then

$$\begin{aligned} A^\times &\cong (\mathbb{R}[x]/(x^2+1))^\times \times (1 + \overline{(x^2+1)}) \\ &\cong (\mathbb{R}[x]/(x^2+1))^\times \times \mathbb{R}[x]/(x^2+1). \end{aligned}$$

Since $\mathbb{R}[x]/(x^2+1) \cong \mathbb{C}$ as $\mathbb{R}$-algebras, we have

$$\mathbb{R}[x]/((x^2+1)^2)^\times \cong \mathbb{C}^\times \times \mathbb{C}.$$

4.2 Real and Complex Normed Algebras

In this section, $\mathbb{k} = \mathbb{R}$ or $\mathbb{C}$. For a detailed look at normed algebras, see [21, 22].

Definition 4.31

Let A be a finite dimensional $\mathbb{k}$-algebra and $\nu\colon A \longrightarrow \mathbb{R}$ be a function. Then ν is a $\mathbb{k}$-*norm* and the pair (A, ν) is called a *normed* $\mathbb{k}$-*algebra*, if $\nu(a) \geqslant 0$ for all $a \in A$ and ν satisfies the conditions

i) for $t \in \mathbb{k}$, $a \in A$, $\nu(ta) = |t|\nu(a)$;
ii) for $a, b \in A$, $\nu(ab) \leqslant \nu(a)\nu(b)$;
iii) for $a, b \in A$, $\nu(a + b) \leqslant \nu(a) + \nu(b)$;
iv) for $a \in A$, $\nu(a) = 0$ if and only if $a = 0$;
v) $\nu(1) = 1$.

Actually, there is a certain amount of variation in the definition of a normed algebra, with some authors using the phrase to denote an algebra with a norm satisfying $\nu(ab) = \nu(a)\nu(b)$ for all $a, b \in A$, rather than the weaker requirement of Definition 4.31(ii), while others do not insist on condition (v).

Example 4.32

For $\mathbb{k} = \mathbb{R}$ or $\mathbb{C}$, let $\mathbb{k}[G]$ be the group algebra of a finite group G. Define $\nu\colon \mathbb{k}[G] \longrightarrow \mathbb{R}$ by

$$\nu(\sum_{g \in G} x_g g) = \sqrt{\sum_{g \in G} |x_g|^2}.$$

Then it is easy to verify that ν is a norm on $\mathbb{k}[G]$, the most interesting condition being the submultiplicative identity of Definition 4.31(ii) which follows from the calculation

$$\begin{aligned}
\nu\left((\sum_{g \in G} x_g g)(\sum_{h \in G} y_h h)\right)^2 &= \nu\left(\sum_{\substack{g \in G \\ h \in G}} x_{gh^{-1}} y_h g\right)^2 \\
&= \sum_{g \in G} |\sum_{h \in G} x_{gh^{-1}} y_h|^2 \\
&\leqslant \sum_{\substack{g \in G \\ h \in G}} |x_{gh^{-1}}|^2 \, |y_h|^2 \\
&= \nu(\sum_{g \in G} x_g g)^2 \nu(\sum_{h \in G} y_h h)^2.
\end{aligned}$$

Notice that each $u \in G$ acts on $\Bbbk[G]$ by left multiplication preserving ν:

$$\begin{aligned}\nu(u \sum_{g \in G} x_g g) &= \nu(\sum_{g \in G} x_g (ug)) \\ &= \nu(\sum_{g \in G} x_{u^{-1}g} g) \\ &= \nu(\sum_{g \in G} x_g g).\end{aligned}$$

Similarly the action by right multiplication preserves ν. Finally, conjugation by an element $u \in G$ preserves ν and also gives a $\Bbbk$-algebra isomorphism.

By forgetting the multiplication, a normed $\Bbbk$-algebra (A, ν) can be viewed as a normed vector space in the sense of Definition 1.51. There is a natural metric ρ_ν on A for which

$$\rho_\nu(a, b) = \nu(a - b)$$

which allows us to introduce topological and analytic ideas when studying (A, ν). In particular, A is complete with respect to this metric, *i.e.*, all Cauchy sequences in A converge. In particular, for each $a \in A$, the exponential series

$$\exp(a) = \sum_{k=0}^{\infty} \frac{1}{k!} a^k$$

converges. Other analytic constructions generalise to this setting. For example, given a function $\alpha\colon (r, s) \longrightarrow A$, where $r, s \in \mathbb{R}$ with $r < s$, we define its derivative at $t \in (r, s)$ by

$$\alpha'(t) = \lim_{h \to 0} \frac{1}{h} \left(\alpha(t + h) - \alpha(t)\right),$$

provided this limit is defined. It is straightforward to carry over the ideas of Chapters 1 and 3 to this situation. In particular, the group of units $A^\times \subseteq A$ is an open subset and the Lie algebra is equal to A with the Lie bracket defined by taking commutators in the algebra A, *i.e.*,

$$[x, y] = xy - yx \quad (x, y \in A).$$

A closed subgroup $G \leqslant A^\times$ has a Lie algebra which is an $\mathbb{R}$-Lie subalgebra of A. We leave the reader to work through the details.

If (A, ν) is a normed $\Bbbk$-algebra then we can define a version of the operator norm $\| \ \|_\nu$ on A by

$$\|a\|_\nu = \sup\{\nu(au) : u \in A,\ \nu(u) = 1\}.$$

In general this does not agree with the norm ν. It does provide another normed algebra $(A, \| \ \|_\nu)$ with a metric defined by

$$\rho'_\nu(a, b) = \|a - b\|_\nu.$$

Proposition 4.33

Let (A, ν) be a normed $\Bbbk$-algebra.
i) The identity map $\mathrm{Id}_A \colon A \longrightarrow A$ is a homeomorphism of metric spaces $(A, \rho_\nu) \longrightarrow (A, \rho'_\nu)$.
ii) The left regular representation $\lambda\colon A \longrightarrow \mathrm{M}_n(\Bbbk)$ with respect to any $\Bbbk$-basis $\{v_1, \ldots, v_n\}$ of A gives rise to continuous injections of metric spaces $(A, \rho_\nu) \longrightarrow (A, \|\ \|)$ and $(A, \rho'_\nu) \longrightarrow (A, \|\ \|)$.

Proof

These results follow from Theorem 1.53 and Corollary 1.54. □

4.3 Linear Algebra over a Division Algebra

Throughout this section, let $\mathbb{D}$ be a finite dimensional division algebra over a field $\Bbbk$.

Definition 4.34

A (*right*) $\mathbb{D}$-*vector space* V is a right $\mathbb{D}$-module, *i.e.*, an abelian group with a right scalar multiplication by elements of $\mathbb{D}$ so that for $u, v \in V$, $x, y \in \mathbb{D}$,

$$\begin{aligned} v(xy) &= (vx)y, \\ v(x+y) &= vx + vy, \\ (u+v)x &= ux + vx, \\ v1 &= v. \end{aligned}$$

All the obvious notions of $\mathbb{D}$-linear transformations, subspaces, kernels and images make sense as do those of spanning set and linear independence over $\mathbb{D}$.

Theorem 4.35

Let V be a $\mathbb{D}$-vector space.
i) V has a $\mathbb{D}$-basis.
ii) If V has a finite spanning set over $\mathbb{D}$ then it has a finite $\mathbb{D}$-basis; furthermore any two such finite bases have the same number of elements.

Definition 4.36

A $\mathbb{D}$-vector space V with a finite basis is called *finite dimensional* and the number of elements in a basis is called the *dimension of V over $\mathbb{D}$*, denoted $\dim_{\mathbb{D}} V$.

For $n \geqslant 1$, we can view $\mathbb{D}^n$ as the set of $n \times 1$ column vectors with entries in $\mathbb{D}$ and this becomes a $\mathbb{D}$-vector space with the obvious scalar multiplication

$$\begin{bmatrix} z_1 \\ z_2 \\ \vdots \\ z_n \end{bmatrix} w = \begin{bmatrix} z_1 w \\ z_2 w \\ \vdots \\ z_n w \end{bmatrix} \qquad (z_1, \ldots, z_n, w \in \mathbb{D}).$$

Proposition 4.37

Let V, W be two finite dimensional vector spaces over $\mathbb{D}$, of dimensions $\dim_{\mathbb{D}} V = m$, $\dim_{\mathbb{D}} W = n$ and with bases $\{v_1, \ldots, v_m\}$, $\{w_1, \ldots, w_n\}$. Then a $\mathbb{D}$-linear transformation $\varphi \colon V \longrightarrow W$ is given by

$$\varphi(v_j) = \sum_{r=1}^{n} w_r a_{rj}$$

for unique elements $a_{ij} \in \mathbb{D}$. Hence if

$$\varphi\left(\sum_{s=1}^{m} v_s x_s\right) = \sum_{r=1}^{n} w_r y_r,$$

then

$$\begin{bmatrix} y_1 \\ y_2 \\ \vdots \\ y_n \end{bmatrix} = \begin{bmatrix} a_{11} & a_{12} & \cdots & a_{1m} \\ a_{21} & a_{22} & \cdots & a_{2m} \\ \vdots & \ddots & \ddots & \vdots \\ a_{n1} & a_{n2} & \cdots & a_{mn} \end{bmatrix} \begin{bmatrix} x_1 \\ x_2 \\ \vdots \\ x_m \end{bmatrix}.$$

In particular, for $V = \mathbb{D}^m$ and $W = \mathbb{D}^n$, every $\mathbb{D}$-linear transformation is obtained in this way from left multiplication by a fixed matrix.

Of course, this is analogous to what happens over a field except that we are careful to keep the scalar action on the right and the matrix action on the left.

Our main interest is in linear transformations which we will identify with the corresponding matrices. If $\theta \colon \mathbb{D}^k \longrightarrow \mathbb{D}^m$ and $\varphi \colon \mathbb{D}^m \longrightarrow \mathbb{D}^n$ are $\mathbb{D}$-linear transformations with corresponding matrices $[\theta], [\varphi]$, then

$$[\theta][\varphi] = [\theta \circ \varphi]. \tag{4.1}$$

Also, the identity and zero functions $\mathrm{Id}, 0\colon \mathbb{D}^m \longrightarrow \mathbb{D}^m$ have $[\mathrm{Id}] = I_m$ and $[0] = O_m$.

Notice that given a $\mathbb{D}$-linear transformation $\varphi\colon V \longrightarrow W$, we can 'forget' the $\mathbb{D}$-structure and just view it as a $\mathbb{k}$-linear transformation. Given $\mathbb{D}$-bases $\{v_1, \ldots, v_m\}$, $\{w_1, \ldots, w_n\}$ and a basis $\{b_1, \ldots, b_d\}$ say for $\mathbb{D}$, the elements

$$v_r b_t \quad (r = 1, \ldots, m,\ t = 1, \ldots, d),$$
$$w_s b_t \quad (s = 1, \ldots, n,\ t = 1, \ldots, d)$$

form $\mathbb{k}$-bases for V, W as $\mathbb{k}$-vector spaces.

We denote the set of all $m \times n$ matrices with entries in $\mathbb{D}$ by $\mathrm{M}_{m,n}(\mathbb{D})$ and $\mathrm{M}_n(\mathbb{D}) = \mathrm{M}_{n,n}(\mathbb{D})$. Then $\mathrm{M}_n(\mathbb{D})$ is a $\mathbb{k}$-algebra of dimension

$$\dim_{\mathbb{k}} \mathrm{M}_n(\mathbb{D}) = n^2 \dim_{\mathbb{k}} \mathbb{D}.$$

The group of units $\mathrm{M}_n(\mathbb{D})^\times$ is usually denoted $\mathrm{GL}_n(\mathbb{D})$. However, for non-commutative $\mathbb{D}$ it turns out that there is no determinant function so we cannot define an analogue of the special linear group. However, we can use the left regular representation to circumvent this problem.

Proposition 4.38

Let A be an algebra over a field $\mathbb{k}$ and $B \subseteq A$ be a finite dimensional subalgebra. If $u \in B$ is a unit in A then $u^{-1} \in B$, hence u is a unit in B.

Proof

Since B is finite dimensional, the powers u^k $(k \geqslant 0)$ are linearly dependent over $\mathbb{k}$, so for some $t_r \in \mathbb{k}$ $(r = 0, \ldots, \ell)$ with $t_\ell \neq 0$ and $\ell \geqslant 1$, there is a non-trivial relation of the form

$$\sum_{r=0}^{\ell} t_r u^r = 0.$$

If we choose k suitably and multiply by a non-zero scalar, then we can assume that

$$u^k - \sum_{r=k+1}^{\ell} t_r u^r = 0.$$

If v is the inverse of u in A, then multiplication by v^{k+1} gives

$$v - \sum_{r=k+1}^{\ell} t_r u^{r-k-1} = 0,$$

from which we obtain

$$v = \sum_{r=k+1}^{\ell} t_r u^{r-k-1} \in B,$$

hence u has an inverse in B. □

For a division algebra $\mathbb{D}$, each matrix $A \in \mathrm{M}_n(\mathbb{D})$ acts by multiplication on the left of $\mathbb{D}^n$. For any subfield $\mathbb{k}_1 \subseteq \mathbb{D}$ containing $\mathbb{k}$, A induces a (right) $\mathbb{k}_1$-linear transformation,

$$\mathbb{D}^n \longrightarrow \mathbb{D}^n; \quad \mathbf{x} \longmapsto A\mathbf{x}.$$

If we choose a $\mathbb{k}_1$-basis for $\mathbb{D}$, A gives rise to a matrix $\Lambda_A \in \mathrm{M}_{nd}(\mathbb{k}_1)$ where $d = \dim_{\mathbb{k}_1} \mathbb{D}_{\mathbb{k}_1}$. It is easy to see that the function

$$\Lambda\colon \mathrm{M}_n(\mathbb{D}) \longrightarrow \mathrm{M}_{nd}(\mathbb{k}_1); \quad \Lambda(A) = \Lambda_A,$$

is a ring homomorphism with $\ker \Lambda = 0$. This allows us to identify $\mathrm{M}_n(\mathbb{D})$ with the subring $\operatorname{im} \Lambda \subseteq \mathrm{M}_{nd}(\mathbb{k}_1)$.

Applying Proposition 4.38 we see that A is invertible in $\mathrm{M}_n(\mathbb{D})$ if and only if Λ_A is invertible in $\mathrm{M}_{nd}(\mathbb{k}_1)$. But the latter is true if and only if $\det \Lambda_A \neq 0$.

Hence to determine invertibility of $A \in \mathrm{M}_n(\mathbb{D})$, it suffices to consider $\det \Lambda_A$ using a subfield $\mathbb{k}_1$. The resulting function

$$\mathrm{Rdet}_{\mathbb{k}_1}\colon \mathrm{M}_n(\mathbb{D}) \longrightarrow \mathbb{k}_1; \quad \mathrm{Rdet}_{\mathbb{k}_1}(A) = \det \Lambda_A,$$

is called the $\mathbb{k}_1$-*reduced determinant* of $\mathrm{M}_n(\mathbb{D})$ and is a group homomorphism. It is actually true that $\det \Lambda_A \in \mathbb{k}$, not just in $\mathbb{k}_1$, although we will not prove this here.

Proposition 4.39

$A \in \mathrm{M}_n(\mathbb{D})$ is invertible if and only if $\mathrm{Rdet}_{\mathbb{k}_1}(A) \neq 0$ for some subfield $\mathbb{k}_1 \subseteq \mathbb{D}$ containing $\mathbb{k}$.

4.4 The Quaternions

Proposition 4.40

If A is a finite dimensional commutative $\mathbb{R}$-division algebra then either $A = \mathbb{R}$ or there is an isomorphism of $\mathbb{R}$-algebras $A \cong \mathbb{C}$.

Proof

Let $\alpha \in A$. Since A is a finite dimensional $\mathbb{R}$-vector space, the powers of α must be linearly dependent, say

$$t_0 + t_1\alpha + \cdots + t_m\alpha^m = 0 \tag{4.2}$$

for some $t_j \in \mathbb{R}$ with $m \geqslant 1$ and $t_m \neq 0$. We can choose m to be minimal with these properties. If $t_0 = 0$, then

$$t_1 + t_2\alpha + t_3\alpha^2 + \cdots + t_m\alpha^{m-1} = 0,$$

contradicting the minimality of m; so $t_0 \neq 0$. In fact, the polynomial

$$p(X) = t_0 + t_1X + \cdots + t_mX^m \in \mathbb{R}[X]$$

is irreducible. To see this, suppose that $p(X) = p_1(X)p_2(X)$; then as A is a division algebra, either $p_1(\alpha) = 0$ or $p_2(\alpha) = 0$, contradicting minimality of m if both $\deg p_1(X) > 0$ and $\deg p_2(X) > 0$.

Consider the $\mathbb{R}$-subspace

$$\mathbb{R}(\alpha) = \{\sum_{j=0}^{k} s_j\alpha^j : s_j \in \mathbb{R}\} \subseteq A.$$

Then $\mathbb{R}(\alpha)$ is easily seen to be an $\mathbb{R}$-subalgebra of A and the elements $1, \alpha, \ldots, \alpha^{m-1}$ form a basis by Equation (4.2), hence $\dim_{\mathbb{R}} \mathbb{R}(\alpha) = m$.

Let $\gamma \in \mathbb{C}$ be any complex root of the irreducible polynomial

$$t_0 + t_1X + \cdots + t_mX^m \in \mathbb{R}[X]$$

(such a root certainly exists by the Fundamental Theorem of Algebra). There is an $\mathbb{R}$-linear transformation which is actually an injection,

$$\varphi\colon \mathbb{R}(\alpha) \longrightarrow \mathbb{C}; \quad \varphi(\sum_{j=0}^{m-1} s_j\alpha^j) = \sum_{j=0}^{m-1} s_j\gamma^j.$$

It is easy to see that this is actually an $\mathbb{R}$-algebra homomorphism. Hence $\varphi\mathbb{R}(\alpha) \subseteq \mathbb{C}$ is a subalgebra. But as $\dim_{\mathbb{R}} \mathbb{C} = 2$, this implies that $m = \dim_{\mathbb{R}} \mathbb{R}(\alpha) \leqslant 2$. If $m = 1$, then by Equation (4.2), $\alpha \in \mathbb{R}$. If $m = 2$, then $\varphi\mathbb{R}(\alpha) = \mathbb{C}$.

So either $\dim_{\mathbb{R}} A = 1$ and $A = \mathbb{R}$, or $\dim_{\mathbb{R}} A > 1$ and we can choose an $\alpha \in A$ with $\mathbb{C} \cong \mathbb{R}(\alpha)$. This means that we can view A as a finite dimensional $\mathbb{C}$-algebra. Now for any $\beta \in A$ there is polynomial

$$q(X) = u_0 + u_1X + \cdots + u_\ell X^\ell \in \mathbb{C}[X]$$

with $\ell \geqslant 1$ and $u_\ell \neq 0$. Again choosing ℓ to be minimal with this property, $q(X)$ is irreducible. But then since $q(X)$ has a root in $\mathbb{C}$, $\ell = 1$ and $\beta \in \mathbb{C}$. This shows that $A = \mathbb{C}$ whenever $\dim_{\mathbb{R}} A > 1$. □

The above proof actually shows that if A is a finite dimensional $\mathbb{R}$-division algebra, then either $A = \mathbb{R}$ or there is a subalgebra isomorphic to $\mathbb{C}$. However, the question of what finite dimensional $\mathbb{R}$-division algebras exist is less easy to decide. In fact, up to isomorphism there is only one other, the skew field of *quaternions* (or *Hamiltonians* after their discoverer William Rowan Hamilton), usually denoted $\mathbb{H}$. We will construct $\mathbb{H}$ as a ring of 2×2 complex matrices. They provide the first example of a non-commutative *Clifford algebra* which we will define later in Chapter 5.

Let

$$\mathbb{H} = \left\{ \begin{bmatrix} z & w \\ -\overline{w} & \overline{z} \end{bmatrix} : z, w \in \mathbb{C} \right\} \subseteq \mathrm{M}_2(\mathbb{C}).$$

It is easy to see that $\mathbb{H}$ is a subring of $\mathrm{M}_2(\mathbb{C})$ and is in fact an $\mathbb{R}$-subalgebra where we view $\mathrm{M}_2(\mathbb{C})$ as an $\mathbb{R}$-algebra of dimension 8. It also contains a copy of $\mathbb{C}$, namely the $\mathbb{R}$-subalgebra

$$\left\{ \begin{bmatrix} z & 0 \\ 0 & \overline{z} \end{bmatrix} : z \in \mathbb{C} \right\} \subseteq \mathbb{H}.$$

However, $\mathbb{H}$ is not a $\mathbb{C}$-algebra since for example

$$\begin{bmatrix} i & 0 \\ 0 & -i \end{bmatrix} \begin{bmatrix} 0 & 1 \\ -1 & 0 \end{bmatrix} = \begin{bmatrix} 0 & i \\ i & 0 \end{bmatrix} = - \begin{bmatrix} 0 & 1 \\ -1 & 0 \end{bmatrix} \begin{bmatrix} i & 0 \\ 0 & -i \end{bmatrix} \neq \begin{bmatrix} 0 & 1 \\ -1 & 0 \end{bmatrix} \begin{bmatrix} i & 0 \\ 0 & -i \end{bmatrix}.$$

Notice that if $z, w \in \mathbb{C}$, then $z = 0 = w$ if and only if $|z|^2 + |w|^2 = 0$. We have

$$\begin{bmatrix} z & w \\ -\overline{w} & \overline{z} \end{bmatrix} \begin{bmatrix} \overline{z} & -w \\ \overline{w} & z \end{bmatrix} = \begin{bmatrix} |z|^2 + |w|^2 & 0 \\ 0 & |z|^2 + |w|^2 \end{bmatrix},$$

hence $\begin{bmatrix} z & w \\ -\overline{w} & \overline{z} \end{bmatrix}$ is invertible if and only if $\begin{bmatrix} z & w \\ -\overline{w} & \overline{z} \end{bmatrix} \neq O$; furthermore in that case,

$$\begin{bmatrix} z & w \\ -\overline{w} & \overline{z} \end{bmatrix}^{-1} = \begin{bmatrix} \dfrac{\overline{z}}{|z|^2 + |w|^2} & \dfrac{-w}{|z|^2 + |w|^2} \\ \dfrac{\overline{w}}{|z|^2 + |w|^2} & \dfrac{z}{|z|^2 + |w|^2} \end{bmatrix}$$

which is in $\mathbb{H}$. So an element of $\mathbb{H}$ is invertible in $\mathbb{H}$ if and only if it is invertible as a matrix. Notice that

$$\mathrm{SU}(2) = \{A \in \mathbb{H} : \det A = 1\} \leqslant \mathbb{H}^{\times}.$$

It is useful to define on $\mathbb{H}$ a norm in the sense of Proposition 1.5:

$$\left\| \begin{bmatrix} z & w \\ -\overline{w} & \overline{z} \end{bmatrix} \right\| = \det \begin{bmatrix} z & w \\ -\overline{w} & \overline{z} \end{bmatrix} = |z|^2 + |w|^2.$$

Then

$$\mathrm{SU}(2) = \{A \in \mathbb{H} : |A| = 1\} \leqslant \mathbb{H}^{\times}.$$

As an $\mathbb{R}$-basis of $\mathbb{H}$ we have the matrices

$$\mathbf{1} = I, \quad \mathbf{i} = \begin{bmatrix} i & 0 \\ 0 & -i \end{bmatrix}, \quad \mathbf{j} = \begin{bmatrix} 0 & 1 \\ -1 & 0 \end{bmatrix}, \quad \mathbf{k} = \begin{bmatrix} 0 & i \\ i & 0 \end{bmatrix}.$$

These satisfy the equations

$$\mathbf{i}^2 = \mathbf{j}^2 = \mathbf{k}^2 = -\mathbf{1}, \quad \mathbf{ij} = \mathbf{k} = -\mathbf{ji}, \quad \mathbf{jk} = \mathbf{i} = -\mathbf{kj}, \quad \mathbf{ki} = \mathbf{j} = -\mathbf{ik}.$$

This should be compared with the vector product on $\mathbb{R}^3$ as discussed in Example 3.2. From now on we will write quaternions in the form

$$q = x\mathbf{i} + y\mathbf{j} + z\mathbf{k} + t\mathbf{1} \quad (x, y, z, t \in \mathbb{R}).$$

q is a *pure quaternion* if and only if $t = 0$; q is a *real quaternion* if and only if $x = y = z = 0$. We can identify the pure quaternion $x\mathbf{i} + y\mathbf{j} + z\mathbf{k}$ with the element $x\mathbf{e}_1 + y\mathbf{e}_2 + z\mathbf{e}_3 \in \mathbb{R}^3$. Using this identification we see that the scalar and vector products on $\mathbb{R}^3$ are related to quaternion multiplication as in the following result.

Proposition 4.41

For two pure quaternions $q_1 = x_1\mathbf{i} + y_1\mathbf{j} + z_1\mathbf{k}$, $q_2 = x_2\mathbf{i} + y_2\mathbf{j} + z_2\mathbf{k}$,

$$q_1q_2 = -(x_1\mathbf{i}+y_1\mathbf{j}+z_1\mathbf{k})\cdot(x_2\mathbf{i}+y_2\mathbf{j}+z_2\mathbf{k})+(x_1\mathbf{i}+y_1\mathbf{j}+z_1\mathbf{k})\times(x_2\mathbf{i}+y_2\mathbf{j}+z_2\mathbf{k}).$$

In particular, q_1q_2 is a pure quaternion if and only if q_1 and q_2 are orthogonal, in which case q_1q_2 is orthogonal to each of them.

The next result describes the general solution of the equation $X^2 + 1 = 0$ in $\mathbb{H}$.

Proposition 4.42

The quaternion $q = x\mathbf{i} + y\mathbf{j} + z\mathbf{k} + t\mathbf{1}$ satisfies $q^2 + \mathbf{1} = \mathbf{0}$ if and only if $t = 0$ and $x^2 + y^2 + z^2 = 1$.

Proof

This follows easily from Proposition 4.41. □

There is a quaternionic analogue of complex conjugation, namely

$$q = x\mathbf{i} + y\mathbf{j} + z\mathbf{k} + t\mathbf{1} \longmapsto \overline{q} = q^* = -x\mathbf{i} - y\mathbf{j} - z\mathbf{k} + t\mathbf{1}.$$

This is 'almost' a ring homomorphism $\mathbb{H} \longrightarrow \mathbb{H}$, in fact it satisfies

$$\overline{(q_1 + q_2)} = \overline{q}_1 + \overline{q}_2; \tag{4.3a}$$

$$\overline{(q_1 q_2)} = \overline{q}_2 \overline{q}_1; \tag{4.3b}$$

$$\overline{q} = q \iff q \text{ is a real quaternion;} \tag{4.3c}$$

$$\overline{q} = -q \iff q \text{ is a pure quaternion.} \tag{4.3d}$$

Because of Equation (4.3b) this is called an *anti-homomorphism of skew rings* or *ring anti-homomorphism*. The inverse of a non-zero quaternion q can be written as

$$q^{-1} = \frac{1}{(q\overline{q})}\overline{q} = \frac{\overline{q}}{(q\overline{q})}. \tag{4.4}$$

The real quantity $q\overline{q}$ is the square of the length of the corresponding vector,

$$|q| = \sqrt{q\overline{q}} = \sqrt{x^2 + y^2 + z^2 + t^2}.$$

For $z = u\mathbf{1} + v\mathbf{i}$ with $u, v \in \mathbb{R}$, $\overline{z} = u\mathbf{1} - v\mathbf{i}$ agrees with the usual complex conjugate.

In terms of the matrix description of $\mathbb{H}$, quaternionic conjugation is given by hermitian conjugation,

$$\begin{bmatrix} z & w \\ -\overline{w} & \overline{z} \end{bmatrix} \longmapsto \begin{bmatrix} z & w \\ -\overline{w} & \overline{z} \end{bmatrix}^* = \begin{bmatrix} \overline{z} & -w \\ \overline{w} & z \end{bmatrix}.$$

To simplify notation, from now on we will write

$$1 = \mathbf{1}, \quad i = \mathbf{i}, \quad j = \mathbf{j}, \quad k = \mathbf{k}.$$

4.5 Quaternionic Matrix Groups

The above norm $|\ |$ on $\mathbb{H}$ extends to a norm on $\mathbb{H}^n$, viewed as a right $\mathbb{H}$-vector space. We can define a quaternionic inner product on $\mathbb{H}$ by

$$\mathbf{x} \cdot \mathbf{y} = \mathbf{x}^* \mathbf{y} = \sum_{r=1}^{n} \overline{x}_r y_r,$$

where we define the *quaternionic conjugate* of a vector by

$$\begin{bmatrix} x_1 \\ x_2 \\ \vdots \\ x_n \end{bmatrix}^* = \begin{bmatrix} \overline{x}_1 & \overline{x}_2 & \cdots & \overline{x}_n \end{bmatrix}.$$

Similarly, for any matrix $[a_{ij}]$ over $\mathbb{H}$ we can define $[a_{ij}]^* = [\overline{a}_{ji}]$.

The length of $\mathbf{x} \in \mathbb{H}^n$ is defined to be

$$|\mathbf{x}| = \sqrt{\mathbf{x}^*\mathbf{x}} = \sqrt{\sum_{r=1}^{n} |x_r|^2}.$$

We can also define a norm on $\mathrm{M}_n(\mathbb{H})$ by the method used in Section 1.2, *i.e.*, for $A \in \mathrm{M}_n(\mathbb{H})$,

$$\|A\| = \sup\left\{ \frac{|A\mathbf{x}|}{|\mathbf{x}|} : \mathbf{0} \neq \mathbf{x} \in \mathbb{H}^n \right\}.$$

The analogue of Proposition 1.5 holds for $\| \;\|$ and the norm $| \;|$ on $\mathbb{H}$, although statements involving scalar multiplication need to be formulated with scalars on the right. There is also a resulting metric on $\mathrm{M}_n(\mathbb{H})$,

$$(A, B) \longmapsto \|A - B\|,$$

and we can use this to do analysis on $\mathrm{M}_n(\mathbb{H})$. The multiplication map

$$\mathrm{M}_n(\mathbb{H}) \times \mathrm{M}_n(\mathbb{H}) \longrightarrow \mathrm{M}_n(\mathbb{H})$$

is again continuous, and the group of invertible elements $\mathrm{GL}_n(\mathbb{H}) \subseteq \mathrm{M}_n(\mathbb{H})$ is actually an open subset. This can be proved using either of the reduced determinants

$$\mathrm{Rdet}_{\mathbb{R}} \colon \mathrm{M}_n(\mathbb{H}) \longrightarrow \mathbb{R}, \quad \mathrm{Rdet}_{\mathbb{C}} \colon \mathrm{M}_n(\mathbb{H}) \longrightarrow \mathbb{C},$$

each of which is continuous. By Proposition 4.39,

$$\mathrm{GL}_n(\mathbb{H}) = \mathrm{M}_n(\mathbb{H}) - \mathrm{Rdet}_{\mathbb{R}}^{-1}\, 0, \tag{4.5a}$$

$$\mathrm{GL}_n(\mathbb{H}) = \mathrm{M}_n(\mathbb{H}) - \mathrm{Rdet}_{\mathbb{C}}^{-1}\, 0. \tag{4.5b}$$

In either case we see that $\mathrm{GL}_n(\mathbb{H})$ is an open subset of $\mathrm{M}_n(\mathbb{H})$. It is also possible to show that the two embeddings

$$\mathrm{GL}_n(\mathbb{H}) \longrightarrow \mathrm{GL}_{4n}(\mathbb{R}), \quad \mathrm{GL}_n(\mathbb{H}) \longrightarrow \mathrm{GL}_{2n}(\mathbb{C}),$$

have closed images. So $\mathrm{GL}_n(\mathbb{H})$ and its closed subgroups are real and complex matrix groups.

Definition 4.43

The $n \times n$ *quaternionic symplectic group* is

$$\mathrm{Sp}(n) = \{A \in \mathrm{GL}_n(\mathbb{H}) : A^*A = I\} \leqslant \mathrm{GL}_n(\mathbb{H}).$$

Then $\mathrm{Sp}(n)$ is easily seen to satisfy

$$\mathrm{Sp}(n) = \{A \in \mathrm{GL}_n(\mathbb{H}) : \forall \mathbf{x}, \mathbf{y} \in \mathbb{H}^n,\ A\mathbf{x} \cdot A\mathbf{y} = \mathbf{x} \cdot \mathbf{y}\}. \tag{4.6}$$

Hence we have the following proposition.

Proposition 4.44

The Lie algebra of $\mathrm{Sp}(n)$ is

$$\mathfrak{sp}(n) = \{Q \in \mathrm{M}_n(\mathbb{H}) : Q^* = -Q\}$$

and this has dimension

$$\dim \mathrm{Sp}(n) = \dim_{\mathbb{R}} \mathfrak{sp}(n) = 2n^2 + n.$$

Proof

In $\mathrm{M}_n(\mathbb{H})$, solution space of the equation $Q^* = -Q$ has dimension

$$3n + 4\binom{n}{2} = 3n + 2(n^2 - n) = 2n^2 + n. \qquad \square$$

These groups $\mathrm{Sp}(n)$ form another infinite family of compact connected matrix groups along with familiar examples such as $\mathrm{SO}(n), \mathrm{U}(n), \mathrm{SU}(n)$. There are further examples, the *spinor groups* $\mathrm{Spin}(n)$ whose description involves the *real Clifford algebras* Cl_n.

4.6 Automorphism Groups of Algebras

Let A be a finite dimensional $\Bbbk$-algebra. We can view A as a $\Bbbk$-vector space and consider its general linear group $\mathrm{GL}_{\Bbbk}(A)$.

Definition 4.45

The group of *$\Bbbk$-algebra automorphisms* of A is

$$\mathrm{Aut}_{\Bbbk}(A) = \{\alpha \in \mathrm{GL}_{\Bbbk}(A) : \alpha \text{ is an algebra automorphism}\}.$$

It is easy to see that $\mathrm{Aut}_{\Bbbk}(A)$ is a group and $\mathrm{Aut}_{\Bbbk}(A) \leqslant \mathrm{GL}_{\Bbbk}(A)$.

If A has an algebra norm ν then by Theorem 1.53, we can give $\mathrm{GL}_{\Bbbk}(A)$ and hence $\mathrm{Aut}_{\Bbbk}(A)$, the metric associated to the operator norm $\| \ \|_\nu$ introduced in Section 1.8. Using these metrics we obtain the following propositions.

Proposition 4.46

$\mathrm{Aut}_{\Bbbk}(A)$ is a closed subgroup of $\mathrm{GL}_{\Bbbk}(A)$, hence is a matrix subgroup.

Proof

Here is one approach. Choose a basis $\{v_1, \dots, v_n\}$ for A and use this to identify A with $\Bbbk^n$ and then produce an isomorphism of matrix groups $\mathrm{GL}_{\Bbbk}(A) \longrightarrow \mathrm{GL}_n(\Bbbk)$ as in Proposition 1.56.

Let $\mu\colon \Bbbk^n \times \Bbbk^n \longrightarrow \Bbbk^n$ be the $\Bbbk$-bilinear map corresponding to the product A. Then for $\alpha \in \mathrm{GL}_n(\Bbbk)$, the condition that it corresponds to an element of $\mathrm{Aut}_{\Bbbk}(A)$ is that

$$\mu \circ (\alpha \times \alpha) = \alpha \circ \mu,$$

and writing $\alpha = [a_{ij}]$, this amounts to n polynomial equations for the a_{ij}. By the sort of considerations we used to identify matrix subgroups in Chapter 1, this shows that $\mathrm{Aut}_{\Bbbk}(A) \leqslant \mathrm{GL}_n(\Bbbk)$ is a closed subgroup. □

Example 4.47

Consider $A = \mathbb{C}$ as an $\mathbb{R}$-algebra. Then $\mathrm{Aut}_{\mathbb{R}}(\mathbb{C}) = \{\mathrm{Id}_{\mathbb{C}}, \overline{(\ )}\}$, where $\overline{(\ )}$ denotes complex conjugation.

Proof

Let $\alpha\colon \mathbb{C} \longrightarrow \mathbb{C}$ be an $\mathbb{R}$-algebra automorphism. Every complex number z has the form $z = x + yi$ with $x, y \in \mathbb{R}$, hence

$$\alpha(z) = \alpha(x + yi) = \alpha(x) + \alpha(y)\alpha(i) = x + y\alpha(i).$$

So it suffices to determine $\alpha(i)$. But applying α to the equation $i^2 + 1 = 0$, we obtain

$$\alpha(i)^2 + 1 = 0,$$

so $\alpha(i) = \pm i$. Hence either $\alpha = \mathrm{Id}_{\mathbb{C}}$ or $\alpha = \overline{(\ )}$. □

This example is rather simple because the algebra is commutative. Non-commutative algebras have lots of automorphisms obtained using conjugation in the group theoretic sense. Let $u \in A^{\times}$ be a unit in the $\Bbbk$-algebra A. The function

$$\chi_u : A \longrightarrow A; \quad \chi_u(a) = uau^{-1}$$

is clearly a $\Bbbk$-linear isomorphism. It is also easy to check that for $a, b \in A$ and $t \in \Bbbk$,

$$\chi_u(ab) = \chi_u(a)\chi_u(b).$$

Hence χ_u is a $\Bbbk$-algebra automorphism of A.

Definition 4.48

An automorphism α of the $\Bbbk$-algebra A is called an *inner automorphism* of A if it has the form χ_u for some unit $u \in A^{\times}$. The inner automorphisms form a subgroup $\mathrm{Inn}_{\Bbbk}(A) \leqslant \mathrm{Aut}_{\Bbbk}(A)$. Automorphisms which are not inner are called *outer automorphisms.*

Of course, if A is commutative, the only inner automorphism is the identity Id_A. For the $\mathbb{R}$-algebra of complex numbers, complex conjugation $\overline{(\ )}$ is an outer automorphism. The construction of inner automorphisms defines a function

$$\chi : A^{\times} \longrightarrow \mathrm{Aut}_{\Bbbk}(A); \quad u \longmapsto \chi_u.$$

The proof of the next result is left as an exercise.

Proposition 4.49

Let A be a finite dimensional $\Bbbk$-algebra.
i) The function $\chi : A^{\times} \longrightarrow \mathrm{Aut}_{\Bbbk}(A)$ is a continuous group homomorphism, with kernel and image

$$\ker \chi = \mathrm{Z}(A^{\times}) = \text{the centre of } A^{\times}, \quad \mathrm{im}\, \chi = \mathrm{Inn}_{\Bbbk}(A) \leqslant \mathrm{Aut}_{\Bbbk}(A).$$

ii) The subgroup of inner automorphisms $\mathrm{Inn}_{\Bbbk}(A) \leqslant \mathrm{Aut}_{\Bbbk}(A)$ is a closed normal subgroup, hence $\mathrm{Inn}_{\Bbbk}(A)$ is a matrix group.

Definition 4.50

For a finite dimensional $\Bbbk$-algebra A, the *outer automorphism group* is the quotient group

$$\mathrm{Out}_{\Bbbk}(A) = \mathrm{Aut}_{\Bbbk}(A)/\,\mathrm{Inn}_{\Bbbk}(A).$$

This has a natural *quotient topology,* which we discuss in Section 8.1.

Example 4.51

Let $A = \mathrm{M}_n(\Bbbk)$. Then

$$\mathrm{Inn}_{\Bbbk}(\mathrm{M}_n(\Bbbk)) = \mathrm{GL}_n(\Bbbk)/\{zI_n : z \in \Bbbk^\times\} = \mathrm{PGL}_n(\Bbbk),$$

the $n \times n$ *projective linear group.*

Example 4.52

For the $\mathbb{R}$-algebra of quaternions $\mathbb{H}$,

$$\mathrm{Aut}_{\mathbb{R}}(\mathbb{H}) = \mathrm{Inn}_{\mathbb{R}}(\mathbb{H}) \cong \mathrm{SO}(3).$$

Proof

Let $\alpha \in \mathrm{Aut}_{\mathbb{R}}(\mathbb{H})$. Then by Propositions 4.41 and 4.42, $\alpha(i), \alpha(j), \alpha(k)$ are pure quaternions satisfying

$$\alpha(i)^2 = \alpha(j)^2 = \alpha(k)^2 = -1, \quad \alpha(i)\alpha(j) = \alpha(k).$$

In particular, viewed as elements of the real vector space $\mathbb{R}^3$ with usual basis $\mathbf{e}_1 = i$, $\mathbf{e}_2 = j$ and $\mathbf{e}_3 = k$, these vectors form an orthonormal basis, hence there is an orthogonal matrix $A \in \mathrm{O}(3)$ for which

$$\alpha(i) = Ai, \quad \alpha(j) = Aj, \quad \alpha(k) = Ak.$$

It is even true that $\det A = 1$ since in terms of the vector product $\times$ we have

$$\alpha(i)\alpha(j) = \alpha(i) \times \alpha(j),$$

and it is standard that for a pair of orthonormal vectors $\mathbf{u}, \mathbf{v}$, the triple $\mathbf{u}, \mathbf{v}, \mathbf{u} \times \mathbf{v}$ is a left-handed orthonormal basis. Thus α is equivalent to a special orthogonal matrix $A \in \mathrm{SO}(3)$ acting on $\mathbb{R}^3$ viewed as the pure quaternions.

To see that such an automorphism is actually inner, recall that in Section 3.5 we defined a surjective group homomorphism $\overline{\mathrm{Ad}}\colon \mathrm{SU}(2) \longrightarrow \mathrm{SO}(3)$. Careful comparison of the definitions of $\overline{\mathrm{Ad}}$ and χ restricted to the subgroup of $\mathbb{H}^\times$ consisting of unit quaternions shows that these are essentially the same mapping. Hence $\alpha \in \mathrm{Inn}_{\mathbb{R}}(\mathbb{H})$.

So $\mathrm{Aut}_{\mathbb{R}}(\mathbb{H}) = \mathrm{Inn}_{\mathbb{R}}(\mathbb{H}) \cong \mathrm{SO}(3)$ as claimed. □

Remark 4.53

Suppose that A is a finite dimensional simple $\Bbbk$-algebra where $\Bbbk$ is actually the centre of A, *i.e.*,

$$\Bbbk = \{z \in A : \forall a \in A,\ az = za\}.$$

Then the important *Noether–Skolem Theorem* (see [15]) asserts that $\mathrm{Aut}_{\Bbbk}(A) = \mathrm{Inn}_{\Bbbk}(A)$, or equivalently that $\mathrm{Out}_{\Bbbk}(A) = \{1\}$. This is true for all division algebras and matrix algebras over a division algebra, which gives a more theoretical explanation for part of Example 4.52 and also shows that in Example 4.51 we actually have $\mathrm{Aut}_{\Bbbk}(\mathrm{M}_n(\Bbbk)) \cong \mathrm{PGL}_n(\Bbbk)$.

EXERCISES

4.1. Recall Example 4.25. Show that in $\mathbb{R}[\mathrm{S}_3]$, the summand

$$A'_3 = \mathbb{R}[\mathrm{S}_3]e_3 = e_3\mathbb{R}[\mathrm{S}_3]$$

contains at least 4 elements of order 2. Deduce that A'_3 cannot be isomorphic to $\mathbb{H}$ as an $\mathbb{R}$-algebra.

4.2. If $\Bbbk = \mathbb{R}$ or $\mathbb{C}$, let (A, ν) be a normed $\Bbbk$-algebra. Consider the set

$$A_1^\times = \{u \in A^\times : \nu(u) = 1 = \nu(u^{-1})\}.$$

a) Show that $A_1^\times \subseteq A$ is a compact subset with respect to the metric associated with ν.
b) Show that $A_1^\times$ is a subgroup of $A^\times$.
c) Find $A_1^\times$ in each of the following cases and determine its Lie algebra as a Lie subalgebra of A with its usual bracket: $A = \mathbb{C}$, $A = \mathbb{H}$, $A = \mathrm{M}_n(\Bbbk)$.

4.3. Show that for $r \in \mathbb{R}$, in the real division algebra of quaternions $\mathbb{H}$, the equation $x^2 = r$ has

- two solutions if $r > 0$;
- infinitely many solutions if $r < 0$ and these are all pure quaternions.

A good question to consider after reading Chapter 5 is whether this result generalises to an arbitrary Clifford algebra Cl_n.

4.4. Using the bases $\{1, i, j, k\}$ of $\mathbb{H}_{\mathbb{R}}$ over $\mathbb{R}$ and $\{1, j\}$ of $\mathbb{H}_{\mathbb{C}}$ over $\mathbb{C}$, determine the reduced determinants $\mathrm{Rdet}_{\mathbb{R}}\colon \mathrm{GL}_n(\mathbb{H}) \longrightarrow \mathbb{R}^\times$ and $\mathrm{Rdet}_{\mathbb{C}}\colon \mathrm{GL}_n(\mathbb{H}) \longrightarrow \mathbb{C}^\times$ for small values of n.

4.5. a) Verify that $\mathrm{M}_n(\mathbb{H})$ is complete with respect to the norm $\|\ \|$. Use this to define an exponential function $\exp\colon \mathrm{M}_n(\mathbb{H}) \longrightarrow \mathrm{GL}_n(\mathbb{H})$ with properties analogous to those for the exponential functions on $\mathrm{M}_n(\mathbb{R})$ and $\mathrm{M}_n(\mathbb{C})$.
b) When $n = 1$, determine $\exp(q)$ using the decomposition $q = r + su$ with $r, s \in \mathbb{R}$ and u a pure quaternion of unit length $|u| = 1$.

4.6. For each of the following matrix groups G, determine the Lie algebra $\mathfrak{g}$ and dimension $\dim G$.
a) $G = \mathrm{GL}_n(\mathbb{H})$.
b) $G = \mathrm{Sp}_n(\mathbb{H})$.
c) $G = \ker \mathrm{Rdet}_{\Bbbk}\colon \mathrm{GL}_n(\mathbb{H}) \longrightarrow \Bbbk^{\times}$, where $\Bbbk = \mathbb{R}, \mathbb{C}$ and n is small.
d) $G = \ker \mathrm{Rdet}_{\Bbbk}\colon \mathrm{Sp}(n) \longrightarrow \Bbbk^{\times}$, where $\Bbbk = \mathbb{R}, \mathbb{C}$ and n is small.

4.7. The group of unit quaternions

$$\mathrm{Sp}(1) = \{q \in \mathbb{H} : |q| = 1\}$$

has an $\mathbb{R}$-linear action on $\mathbb{H}$ given by

$$q \cdot x = qxq^{-1} = qx\overline{q} \quad (x \in \mathbb{H}).$$

a) By identifying $\mathbb{H}$ with $\mathbb{R}^4$ using the basis $\{i, j, k, 1\}$, show that this defines a Lie homomorphism $\mathrm{Sp}(1) \longrightarrow \mathrm{SO}(4)$.
b) Show that this action restricts to an action of $\mathrm{Sp}(1)$ on the space of pure quaternions and by identifying this with $\mathbb{R}^3$ using the basis $\{i, j, k\}$, show that this defines a surjective Lie homomorphism $\alpha\colon \mathrm{Sp}(1) \longrightarrow \mathrm{SO}(3)$. Show that $\ker \alpha = \{1, -1\}$.

4.8. Using the surjective homomorphism $\alpha\colon \mathrm{Sp}(1) \longrightarrow \mathrm{SO}(3)$ of the previous question, for a subgroup $G \leqslant \mathrm{SO}(3)$ set

$$\widetilde{G} = \alpha^{-1}G = \{g \in \mathrm{Sp}(1) : \alpha(g) \in G\} \leqslant \mathrm{Sp}(1).$$

From now on assume that G is *finite*.
a) Determine the order of $\widetilde{G}$.
b) Show that the order of the centre of $\widetilde{G}$, $\mathrm{Z}(\widetilde{G})$, is even.
c) If G contains an element of order 2, show that the group homomorphism $\alpha\colon \widetilde{G} \longrightarrow G$ is not *split* in the sense that there is no group homomorphism $\beta\colon \widetilde{G} \longrightarrow G$ for which $\alpha \circ \beta = \mathrm{Id}_G$.

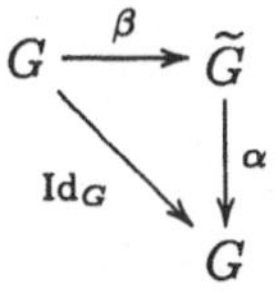

d) Show that $\mathrm{Q}_8 = \{\pm 1, \pm i, \pm j, \pm k\}$ is a subgroup of $\mathrm{Sp}(1)$ and find a geometric interpretation as a group of symmetries for $\alpha \mathrm{Q}_8 \leqslant \mathrm{SO}(3)$. Generalise this by considering for each $n \geqslant 2$,

$$\mathrm{Q}_{2n} = \left\{e^{2\pi i r/n} : r = 0, \ldots, n-1\right\} \cup \left\{e^{2\pi i r/n} j : r = 0, \ldots, n-1\right\}.$$

e) Show that the set T_{24} consisting of the 24 elements

$$\pm 1,\ \pm i,\ \pm j,\ \pm k,\ \frac{1}{2}(\pm 1 \pm i \pm j \pm k),$$

is a subgroup of $\mathrm{Sp}(1)$ and find a geometric interpretation for the group $\alpha\mathrm{T}_{24} \leqslant \mathrm{SO}(3)$.

f) ⚠⚠ Let **Icos** be a regular icosahedron in $\mathbb{R}^3$ centred at the origin. The group of direct symmetries of **Icos** is known to be isomorphic to the alternating group, $\mathrm{Symm}^+(\mathbf{Icos}) \cong \mathrm{A}_5$. Determine $\alpha^{-1}\,\mathrm{Symm}^+(\mathbf{Icos}) \leqslant \mathrm{Sp}(1)$.

[This requires a good way to view the icosahedron relative to the x, y, z-axes. The resulting subgroup of $\mathrm{Sp}(1)$ is called the *binary icosahedral group* since it provides a double covering of the symmetry group $\mathrm{Symm}^+(\mathbf{Icos})$. It also provides a non-split double covering $\widetilde{\mathrm{A}}_5 \longrightarrow \mathrm{A}_5$ of the simple group A_5.]

4.9. ⚠⚠ a) Show that for each $n \geqslant 1$,

$$\dim \mathrm{Sp}(n) = \dim \mathrm{Symp}_{2n}(\mathbb{R}).$$

b) Notice that there are embeddings $\mathrm{Symp}_{2n}(\mathbb{R}) \leqslant \mathrm{GL}_{2n}(\mathbb{C})$ and $\mathrm{Sp}(n) \leqslant \mathrm{GL}_{2n}(\mathbb{C})$, hence $\mathfrak{sp}(n)_{\mathbb{C}} \leqslant \mathfrak{gl}_{2n}(\mathbb{C})$ and $\mathfrak{symp}_{2n}(\mathbb{R})_{\mathbb{C}} \leqslant \mathfrak{gl}_{2n}(\mathbb{C})$. Show that as $\mathbb{C}$-Lie algebras

$$\mathfrak{sp}(n)_{\mathbb{C}} \cong \mathfrak{symp}_{2n}(\mathbb{R})_{\mathbb{C}}.$$

4.10. Prove Proposition 4.49.

5
Clifford Algebras and Spinor Groups

In this chapter, we generalise the quaternions by studying the *real Clifford algebras*, and our account of these is heavily influenced by the classic paper of Atiyah, Bott & Shapiro [3]; Porteous [23, 24] also provides an accessible description, as does Curtis [7] but there are some errors and omissions in that account. Lawson & Michelsohn [19] provides a more sophisticated introduction which shows how central Clifford algebras have become to modern geometry and topology; they also appear in Quantum Theory in connection with the Dirac operator. There is also a theory of *Clifford Analysis* in which the field of complex numbers is replaced by a Clifford algebra and a suitable class of *Clifford analytic functions* generalising complex analytic functions is studied; motivation for this is provided by the above applications. The groups of units in Clifford algebras contain the *spinor groups* which we define and also show how they provide double coverings of the special orthogonal groups.

We restrict attention to real Clifford algebras associated to positive definite inner products on $\mathbb{R}^n$. There are also complex and indefinite Clifford algebras discussed in [3, 19, 23, 24].

5.1 Real Clifford Algebras

The sequence of real division algebras $\mathbb{R}, \mathbb{C}, \mathbb{H}$ can be extended by introducing the (*real*) *Clifford algebras* Cl_n for $n \geqslant 0$, where the first few satisfy

$$\mathrm{Cl}_0 = \mathbb{R}, \quad \mathrm{Cl}_1 \cong \mathbb{C}, \quad \mathrm{Cl}_2 \cong \mathbb{H},$$

as $\mathbb{R}$-algebras.

We begin by describing Cl_n as an $\mathbb{R}$-vector space and define the product in terms of a particular basis. There are elements $e_1, e_2, \dots, e_n \in \mathrm{Cl}_n$ for which

$$\begin{cases} e_s e_r = -e_r e_s & \text{if } s \neq r, \\ e_r^2 = -1. \end{cases} \tag{5.1}$$

Moreover, Cl_n has an $\mathbb{R}$-basis consisting of the elements $e_{i_1} e_{i_2} \cdots e_{i_r}$ corresponding to increasing sequences $1 \leqslant i_1 < i_2 < \cdots < i_r \leqslant n$ with $0 \leqslant r \leqslant n$. Thus

$$\dim_{\mathbb{R}} \mathrm{Cl}_n = 2^n. \tag{5.2}$$

When $r = 0$, the element $e_{i_1} e_{i_2} \cdots e_{i_r}$ is taken to be 1.

Doing calculations with these basis elements is easy as the following example illustrates.

Example 5.1

Let i, j, k, ℓ be distinct numbers in the range 1 to n. Then

$$e_i e_j e_k e_\ell = e_k e_\ell e_i e_j.$$

Proof

Repeatedly using the relations of (5.1) we obtain

$$e_i e_j e_k e_\ell = -e_i e_k e_j e_\ell = e_k e_i e_j e_\ell = -e_k e_i e_\ell e_j = e_k e_\ell e_i e_j. \qquad \square$$

Proposition 5.2

There are isomorphisms of $\mathbb{R}$-algebras

$$\mathrm{Cl}_1 \cong \mathbb{C}, \quad \mathrm{Cl}_2 \cong \mathbb{H}.$$

Proof

For Cl_1, the function

$$\mathrm{Cl}_1 \longrightarrow \mathbb{C}; \quad x + ye_1 \longmapsto x + yi \quad (x, y \in \mathbb{R}),$$

is an $\mathbb{R}$-linear ring isomorphism.

Similarly, for Cl_2, the function

$$\mathrm{Cl}_2 \longrightarrow \mathbb{H}; \quad t1 + xe_1 + ye_2 + ze_1e_2 \longmapsto t1 + xi + yj + zk \quad (t, x, y, z \in \mathbb{R}),$$

is an $\mathbb{R}$-linear ring isomorphism. □

We can order the basis monomials in the e_r by declaring $e_{i_1}e_{i_2}\cdots e_{i_r}$ to be numbered

$$1 + 2^{i_1-1} + 2^{i_2-1} + \cdots + 2^{i_r-1},$$

which should be interpreted as 1 when $r = 0$. Every integer k for which $1 \leqslant k \leqslant 2^n$ has a unique *binary expansion*

$$k = k_0 + 2k_1 + \cdots + 2^j k_j + \cdots + 2^n k_n,$$

where each $k_j = 0, 1$. This provides a one to one correspondence between such numbers k and the basis monomials of Cl_n. Here are the orderings of the bases for the first few Clifford algebras.

Cl_1	$1, e_1$
Cl_2	$1, e_1, e_2, e_1e_2$
Cl_3	$1, e_1, e_2, e_1e_2, e_3, e_1e_3, e_2e_3, e_1e_2e_3$

Using the left regular representation over $\mathbb{R}$ associated with this basis of Cl_n, we can realise Cl_n as a subalgebra of $\mathrm{M}_{2^n}(\mathbb{R})$.

Example 5.3

For Cl_1 we have the basis $\{1, e_1\}$ and find that

$$\rho(1) = I_2, \quad \rho(e_1) = \begin{bmatrix} 0 & -1 \\ 1 & 0 \end{bmatrix},$$

while the general formula is

$$\rho(x + ye_1) = \begin{bmatrix} x & -y \\ y & x \end{bmatrix} \quad (x, y \in \mathbb{R}).$$

Applying ρ to the basis $\{1, e_1, e_2, e_1e_2\}$ we obtain the following elements in $M_4(\mathbb{R})$:

$$\rho(1) = I_4, \qquad \rho(e_1) = \begin{bmatrix} 0 & -1 & 0 & 0 \\ 1 & 0 & 0 & 0 \\ 0 & 0 & 0 & -1 \\ 0 & 0 & 1 & 0 \end{bmatrix},$$

$$\rho(e_2) = \begin{bmatrix} 0 & 0 & -1 & 0 \\ 0 & 0 & 0 & 1 \\ 1 & 0 & 0 & 0 \\ 0 & -1 & 0 & 0 \end{bmatrix}, \qquad \rho(e_1e_2) = \begin{bmatrix} 0 & 0 & 0 & -1 \\ 0 & 0 & -1 & 0 \\ 0 & 1 & 0 & 0 \\ 1 & 0 & 0 & 0 \end{bmatrix}.$$

In all cases the matrices $\rho(e_{i_1}e_{i_2}\cdots e_{i_r})$ are *generalised permutation matrices* all of whose entries are $0, \pm 1$ and exactly one non-zero entry in each row and column. These are always orthogonal matrices of determinant 1.

The Clifford algebras Cl_n are characterised by an important *universal property*. First notice that there is an $\mathbb{R}$-linear transformation

$$j_n \colon \mathbb{R}^n \longrightarrow \mathrm{Cl}_n; \quad j_n\left(\sum_{r=1}^{n} x_r \mathbf{e}_r\right) = \sum_{r=1}^{n} x_r e_r,$$

for which

$$j_n\left(\sum_{r=1}^{n} x_r \mathbf{e}_r\right)^2 = -\sum_{r=1}^{n} x_r^2 = -\left|\sum_{r=1}^{n} x_r \mathbf{e}_r\right|^2. \tag{5.3}$$

Theorem 5.4 (Universal property of a Clifford algebra)

Let A be an $\mathbb{R}$-algebra. If $f \colon \mathbb{R}^n \longrightarrow A$ is an $\mathbb{R}$-linear transformation for which

$$f(\mathbf{x})^2 = -|\mathbf{x}|^2 1 \quad (\mathbf{x} \in \mathbb{R}^n),$$

then there is a unique homomorphism of $\mathbb{R}$-algebras $F \colon \mathrm{Cl}_n \longrightarrow A$ for which $F \circ j_n = f$, *i.e.*, for all $\mathbf{x} \in \mathbb{R}^n$, $F(j_n(\mathbf{x})) = f(\mathbf{x})$.

This can be indicated in the following commutative diagram

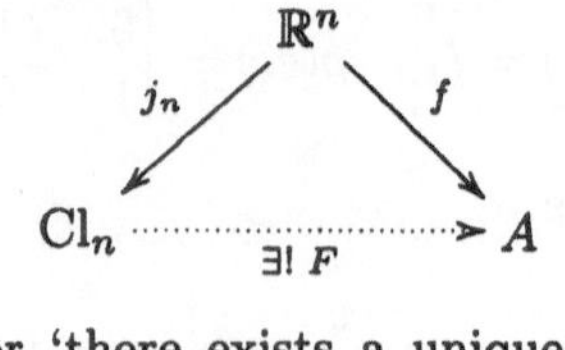

in which '$\exists!\ F$' stands for 'there exists a unique F' and the dotted arrow indicates a function F which solves the following equation amongst functions, $F \circ j_n = f$.

Proof

The homomorphism $F\colon \mathrm{Cl}_n \longrightarrow A$ is defined by setting $F(e_r) = f(\mathbf{e}_r)$ and showing that it extends it to a ring homomorphism given on basis monomials by

$$F(e_{i_1}\cdots e_{i_r}) = f(\mathbf{e}_{i_1})\cdots f(\mathbf{e}_{i_r}).$$

The details are left as an exercise. □

Corollary 5.5

Let U be an $\mathbb{R}$-algebra and $j\colon \mathbb{R}^n \longrightarrow U$ be an $\mathbb{R}$-linear transformation for which

$$j(\mathbf{x})^2 = -|\mathbf{x}|^2 1 \quad (\mathbf{x} \in \mathbb{R}^n).$$

Suppose that U and j have the universal property enjoyed by Cl_n.

If A is an $\mathbb{R}$-algebra and $h\colon \mathbb{R}^n \longrightarrow A$ is an $\mathbb{R}$-linear transformation for which

$$h(\mathbf{x})^2 = -|\mathbf{x}|^2 1 \quad (\mathbf{x} \in \mathbb{R}^n),$$

then there is a unique homomorphism of $\mathbb{R}$-algebras $H\colon U \longrightarrow A$ satisfying $H \circ j = f$, i.e., *for all $\mathbf{x} \in \mathbb{R}^n$, $H(j(\mathbf{x})) = h(\mathbf{x})$.*

Then there is a unique $\mathbb{R}$-algebra isomorphism $\Phi\colon \mathrm{Cl}_n \longrightarrow U$ which satisfies $\Phi \circ j_n = j$.

Proof

The algebra homomorphism $\Phi\colon \mathrm{Cl}_n \longrightarrow U$ is constructed using the universal property Cl_n applied to the linear transformation $j\colon \mathbb{R}^n \longrightarrow U$. On the other hand, applying the universal property of U to $j_n\colon \mathbb{R}^n \longrightarrow \mathrm{Cl}_n$, we obtain an algebra homomorphism $\Theta\colon U \longrightarrow \mathrm{Cl}_n$. These homomorphisms satisfy the equations

$$\Phi \circ j_n = j, \quad \Theta \circ j = j_n.$$

Notice that the following equations also hold:

$$\Theta \circ \Phi \circ j_n = \Theta \circ j = j_n, \quad \Phi \circ \Theta \circ j = \Phi \circ j_n = j.$$

By the uniqueness part of the universality of Cl_n applied to the linear transformation $j_n\colon \mathbb{R}^n \longrightarrow \mathrm{Cl}_n$ together with the identity $\mathrm{Id}_{\mathrm{Cl}_n} \circ j_n = j_n$, we have $\Theta \circ \Phi = \mathrm{Id}_{\mathrm{Cl}_n}$. Similarly, combining the uniqueness part of the universality of U

applied to the linear transformation $j\colon \mathbb{R}^n \longrightarrow U$ with the identity $\mathrm{Id}_U \circ j = j$, we obtain $\Phi \circ \Theta = \mathrm{Id}_U$.

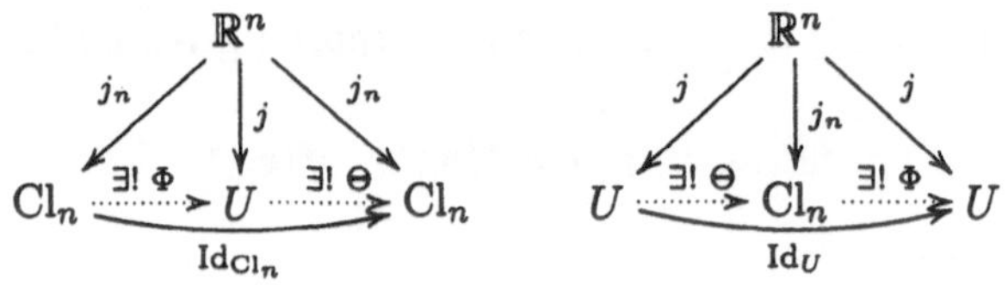

Hence Φ is an algebra isomorphism with inverse $\Phi^{-1} = \Theta$. □

This uniqueness up to isomorphism is an important characteristic of objects satisfying such universal properties. Another example we discuss is that of the quotient topology, see Proposition 8.3 and Corollary 8.4.

Example 5.6

There is an $\mathbb{R}$-linear transformation

$$\alpha_0\colon \mathbb{R}^n \longrightarrow \mathrm{Cl}_n; \quad \alpha_0(\mathbf{x}) = -j_n(\mathbf{x}) = j_n(-\mathbf{x}).$$

Then in Cl_n,

$$\alpha_0(\mathbf{x})^2 = j_n(-\mathbf{x})^2 = -|\mathbf{x}|^2,$$

so by Theorem 5.4 there is a unique algebra homomorphism $\alpha\colon \mathrm{Cl}_n \longrightarrow \mathrm{Cl}_n$ for which

$$\alpha(j_n(\mathbf{x})) = \alpha_0(\mathbf{x}).$$

Since $j_n(\mathbf{e}_r) = e_r$, this implies that

$$\alpha(e_r) = -e_r.$$

Notice that for $1 \leqslant i_1 < i_2 < \cdots < i_k \leqslant n$,

$$\alpha(e_{i_1}e_{i_2}\cdots e_{i_k}) = (-1)^k e_{i_1}e_{i_2}\cdots e_{i_k} = \begin{cases} e_{i_1}e_{i_2}\cdots e_{i_k} & \text{if } k \text{ is even,} \\ -e_{i_1}e_{i_2}\cdots e_{i_k} & \text{if } k \text{ is odd.} \end{cases}$$

It is easy to see that α is an isomorphism and hence an automorphism, often called the *canonical automorphism* of Cl_n.

We record the explicit forms of the next few Clifford algebras. Consider the $\mathbb{R}$-algebra $\mathrm{M}_2(\mathbb{H})$ of dimension 16. There is an $\mathbb{R}$-linear transformation

$$\theta_4\colon \mathbb{R}^4 \longrightarrow \mathrm{M}_2(\mathbb{H});$$

$$\theta_4(x_1\mathbf{e}_1 + x_2\mathbf{e}_2 + x_3\mathbf{e}_3 + x_4\mathbf{e}_4) = \begin{bmatrix} x_1 i + x_2 j + x_3 k & x_4 k \\ x_4 k & x_1 i + x_2 j - x_3 k \end{bmatrix}.$$

Direct calculation shows that θ_4 satisfies the condition of Theorem 5.4, hence there is a unique $\mathbb{R}$-algebra homomorphism $\Theta_4 \colon \mathrm{Cl}_4 \longrightarrow \mathrm{M}_2(\mathbb{H})$ for which $\Theta_4 \circ j_4 = \theta_4$. It can be shown that this is an isomorphism of $\mathbb{R}$-algebras, so

$$\mathrm{Cl}_4 \cong \mathrm{M}_2(\mathbb{H}).$$

Since $\mathbb{R} \subseteq \mathbb{R}^2 \subseteq \mathbb{R}^3 \subseteq \mathbb{R}^4$, we obtain compatible injective homomorphisms

$$\Theta_1 \colon \mathrm{Cl}_1 \longrightarrow \mathrm{M}_2(\mathbb{H}), \quad \Theta_2 \colon \mathrm{Cl}_2 \longrightarrow \mathrm{M}_2(\mathbb{H}), \quad \Theta_3 \colon \mathrm{Cl}_3 \longrightarrow \mathrm{M}_2(\mathbb{H}),$$

which turn out to have images

$$\begin{aligned} \operatorname{im}\Theta_1 &= \{zI_2 : z \in \mathbb{C}\}, \\ \operatorname{im}\Theta_2 &= \{qI_2 : q \in \mathbb{H}\}, \\ \operatorname{im}\Theta_3 &= \left\{ \begin{bmatrix} q_1 & 0 \\ 0 & q_2 \end{bmatrix} : q_1, q_2 \in \mathbb{H} \right\}. \end{aligned}$$

This shows that there is an isomorphism of $\mathbb{R}$-algebras

$$\mathrm{Cl}_3 \cong \mathbb{H} \times \mathbb{H},$$

where the latter is the direct product of Definition 4.8. By direct calculation, we also obtain

$$\mathrm{Cl}_5 \cong \mathrm{M}_4(\mathbb{C}), \quad \mathrm{Cl}_6 \cong \mathrm{M}_8(\mathbb{R}), \quad \mathrm{Cl}_7 \cong \mathrm{M}_8(\mathbb{R}) \times \mathrm{M}_8(\mathbb{R}).$$

These results are summarised in Table 5.1, whose last column gives their dimensions.

Cl_0	$\mathbb{R}$	1
Cl_1	$\mathbb{C}$	2
Cl_2	$\mathbb{H}$	4
Cl_3	$\mathbb{H} \times \mathbb{H}$	8
Cl_4	$\mathrm{M}_2(\mathbb{H})$	16
Cl_5	$\mathrm{M}_4(\mathbb{C})$	32
Cl_6	$\mathrm{M}_8(\mathbb{R})$	64
Cl_7	$\mathrm{M}_8(\mathbb{R}) \times \mathrm{M}_8(\mathbb{R})$	128
Cl_8	$\mathrm{M}_{16}(\mathbb{R})$	256

Table 5.1 The first 8 Clifford algebras

To go beyond this we use the following periodicity result, in which $\mathrm{M}_m(\mathrm{Cl}_n)$ denotes the ring of $m \times m$ matrices with entries in Cl_n.

Theorem 5.7

For $n \geqslant 0$,

$$\mathrm{Cl}_{n+8} \cong \mathrm{M}_{16}(\mathrm{Cl}_n).$$

We will not give the proof (which can be found in [3, 23, 24]) but note that it involves the useful observation that for a unital ring R, there is an isomorphism of rings

$$\mathrm{M}_m(\mathrm{M}_n(R)) \cong \mathrm{M}_{mn}(R), \tag{5.4}$$

obtained by expressing each $m \times m$ matrix of $n \times n$ matrices as an $mn \times mn$ matrix. Thus we obtain for example

$$\begin{aligned}
\mathrm{Cl}_{10} &\cong \mathrm{M}_{16}(\mathbb{H}), \\
\mathrm{Cl}_{12} &\cong \mathrm{M}_{16}(\mathrm{M}_2(\mathbb{H})) \cong \mathrm{M}_{32}(\mathbb{H}), \\
\mathrm{Cl}_{14} &\cong \mathrm{M}_{16}(\mathrm{M}_8(\mathbb{R})) \cong \mathrm{M}_{128}(\mathbb{R}).
\end{aligned}$$

In the next section we will make use of some more structure in Cl_n. First there is a *conjugation* $\overline{(\ )}\colon \mathrm{Cl}_n \longrightarrow \mathrm{Cl}_n$ defined by

$$\overline{e_{i_1} e_{i_2} \cdots e_{i_k}} = (-1)^k e_{i_k} e_{i_{k-1}} \cdots e_{i_1}$$

whenever $1 \leqslant i_1 < i_2 < \cdots < i_k \leqslant n$, and satisfying

$$\begin{aligned}
\overline{x + y} &= \overline{x} + \overline{y}, \\
\overline{tx} &= t\overline{x},
\end{aligned}$$

for $x, y \in \mathrm{Cl}_n$ and $t \in \mathbb{R}$. Notice that if $n > 1$, $\overline{(\ )}$ is not a ring homomorphism since whenever $r < s$,

$$\overline{e_r e_s} = e_s e_r = -e_r e_s = -\overline{e_r}\,\overline{e_s} \neq \overline{e_r}\,\overline{e_s}.$$

However, it is a ring *anti-homomorphism* in the sense that for all $x, y \in \mathrm{Cl}_n$,

$$\overline{xy} = \overline{y}\,\overline{x} \quad (x, y \in \mathrm{Cl}_n). \tag{5.5}$$

When $n = 1$ or 2, $\overline{(\ )}$ agrees with the conjugations already defined in $\mathbb{C}$ and $\mathbb{H}$.

Second there is the canonical automorphism $\alpha\colon \mathrm{Cl}_n \longrightarrow \mathrm{Cl}_n$ defined in Example 5.6. We can use α to define a $\pm$-*grading* on Cl_n:

$$\mathrm{Cl}_n^+ = \{u \in \mathrm{Cl}_n : \alpha(u) = u\}, \quad \mathrm{Cl}_n^- = \{u \in \mathrm{Cl}_n : \alpha(u) = -u\}.$$

Proposition 5.8

i) Every element $v \in \mathrm{Cl}_n$ can be uniquely expressed as $v = v^+ + v^-$ with $v^+ \in \mathrm{Cl}_n^+$ and $v^- \in \mathrm{Cl}_n^-$. Hence $\mathrm{Cl}_n = \mathrm{Cl}_n^+ \oplus \mathrm{Cl}_n^-$ as an $\mathbb{R}$-vector space.
ii) This decomposition is multiplicative in the sense that

$$\begin{cases} uv \in \mathrm{Cl}_n^+ & \text{if } u, v \in \mathrm{Cl}_n^+ \text{ or } u, v \in \mathrm{Cl}_n^-, \\ uv, vu \in \mathrm{Cl}_n^+ & \text{if } u \in \mathrm{Cl}_n^+ \text{ and } v \in \mathrm{Cl}_n^-. \end{cases}$$

Proof

(i) The elements

$$v^+ = \frac{1}{2}(v + \alpha(v)), \quad v^- = \frac{1}{2}(v - \alpha(v)), \tag{5.6}$$

satisfy $\alpha(v^+) = v^+$, $\alpha(v^-) = -v^-$ and $v = v^+ + v^-$, hence $v^+ \in \mathrm{Cl}_n^+$ and $v^- \in \mathrm{Cl}_n^-$. On the other hand, if $v = v' + v''$ with $v' \in \mathrm{Cl}_n^+$ and $v'' \in \mathrm{Cl}_n^-$, then $\alpha(v) = v' - v''$ and so

$$\frac{1}{2}(v + \alpha(v)) = v', \quad \frac{1}{2}(v - \alpha(v)) = v''.$$

So this expression is the only one with such properties and therefore defines the stated vector space direct sum decomposition.
(ii) This is easily checked using the fact that α is a ring homomorphism. □

Notice that for bases of $\mathrm{Cl}_n^\pm$ we have the monomials

$$\begin{cases} e_{j_1} \cdots e_{j_{2m}} \in \mathrm{Cl}_n^+ & (1 \leqslant j_1 < \cdots < j_{2m} \leqslant n), \\ e_{j_1} \cdots e_{j_{2m+1}} \in \mathrm{Cl}_n^- & (1 \leqslant j_1 < \cdots < j_{2m+1} \leqslant n). \end{cases} \tag{5.7}$$

For later use, it is useful to record the following identity which holds for all $u \in \mathrm{Cl}_n$:

$$\alpha(\overline{u}) = \overline{\alpha(u)}. \tag{5.8}$$

This is verified by checking that it holds on the monomial basis in the e_i.

Remark 5.9

It is worth noting that the composite of α and $\overline{(\ )}$ (in either order) is another anti-homomorphism of Cl_n; then Id, α, $\overline{(\ )}$ and $\overline{\alpha} = \alpha \circ \overline{(\ )}$ form a finite group of order 4 which is not cyclic since the square of every element is the identity.

Finally, we introduce an inner product $\cdot$ and a norm $|\ |$ on Cl_n by defining the distinct monomials $e_{i_1}e_{i_2}\cdots e_{i_k}$ with $1 \leqslant i_1 < i_2 < \cdots < i_k \leqslant n$ to be an orthonormal basis, *i.e.*,

$$(e_{i_1}e_{i_2}\cdots e_{i_k})\cdot(e_{j_1}e_{j_2}\cdots e_{j_\ell}) = \begin{cases} 1 & \text{if } \ell = k \text{ and } i_r = j_r \text{ for all } r, \\ 0 & \text{otherwise,} \end{cases}$$

and

$$|x| = \sqrt{x \cdot x}.$$

We leave it to the reader to check that $|\ |$ is a norm. Perhaps a more illuminating way to define the inner product $\cdot$ is by using the formula

$$u \cdot v = \frac{1}{2}\operatorname{Re}(\overline{u}\,v + \overline{v}\,u), \tag{5.9}$$

where for $w \in \mathrm{Cl}_n$ we define its *real part* $\operatorname{Re} w$ to be the coefficient of 1 when w is expanded as an $\mathbb{R}$-linear combination of the basis monomials $e_{i_1}\cdots e_{i_r}$. It is easily verified that for any $u, v \in \mathrm{Cl}_n$ and $w \in j_n\mathbb{R}^n$,

$$(wu)\cdot(wv) = |w|^2(u \cdot v). \tag{5.10}$$

In particular, when $|w| = 1$ left multiplication by w defines an $\mathbb{R}$-linear transformation on Cl_n which is an isometry. The norm $|\ |$ gives rise to a metric on Cl_n which makes the group of units $\mathrm{Cl}_n^{\times}$ into a topological group and the above embeddings of Cl_n into matrix rings are then continuous, so $\mathrm{Cl}_n^{\times}$ is a matrix group. Unfortunately, these embeddings are not norm preserving in general. For example, $2 + e_1e_2e_3 \in \mathrm{Cl}_3$ has $|2 + e_1e_2e_3| = \sqrt{5}$, but the corresponding matrix in $\mathrm{M}_8(\mathbb{R})$ has norm $\sqrt{3}$. However, by defining for each $w \in \mathrm{Cl}_n$

$$\|w\| = \sup\{|wx| : x \in \mathrm{Cl}_n,\ |x| = 1\} = \max\{|wx| : x \in \mathrm{Cl}_n,\ |x| = 1\},$$

we obtain another equivalent norm on Cl_n for which the above embedding $\mathrm{Cl}_n \longrightarrow \mathrm{M}_{2^n}(\mathbb{R})$ does preserve norms. For $w \in j_n\mathbb{R}^n$ we do have $\|w\| = |w|$ and more generally, for $w_1, \ldots, w_k \in j_n\mathbb{R}^n$,

$$\|w_1\cdots w_k\| = |w_1\cdots w_k| = |w_1|\cdots|w_k|.$$

For $x, y \in \mathrm{Cl}_n$,

$$\|xy\| \leqslant \|x\|\,\|y\|,$$

without equality in general.

5.2 Clifford Groups

Using the injective linear transformation $j_n\colon \mathbb{R}^n \longrightarrow \mathrm{Cl}_n$ we can identify $\mathbb{R}^n$ with a subspace of Cl_n, *i.e.*,

$$\sum_{r=1}^{n} x_r \mathbf{e}_r \longleftrightarrow j_n(\sum_{r=1}^{n} x_r \mathbf{e}_r) = \sum_{r=1}^{n} x_r e_r.$$

From now on we do this without further comment, writing elements as $x \in \mathbb{R}^n$ rather than $\mathbf{x}$.

Notice that $\mathbb{R}^n \subseteq \mathrm{Cl}_n^-$, so for $x \in \mathbb{R}^n$, $u \in \mathrm{Cl}_n^+$ and $v \in \mathrm{Cl}_n^-$,

$$xu, ux \in \mathrm{Cl}_n^-, \quad xv, vx \in \mathrm{Cl}_n^+. \tag{5.11}$$

Definition 5.10

For $n \geqslant 1$, the *Clifford group* Γ_n is the subgroup

$$\Gamma_n = \{u \in \mathrm{Cl}_n^\times : \forall x \in \mathbb{R}^n,\ \alpha(u)xu^{-1} \in \mathbb{R}^n\} \leqslant \mathrm{Cl}_n^\times.$$

Notice that $\mathbb{R}^\times \leqslant \Gamma_n$ is a normal subgroup and is the centre of Γ_n. Also for any non-zero $u \in \mathbb{R}^n$ and $x \in \mathbb{R}^n$ it is easily seen that $\alpha(u)xu^{-1} \in \mathbb{R}^n$, hence $u \in \Gamma_n$.

Proposition 5.11

Γ_n is a closed subgroup of $\mathrm{Cl}_n^\times$.

Proof

There is a continuous action

$$\mathrm{Cl}_n^\times \times \mathrm{Cl}_n \longrightarrow \mathrm{Cl}_n; \quad (u, v) = \alpha(u)vu^{-1}.$$

Notice that for each $u \in \mathrm{Cl}_n^\times$, the function

$$\mathrm{Cl}_n \longrightarrow \mathrm{Cl}_n; \quad v \longmapsto \alpha(u)vu^{-1},$$

is a linear isomorphism. Since $\mathbb{R}^n \subseteq \mathrm{Cl}_n$ is an $\mathbb{R}$-subspace, it is closed and so by Section 1.9, its stabiliser

$$\mathrm{Stab}_{\mathrm{Cl}_n^\times}(\mathbb{R}^n) \leqslant \mathrm{Cl}_n^\times$$

is a closed subgroup of $\mathrm{Cl}_n^\times$. But this stabiliser is clearly Γ_n. □

For each $u \in \mathrm{Cl}_n^{\times}$, we can restrict the function $\mathrm{Cl}_n \longrightarrow \mathrm{Cl}_n$ of this proof to

$$\rho_u : \mathbb{R}^n \longrightarrow \mathbb{R}^n; \quad \rho_u(v) = \alpha(u)vu^{-1}.$$

Since we already know that $\mathrm{Cl}_n^{\times}$ is a matrix group we have the following corollary.

Corollary 5.12

For $n \geqslant 1$, Γ_n is a matrix group.

Proposition 5.13

For $u \in \Gamma_n$, $\alpha(u)$ and $\overline{u}$ are also in Γ_n.

Proof

For $u \in \Gamma_n$ and $x \in \mathbb{R}^n$, $\alpha(x) = -x = \overline{x} \in \mathbb{R}^n$. By Equations (5.8) and (5.5) together with the fact that $\alpha \circ \alpha = \mathrm{Id}$, we have

$$\begin{aligned}\alpha(\alpha(u))x\alpha(u)^{-1} &= \alpha(\alpha(u))\alpha(-x)\alpha(u)^{-1} \\ &= \alpha(\alpha(u)(-x)u^{-1}) \in \mathbb{R}^n,\end{aligned}$$

and

$$\begin{aligned}\alpha(\overline{u})x\overline{u}^{-1} &= \alpha(\overline{u})\overline{(-x)}\,\overline{u^{-1}} \\ &= \overline{\alpha(u)}\,\overline{(-x)}\,\overline{u^{-1}} \\ &= \overline{u^{-1}(-x)\alpha(u)} \\ &= \overline{\alpha(\alpha(u^{-1})\alpha(-x)u)} \in \mathbb{R}^n,\end{aligned}$$

since $\mathbb{R}^n$ is closed under α, $\overline{(\ )}$ and $\rho_{\alpha(u^{-1})}$. □

Proposition 5.14

For each $u \in \Gamma_n$, the function $\rho_u : \mathbb{R}^n \longrightarrow \mathbb{R}^n$ is an $\mathbb{R}$-linear isometry.

Proof

For any $x \in \mathbb{R}^n$,

$$\begin{aligned}|\rho_u(x)|^2 &= -\rho_u(x)^2 \\ &= -((-u)xu^{-1})^2 \\ &= -(-u)xu^{-1}(-u)xu^{-1} \\ &= -ux^2u^{-1} \\ &= u|x|^2u^{-1} \\ &= |x|^2.\end{aligned}$$

Hence $|\rho_u(x)| = |x|$ for all $x \in \mathbb{R}^n$. □

This means that for $u \in \Gamma_n$, if we express ρ_u in terms of the standard basis $\{e_1, \ldots, e_n\}$, then we have $\rho_u \in \mathrm{O}(n)$. Hence there is a group homomorphism

$$\rho\colon \Gamma_n \longrightarrow \mathrm{O}(n); \quad \rho(u) = \rho_u.$$

In fact, ρ is also continuous. We also need to identify the kernel of ρ.

Proposition 5.15

$$\ker \rho = \mathbb{R}^\times = \{t1 : t \in \mathbb{R},\ t \neq 0\}.$$

Proof

Suppose that $u \in \ker \rho$. Then for every $x \in \mathbb{R}^n$, $\alpha(u)xu^{-1} = x$, *i.e.*, $\alpha(u)x = xu$. Writing $u = u^+ + u^-$ with $u^\pm \in \mathrm{Cl}_n^\pm$, we have

$$u^+x = xu^+, \tag{5.12a}$$

$$-u^-x = xu^-. \tag{5.12b}$$

For each $r = 1, \ldots, n$, we can write

$$u^+ = a_r^+ + e_rb_r^-, \quad u^- = a_r^- + e_rb_r^+,$$

where $a_r^\pm, b_r^\pm \in \mathrm{Cl}_n^\pm$ do not involve e_r in their expansions in terms of the monomial bases of Equation (5.7).

On taking $x = e_r$, by Equation (5.12a) together with the identities

$$e_ra_r^+ = a_r^+e_r, \quad e_rb_r^- = -b_r^-e_r,$$

we obtain

$$a_r^+ e_r + b_r^- = a_r^+ e_r - b_r^- .$$

Now by comparing the parts involving or not involving e_r we find that $b_r^- = 0$. Similarly, Equation (5.12b) gives

$$-a_r^- e_r + b_r^+ = -a_r^- e_r - b_r^+ ,$$

from which we obtain $b_r^+ = 0$. So we have $u^+ = a_r^+$ and $u^- = a_r^-$, showing that neither of these involves e_r. But as this is true for *all* values of r, we must have $u^- = 0$ and $u^+ = t1$ for some $t \in \mathbb{R}$, hence $u = t1$. □

Proposition 5.16

For $u \in \Gamma_n$, $u\overline{u} \in \mathbb{R}^\times$ and $\overline{u}u = u\overline{u}$. If $v \in \Gamma_n$ as well, then

$$uv\,\overline{uv} = u\overline{u}v\overline{v}.$$

Proof

For $u \in \Gamma_n$, $u\overline{u} \in \Gamma_n$ by Proposition 5.13. We will show that $u\overline{u} \in \ker \rho$.

Let $x \in \mathbb{R}^n$. Then using the notation $\overline{\alpha} = \alpha \circ \overline{(\ )} = \overline{(\ )} \circ \alpha$ of Remark 5.9 we have

$$\begin{aligned}
\rho_{u\overline{u}}(x) &= \alpha(u\overline{u})x(u\overline{u})^{-1} \\
&= \alpha(u)\alpha(\overline{u})x\overline{u^{-1}}u^{-1} \\
&= \alpha(u)\left(\alpha(\overline{u})x\overline{u^{-1}}\right)u^{-1} \\
&= \alpha(u)\alpha\left(\overline{u}\,\overline{x}\alpha(\overline{u^{-1}})\right)u^{-1},
\end{aligned}$$

since $\alpha(\overline{x}) = x$. Hence

$$\begin{aligned}
\rho_{u\overline{u}}(x) &= \alpha(u)\alpha\left(\overline{\alpha(u^{-1})xu}\right)u^{-1} \\
&= \alpha(u)\alpha\left(-\alpha(u^{-1})xu\right)u^{-1}
\end{aligned}$$

[since $\overline{y} = -y$ for $y \in \mathbb{R}^n$]

$$= \alpha(u)\left(\alpha(u^{-1})xu\right)u^{-1}$$

[since $\alpha(y) = -y$ for $y \in \mathbb{R}^n$]

$$= \alpha(uu^{-1})x(uu^{-1}) = x.$$

Thus $\rho_{u\overline{u}}(x) = x$. We also have

$$\overline{u}u = u^{-1}u\overline{u}u = u\overline{u}u^{-1}u$$

since $u\overline{u} \in \mathbb{R}^\times$, hence $\overline{u}u = u\overline{u}$.

If $v \in \Gamma_n$, then

$$uv\,\overline{uv} = uv\overline{v}\,\overline{u} = u\overline{u}v\overline{v},$$

since $v\overline{v} \in \mathbb{R}^\times$ and so commutes with every element of Cl_n. □

Proposition 5.17

The function

$$\nu\colon \Gamma_n \longrightarrow \mathbb{R}^\times; \quad \nu(u) = u\overline{u}$$

is a continuous group homomorphism.

Proof

Notice that if $u \in \Gamma_n$,

$$|u|^2 = \mathrm{Re}(\overline{u}u) = \mathrm{Re}(u\overline{u}) = u\overline{u} = \nu(u),$$

hence ν is continuous and always takes *positive* values. Proposition 5.16 implies that it is a group homomorphism. □

We will study the kernel of ν more thoroughly in the next section.

5.3 Pinor and Spinor Groups

In this section we will describe for each $n \geqslant 1$, the compact connected *spinor group* $\mathrm{Spin}(n)$ which is a group of units in the Clifford algebra Cl_n. Moreover, there is a surjective Lie homomorphism $\mathrm{Spin}(n) \longrightarrow \mathrm{SO}(n)$ whose kernel has two elements. We will first introduce the *pinor group* $\mathrm{Pin}(n)$ which has two connected components, one of them being $\mathrm{Spin}(n)$.

Definition 5.18

For $n \geqslant 1$, the *pinor group* $\mathrm{Pin}(n)$ is the kernel of $\nu\colon \Gamma_n \longrightarrow \mathbb{R}^\times$.

Notice that $\mathrm{Pin}(n)$ is a closed subgroup of $\Gamma_n \subseteq \mathrm{Cl}_n$ and is a bounded subset with respect to the metric induced from the norm $|\ |$, hence it is compact.

Definition 5.19

For $n \geqslant 1$, the *spinor group* $\mathrm{Spin}(n)$ is

$$\mathrm{Spin}(n) = \mathrm{Pin}(n) \cap \mathrm{Cl}_n^+ \leqslant \mathrm{Pin}(n).$$

The restriction of α to a function $\alpha\colon \mathrm{Pin}(n) \longrightarrow \mathrm{Pin}(n)$ is a continuous group homomorphism for which

$$\{u \in \mathrm{Pin}(n) : \alpha(u) = u\} = \mathrm{Pin}(n) \cap \mathrm{Cl}_n^+,$$

hence $\mathrm{Spin}(n) \leqslant \mathrm{Pin}(n)$ is a closed subgroup. Later we will see that it is also a normal subgroup.

Our principal goal in this section is to show that the restricted homomorphism $\rho\colon \mathrm{Pin}(n) \longrightarrow \mathrm{O}(n)$ is surjective and $\mathrm{Spin}(n) = \rho^{-1}\,\mathrm{SO}(n)$. We will do this by showing that $\mathrm{Pin}(n)$ is generated by certain elements $u \in \mathbb{R}^n$ for which ρ_u is a reflection.

Within $\mathbb{R}^n \subseteq \mathrm{Cl}_n$ is the *unit sphere*

$$\mathbb{S}^{n-1} = \{\mathbf{x} \in \mathbb{R}^n : |x| = 1\} = \{\sum_{r=1}^{n} x_r e_r : \sum_{r=1}^{n} x_r^2 = 1\}.$$

Lemma 5.20

Let $u \in \mathbb{S}^{n-1} \subseteq \mathrm{Cl}_n$. Then u is a unit in Cl_n, *i.e.*, $u \in \mathrm{Cl}_n^\times$, and $u^{-1} \in \mathbb{S}^{n-1}$.

Proof

Since $u \in \mathbb{R}^n$,

$$(-u)u = u(-u) = -u^2 = -(-|u|^2) = 1,$$

so $-u$ is the inverse of u. Of course $-u \in \mathbb{S}^{n-1}$. □

More generally, for $u_1, \ldots, u_k \in \mathbb{S}^{n-1}$ we have

$$(u_1 \cdots u_k)^{-1} = (-1)^k u_k \cdots u_1 = \overline{u_1 \cdots u_k}. \tag{5.13}$$

Recall that in a group G, the subgroup *generated* by the subset $S \subseteq G$ is the smallest subgroup of G containing S, often denoted $\langle S\rangle \leqslant G$. A typical element of $\langle S\rangle$ is a product of elements of the form s or s^{-1} where $s \in S$, *i.e.*,

$$s_1^{\varepsilon_1} \cdots s_k^{\varepsilon_k} \quad (\varepsilon_1, \ldots, \varepsilon_k = \pm 1).$$

Furthermore, if $H \leqslant G$ and $S \subseteq H$ then $\langle S\rangle \leqslant H$.

We will show that $\mathbb{S}^{n-1} \subseteq \mathrm{Pin}(n)$ and therefore $\langle \mathbb{S}^{n-1}\rangle \leqslant \mathrm{Pin}(n)$. We begin with the following useful result.

Lemma 5.21

Let $u, v \in \mathbb{R}^n \subseteq \mathrm{Cl}_n$. If $u \cdot v = 0$, then $vu = -uv$.

Proof

Writing $u = \sum_{r=1}^n x_r e_r$ and $v = \sum_{s=1}^n y_s e_s$ with $x_r, y_s \in \mathbb{R}$, we obtain

$$\begin{aligned} vu = \sum_{s=1}^n \sum_{r=1}^n y_s x_r e_s e_r &= \sum_{r=1}^n y_r x_r e_r^2 + \sum_{r<s} (x_s y_r - x_r y_s) e_r e_s \\ &= -\sum_{r=1}^n y_r x_r - \sum_{r<s} (x_r y_s - x_s y_r) e_r e_s \\ &= -u.v - \sum_{r<s} (x_r y_s - x_s y_r) e_r e_s \\ &= -\sum_{r<s} (x_r y_s - x_s y_r) e_r e_s \\ &= v.u - \sum_{r<s} (x_r y_s - x_s y_r) e_r e_s \\ &= -\sum_{r=1}^n \sum_{s=1}^n x_r y_s e_r e_s \\ &= -uv. \end{aligned}$$

□

Notice that for $u \in \mathbb{S}^{n-1}$ and $x \in \mathbb{R}^n$, by Equation (5.13),

$$\alpha(u) x u^{-1} = (-u) x (-u) = uxu. \tag{5.14}$$

If $u \cdot x = 0$, then Lemma 5.21 implies that

$$\alpha(u) x u^{-1} = -u^2 x = -(-1)x = x, \tag{5.15a}$$

since $u^2 = -|u|^2 = -1$. On the other hand, if $x = tu$ for some $t \in \mathbb{R}$, then

$$\alpha(u) x u^{-1} = tu^3 = -tu. \tag{5.15b}$$

So in particular $\alpha(u) x^{-1} \in \mathbb{R}^n$ which yields the following result.

Proposition 5.22

The subgroup of $\mathrm{Cl}_n^\times$ generated by $\mathbb{S}^{n-1}$ is contained in $\mathrm{Pin}(n)$, *i.e.*, $\langle \mathbb{S}^{n-1} \rangle \leqslant \mathrm{Pin}(n)$.

Remark 5.23

By Lemma 5.20, every element of $\langle \mathbb{S}^{n-1} \rangle$ is a product of elements of $\mathbb{S}^{n-1}$.

Next we will investigate the restriction of the homomorphism $\rho\colon \operatorname{Pin}(n) \longrightarrow \operatorname{O}(n)$ to $\rho\colon \langle \mathbb{S}^{n-1} \rangle \longrightarrow \operatorname{O}(n)$. For each $u \in \mathbb{S}^{n-1}$, we have an $\mathbb{R}$-linear isometry

$$\rho_u : \mathbb{R}^n \longrightarrow \mathbb{R}^n; \quad \rho_u(x) = \alpha(u)xu^{-1} = uxu.$$

More generally, for $u \in \langle \mathbb{S}^{n-1} \rangle$, suppose that $u = u_1 \cdots u_r$ for $u_1, \ldots, u_r \in \mathbb{S}^{n-1}$. Then

$$\begin{aligned} \rho_u(x) &= \alpha(u_1 \cdots u_r)x(u_1 \cdots u_r)^{-1} \\ &= ((-1)^r u_1 \cdots u_r)x((-1)^r u_r \cdots u_1) \\ &= u_1 \cdots u_r x u_r \cdots u_1. \end{aligned} \tag{5.16}$$

Proposition 5.24

For $u \in \mathbb{S}^{n-1}$, $\rho_u : \mathbb{R}^n \longrightarrow \mathbb{R}^n$ is reflection in the hyperplane orthogonal to u.

Proof

Recalling the defining equation (1.4) of a hyperplane reflection, Equations (5.15) show that

$$\rho_u(x) = \begin{cases} x & \text{if } u \cdot x = 0, \\ -x & \text{if } x = tu \text{ for some } t \in \mathbb{R}. \end{cases}$$

So $\rho_u : \mathbb{R}^n \longrightarrow \mathbb{R}^n$ is indeed reflection in the hyperplane orthogonal to u. □

Proposition 5.25

i) $\rho\colon \langle \mathbb{S}^{n-1} \rangle \longrightarrow \operatorname{O}(n)$ is surjective and $\ker \rho = \{1, -1\}$.
ii) $\rho\colon \operatorname{Pin}(n) \longrightarrow \operatorname{O}(n)$ is surjective with $\ker \rho = \{1, -1\}$.
iii) $\langle \mathbb{S}^{n-1} \rangle = \operatorname{Pin}(n)$.

Proof

(i) The observation in the proof of Proposition 5.24 shows that reflection in the hyperplane orthogonal to $u \in \mathbb{S}^{n-1}$ has the form ρ_u. Surjectivity is a consequence of Proposition 1.41.
(ii) If $t \in \ker \rho = \operatorname{Pin}(n) \cap \mathbb{R}^{\times}$, then $|t| = 1$ then $|t|^2 = t\bar{t} = t^2$, so $t = \pm 1$. On the other hand, $\pm 1 \in \ker \rho$.

(iii) Suppose that $v \in \mathrm{Pin}(n)$. Then $\rho_v \in \mathrm{O}(n)$ can be expressed as a product of hyperplane reflections, each of which has the form ρ_w for some vector $w \in \mathbb{S}^{n-1}$. Hence $\rho_v = \rho_{w_1 \cdots w_k}$ for suitable $w_1, \ldots, w_k \in \mathbb{S}^{n-1}$. But then

$$(-1)^k v w_k \cdots w_1 \in \ker \rho = \{\pm 1\}.$$

Thus $v = \pm w_1 \cdots w_k \in \langle \mathbb{S}^{n-1} \rangle$. □

At this point we have a great deal of information about elements of $\mathrm{Pin}(n)$ which we collect into a theorem.

Theorem 5.26

i) $\mathrm{Pin}(n)$ is the disjoint union of open subsets

$$\mathrm{Pin}(n) = \mathrm{Pin}(n) \cap \mathrm{Cl}_n^+ \cup \mathrm{Pin}(n) \cap \mathrm{Cl}_n^- = \mathrm{Spin}(n) \cup \mathrm{Pin}(n) \cap \mathrm{Cl}_n^-.$$

ii) $\mathrm{Spin}(n)$ is a normal subgroup of $\mathrm{Pin}(n)$ and every element $v \in \mathrm{Spin}(n)$ can be expressed as an even-length product $v = v_1 \cdots u_{2\ell}$ where $v_1, \ldots, v_{2\ell} \in \mathbb{S}^{n-1}$.
iii) For any $w \in \mathbb{S}^{n-1}$,

$$\mathrm{Pin}(n) \cap \mathrm{Cl}_n^- = w\, \mathrm{Spin}(n),$$

and every element $v \in \mathrm{Pin}(n) \cap \mathrm{Cl}_n^-$ can be expressed as an odd-length product $v = v_1 \cdots u_{2\ell+1}$ where $v_1, \ldots, v_{2\ell+1} \in \mathbb{S}^{n-1}$.

Proof

(i) Since $\mathrm{Pin}(n) = \langle \mathbb{S}^{n-1} \rangle$, every element $u \in \mathrm{Pin}(n)$ can be expressed as $u = u_1 \cdots u_k$ for some $u_1, \ldots, u_k \in \mathbb{S}^{n-1}$. Since

$$\alpha(u_1 \cdots u_k) = (-1)^k u_1 \cdots u_k,$$

we find that $u \in \mathrm{Pin}(n) \cap \mathrm{Cl}_n^+$ if k is even, while $u \in \mathrm{Pin}(n) \cap \mathrm{Cl}_n^-$ if k is odd. The function

$$\widetilde{\alpha}\colon \mathrm{Pin}(n) \longrightarrow \{+1, -1\}; \quad \widetilde{\alpha}(u) = \alpha(u)u^{-1},$$

is continuous and

$$\widetilde{\alpha}^{-1}\{1\} = \mathrm{Pin}(n) \cap \mathrm{Cl}_n^+, \quad \widetilde{\alpha}^{-1}\{-1\} = \mathrm{Pin}(n) \cap \mathrm{Cl}_n^-,$$

showing that these are disjoint open subsets.
(ii) If $u \in \mathrm{Spin}(n) = \mathrm{Pin}(n) \cap \mathrm{Cl}_n^+$ and $v \in \mathrm{Pin}(n)$, then writing $u = u_1 \cdots u_{2k}$ and $v = v_1 \cdots v_{2\ell+1}$ with $u_i, v_j \in \mathbb{S}^{n-1}$ we find

$$vuv^{-1} = -v_1 \cdots u_{2\ell+1} u_1 \cdots u_{2k} v_{2\ell+1} \cdots v_1 \in \mathrm{Pin}(n) \cap \mathrm{Cl}_n^+.$$

(iii) If $u \in \mathrm{Pin}(n) \cap \mathrm{Cl}_n^-$ then for each $w \in \mathbb{S}^{n-1}$ we have

$$w^{-1}u = (-w)u \in \mathrm{Pin}(n) \cap \mathrm{Cl}_n^+. \qquad \square$$

From now on we will make use of the notation

$$\mathrm{Pin}(n)^+ = \mathrm{Spin}(n) = \mathrm{Pin}(n) \cap \mathrm{Cl}_n^+, \quad \mathrm{Pin}(n)^- = \mathrm{Pin}(n) \cap \mathrm{Cl}_n^-.$$

We can also characterise $\mathrm{Spin}(n)$ and $\mathrm{Pin}(n)^-$ using the surjective homomorphism $\rho\colon \mathrm{Pin}(n) \longrightarrow \mathrm{O}(n)$. If $u_1, \ldots, u_k \in \mathbb{S}^{n-1}$, then

$$\det \rho_{u_1 \cdots u_k} = \det \rho_{u_1} \cdots \det \rho_{u_k} = (-1)^k,$$

since each ρ_{u_r} is a hyperplane reflection for which $\det \rho_{u_r} = -1$. Since

$$\begin{aligned} \mathrm{SO}(n) = \mathrm{O}(n)^+ &= \{A \in \mathrm{O}(n) : \det A = 1\}, \\ \mathrm{O}(n)^- &= \{A \in \mathrm{O}(n) : \det A = -1\}, \end{aligned}$$

this means that for $u \in \mathrm{Pin}(n)$,

$$\begin{aligned} u \in \mathrm{Spin}(n) &\iff \rho_u \in \mathrm{SO}(n), \\ u \in \mathrm{Pin}(n)^- &\iff \rho_u \in \mathrm{O}(n)^-. \end{aligned}$$

Theorem 5.27

The continuous homomorphism $\rho\colon \mathrm{Pin}(n) \longrightarrow \mathrm{O}(n)$ is surjective and

$$\rho^{-1}\,\mathrm{SO}(n) = \mathrm{Spin}(n), \quad \rho^{-1}\,\mathrm{O}(n)^- = \mathrm{Pin}(n)^-.$$

Hence the restriction of ρ to $\rho^+\colon \mathrm{Spin}(n) \longrightarrow \mathrm{SO}(n)$ is also surjective and the kernels of these homomorphisms are $\ker \rho = \rho^+ = \{+1, -1\}$.

We can also say more about the topology of $\mathrm{Spin}(n)$ and $\mathrm{Pin}(n)$.

Theorem 5.28

$\mathrm{Spin}(n)$ is a compact, path connected, closed normal subgroup of $\mathrm{Pin}(n)$. Furthermore, if $n \geqslant 3$ the fundamental group of $\mathrm{Spin}(n)$ is trivial, $\pi_1 \mathrm{Spin}(n) = \{1\}$.

Proof

We only discuss connectivity. Recall that the sphere $\mathbb{S}^{n-1} \subseteq \mathbb{R}^n \subseteq \mathrm{Cl}_n$ is path connected. Choose a base point $u_0 \in \mathbb{S}^{n-1}$. Now for an element $u = u_1 \cdots u_k \in \mathrm{Spin}(n)$ with $u_1, \ldots, u_k \in \mathbb{S}^{n-1}$, as noted in the proof of Proposition 5.25, we must have k even, say $k = 2m$. In fact, we might as well take m to be even since $u = u(-w)w$ for any $w \in \mathbb{S}^{n-1}$. Then there are continuous paths

$$p_r \colon [0,1] \longrightarrow \mathbb{S}^{n-1} \quad (r = 1, \ldots, 2m),$$

for which $p_r(0) = u_0$ and $p_r(1) = u_r$. Then

$$p \colon [0,1] \longrightarrow \mathbb{S}^{n-1}; \quad p(t) = p_1(t) \cdots p_{2m}(t)$$

is a continuous path in $\operatorname{Pin}(n)$ with

$$p(0) = u_0^{2m} = (-1)^m = 1, \quad p(1) = u.$$

But $t \mapsto \rho(p(t))$ is a continuous path in $\mathrm{O}(n)$ with $\rho(p(0)) \in \mathrm{SO}(n)$, hence $\rho(p(t)) \in \mathrm{SO}(n)$ for all t. This shows that p is a path in $\operatorname{Spin}(n)$. So every element $u \in \operatorname{Spin}(n)$ can be connected to 1 and therefore $\operatorname{Spin}(n)$ is path connected.

The final statement involves homotopy theory and is not proved here. It should be compared with the fact that for $n \geqslant 3$, $\pi_1 \mathrm{SO}(n) \cong \{1, -1\}$ and in fact the map is an example of a *universal covering*. □

The double covering maps $\rho \colon \operatorname{Spin}(n) \longrightarrow \mathrm{SO}(n)$ generalise the case of $\mathrm{SU}(2) \longrightarrow \mathrm{SO}(3)$ discussed in Section 3.5. In fact, around each element $u \in$ there is an open neighbourhood $N_u \subseteq \operatorname{Spin}(n)$ for which $\rho \colon N_u \longrightarrow \rho N_u$ is a homeomorphism, and actually a diffeomorphism. This implies the following.

Proposition 5.29

The derivative $\mathrm{d}\rho \colon \mathfrak{spin}(n) \longrightarrow \mathfrak{so}(n)$ is an isomorphism of $\mathbb{R}$-Lie algebras and

$$\dim \operatorname{Spin}(n) = \dim \mathrm{SO}(n) = \binom{n}{2}.$$

The Lie algebra of $\operatorname{Spin}(n)$ can be described as a Lie subalgebra of Cl_n.

Proposition 5.30

For $n \geqslant 2$, the Lie algebra of $\operatorname{Spin}(n)$ is

$$\mathfrak{spin}(n) = \left\{ \sum_{1 \leqslant i < j \leqslant n} t_{ij} e_i e_j : t_{ij} \in \mathbb{R} \right\} \subseteq \mathrm{Cl}_n.$$

Proof

For $1 \leqslant i < j \leqslant n$ and $t \in \mathbb{R}$, consider

$$\exp(te_ie_j) = \sum_{r=0}^{\infty} \frac{t^r}{r!}(e_ie_j)^r,$$

which converges to an element of $\mathrm{Cl}_n^\times$. Notice that

$$(e_ie_j)^r = (-1)^{\binom{r}{2}} e_i^r e_j^r = \begin{cases} 1 & \text{if } r \equiv 0 \bmod 4, \\ -1 & \text{if } r \equiv 2 \bmod 4, \\ e_ie_j & \text{if } r \equiv 1 \bmod 4, \\ -e_ie_j & \text{if } r \equiv 3 \bmod 4. \end{cases}$$

Hence we have

$$\exp(te_ie_j) = \cos t + \sin t\, e_ie_j. \tag{5.17}$$

We also obtain

$$e_i(e_ie_j)^r = (-e_ie_j)^r e_i,$$
$$e_j(e_ie_j)^r = (-e_ie_j)^r e_j,$$

and if $k \neq i, j$,

$$e_k(e_ie_j)^r = (e_ie_j)^r e_k.$$

Then since $(e_ie_j)^2 = -1$,

$$\begin{aligned} \alpha(\exp(te_ie_j))e_i \exp(te_ie_j)^{-1} &= \exp(te_ie_j)e_i \exp(-te_ie_j) \\ &= (\cos t + \sin t\, e_ie_j)e_i(\cos t - \sin t\, e_ie_j) \\ &= (\cos t + \sin t\, e_ie_j)^2 e_i \\ &= (\cos 2t + \sin 2t\, e_ie_j)e_i \\ &= \cos 2t\, e_i + \sin 2t\, e_j, \\ \alpha(\exp(te_ie_j))e_j \exp(te_ie_j)^{-1} &= (\cos t + \sin t\, e_ie_j)e_j(\cos t - \sin t\, e_ie_j) \\ &= (\cos t + \sin t\, e_ie_j)^2 e_j \\ &= (\cos 2t + \sin 2t\, e_ie_j)e_j \\ &= \cos 2t\, e_j - \sin 2t\, e_i, \end{aligned}$$

and if $k \neq i, j$,

$$\begin{aligned} \alpha(\exp(te_ie_j))e_k \exp(te_ie_j)^{-1} &= \exp(te_ie_j)\exp(te_ie_j)^{-1} e_k \\ &= e_k. \end{aligned}$$

Since the e_ℓ form a basis of $\mathbb{R}^n$, this shows that $\exp(te_ie_j) \in \Gamma_n \cap \mathrm{Cl}_n^+$. But also

$$\begin{aligned}\exp(te_ie_j)\overline{\exp(te_ie_j)} &= \exp(te_ie_j)\exp((-1)^2te_je_i)\\ &= \exp(te_ie_j)\exp(-te_ie_j)\\ &= 1,\end{aligned}$$

hence $\exp(te_ie_j) \in \mathrm{Spin}(n)$.

Taking the derivative at 0 of the curve

$$\alpha_{ij} \colon \mathbb{R} \longrightarrow \mathrm{Spin}(n); \quad \alpha_{ij}(t) = \exp(te_ie_j),$$

we obtain e_ie_j, so $e_ie_j \in \mathfrak{spin}(n)$. As these $\binom{n}{2}$ elements e_ie_j are clearly linearly independent, by Proposition 5.29 they must form a basis. □

With the aid of the calculations in this proof, the action of the derivative $\mathrm{d}\,\rho\colon \mathfrak{spin}(n) \longrightarrow \mathfrak{so}(n)$ on the basis elements e_ie_j can be determined, namely

$$\mathrm{d}\,\rho(e_ie_j) = 2E^{ji} - 2E^{ij}, \tag{5.18a}$$

where we use the basis for $\mathrm{M}_n(\mathbb{R})$ provided by Equation (3.24). Similarly, the effect of the homomorphism $\rho\colon \mathrm{Spin}(n) \longrightarrow \mathrm{SO}(n)$ on $\exp(te_ie_j)$ is

$$\rho(\exp(te_ie_j)) = \cos 2t\, I_n + \sin 2t\,(E^{ji} - E^{ij}). \tag{5.18b}$$

5.4 The Centres of Spinor Groups

Recall that for a group G the *centre* of G is

$$\mathrm{Z}(G) = \{c \in G : \forall g \in G,\ gc = cg\}.$$

Then $\mathrm{Z}(G) \lhd G$. The centres of the special orthogonal groups are easily found and described.

Proposition 5.31

For $n \geqslant 3$,

$$\mathrm{Z}(\mathrm{SO}(n)) = \{tI_n : t = \pm 1,\ t^n = 1\} = \begin{cases}\{I_n\} & \text{if } n \text{ is odd},\\ \{\pm I_n\} & \text{if } n \text{ is even}.\end{cases}$$

Before stating our main result on the centres of spinor groups, we note that $\mathrm{Spin}(1)$ and $\mathrm{Spin}(2)$ are both abelian.

Proposition 5.32

For $n \geqslant 3$,

$$\mathrm{Z}(\mathrm{Spin}(n)) = \begin{cases} \{\pm 1\} & \text{if } n \text{ is odd,} \\ \{\pm 1, \pm e_1 \cdots e_n\} & \text{if } n \equiv 2 \bmod 4, \\ \{\pm 1, \pm e_1 \cdots e_n\} & \text{if } n \equiv 0 \bmod 4. \end{cases}$$

$$\cong \begin{cases} \mathbb{Z}/2 & \text{if } n \text{ is odd,} \\ \mathbb{Z}/4 & \text{if } n \equiv 2 \bmod 4, \\ \mathbb{Z}/2 \times \mathbb{Z}/2 & \text{if } n \equiv 0 \bmod 4. \end{cases}$$

Proof

If $g \in \mathrm{Z}(\mathrm{Spin}(n))$, then since $\rho\colon \mathrm{Spin}(n) \longrightarrow \mathrm{SO}(n)$, $\rho(g) \in \mathrm{Z}(\mathrm{SO}(n))$. As $\pm 1 \in \mathrm{Z}(\mathrm{Spin}(n))$, this gives $|\,\mathrm{Z}(\mathrm{Spin}(n))| = 2|\,\mathrm{Z}(\mathrm{SO}(n))|$ and indeed

$$\mathrm{Z}(\mathrm{Spin}(n)) = \rho^{-1}\,\mathrm{Z}(\mathrm{SO}(n)).$$

For n even,

$$\begin{aligned}(\pm e_1 \cdots e_n)^2 &= e_1 \cdots e_n e_1 \cdots e_n \\ &= (-1)^{\binom{n}{2}} e_1^2 \cdots e_n^2 \\ &= (-1)^{\binom{n}{2}+n} = (-1)^{\binom{n+1}{2}}.\end{aligned}$$

Since

$$\binom{n+1}{2} = \frac{(n+1)n}{2} \equiv \begin{cases} 1 \bmod 2 & \text{if } n \equiv 2 \bmod 4, \\ 0 \bmod 2 & \text{if } n \equiv 0 \bmod 4, \end{cases}$$

this implies that

$$(\pm e_1 \cdots e_n)^2 = \begin{cases} -1 & \text{if } n \equiv 2 \bmod 4, \\ \ \ 1 & \text{if } n \equiv 0 \bmod 4. \end{cases}$$

So for n even, the multiplicative order of $\pm e_1 \cdots e_n$ is 2 or 4 depending on the congruence class of n modulo 4. This gives the stated groups. □

5.5 Finite Subgroups of Spinor Groups

Each orthogonal group $\mathrm{O}(n)$ and $\mathrm{SO}(n)$ contains finite subgroups. For example, when $n = 2, 3$, these correspond to symmetry groups of compact plane figures and solids. The case of $n = 3$ was explored in the exercises of Chapter 4. Here we make some remarks about the symmetric and alternating groups.

Recall that for each $n \geqslant 1$ the *symmetric group* S_n is the group of all permutations of the set $\mathbf{n} = \{1, \dots, n\}$. The corresponding *alternating group* $A_n \leqslant S_n$ is the subgroup consisting of all *even* permutations, *i.e.*, the elements $\sigma \in S_n$ for which $\operatorname{sgn}(\sigma) = 1$ where $\operatorname{sgn}\colon S_n \longrightarrow \{\pm 1\}$ is the sign homomorphism.

For a field $\Bbbk$, we can make S_n act on $\Bbbk^n$ by linear transformations:

$$\sigma \cdot \begin{bmatrix} x_1 \\ x_2 \\ \vdots \\ x_n \end{bmatrix} = \begin{bmatrix} x_{\sigma^{-1}(1)} \\ x_{\sigma^{-1}(2)} \\ \vdots \\ x_{\sigma^{-1}(n)} \end{bmatrix}.$$

Notice that $\sigma(\mathbf{e}_r) = \mathbf{e}_{\sigma(r)}$. The matrix $[\sigma]$ of the linear transformation induced by σ with respect to the basis of $\mathbf{e}_r$'s has all its entries 0 or 1, with exactly one 1 in each row and column. For example, when $n = 3$,

$$[(1\,2\,3)] = \begin{bmatrix} 0 & 0 & 1 \\ 1 & 0 & 0 \\ 0 & 1 & 0 \end{bmatrix}, \quad [(1\,3)] = \begin{bmatrix} 0 & 0 & 1 \\ 0 & 1 & 0 \\ 1 & 0 & 0 \end{bmatrix}.$$

When $\Bbbk = \mathbb{R}$ each of these matrices is orthogonal, while when $\Bbbk = \mathbb{C}$ it is unitary. For a given n we can view S_n as the subgroup of $O(n)$ or $U(n)$ consisting of all such matrices which are usually called *permutation matrices.*

Proposition 5.33

For each $\sigma \in S_n$ we have $\operatorname{sgn}(\sigma) = \det([\sigma])$. Hence

$$A_n = \begin{cases} \mathrm{SO}(n) \cap S_n & \text{if } \Bbbk = \mathbb{R}, \\ \mathrm{SU}(n) \cap S_n & \text{if } \Bbbk = \mathbb{C}. \end{cases}$$

Recall that if $n \geqslant 5$, A_n is a simple group.

As $\rho\colon \mathrm{Pin}(n) \longrightarrow O(n)$ is onto, there are finite subgroups $\widetilde{S}_n = \rho^{-1} S_n \leqslant \mathrm{Pin}(n)$ and $\widetilde{A}_n = \rho^{-1} A_n \leqslant \mathrm{Spin}(n)$ for which there are surjective homomorphisms $\rho\colon \widetilde{S}_n \longrightarrow S_n$ and $\rho\colon \widetilde{A}_n \longrightarrow A_n$ whose kernels contain the elements ± 1. Note that $|\widetilde{S}_n| = 2 \cdot n!$, while $|\widetilde{A}_n| = n!$. However, for $n \geqslant 4$, there are no homomorphisms $\tau\colon S_n \longrightarrow \widetilde{S}_n$, $\tau\colon A_n \longrightarrow \widetilde{A}_n$ for which $\rho \circ \tau = \mathrm{Id}$.

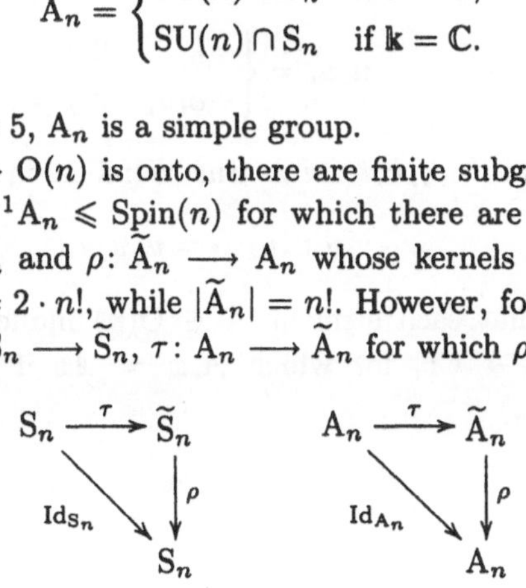

Similar considerations apply to other finite subgroups of $\mathrm{O}(n)$.

In $\mathrm{Cl}_n^\times$ we have a subgroup E_n consisting of all the elements

$$\pm e_{i_1} \cdots e_{i_r} \quad (1 \leqslant i_1 < \cdots < i_r \leqslant n,\ 0 \leqslant r).$$

The order of this group is $|E_n| = 2^{n+1}$ and as it contains ± 1, its image under $\rho\colon \mathrm{Pin}(n) \longrightarrow \mathrm{O}(n)$ is $\overline{E}_n = \rho E_n$ of order $|\overline{E}_n| = 2^n$. In fact, $\{\pm 1\} = \mathrm{Z}(E_n)$ is also the commutator subgroup since $e_i e_j e_i^{-1} e_j^{-1} = -1$ and so $\overline{E}_n$ is abelian. Every non-trivial element in $\overline{E}_n$ has order 2 since $e_i^2 = -1$, hence $\overline{E}_n \leqslant \mathrm{O}(n)$ is an *elementary* 2*-group*, *i.e.*, it is isomorphic to $(\mathbb{Z}/2)^n$. Each element $\rho(e_r) \in \mathrm{O}(n)$ is a generalised permutation matrix with all its non-zero entries on the main diagonal. There is also a subgroup $\overline{E}_n^0 = \rho E_n^0 \leqslant \mathrm{SO}(n)$ of order 2^{n-1}, where

$$E_n^0 = E_n \cap \mathrm{Spin}(n).$$

In fact $\overline{E}_n^0$ is isomorphic to $(\mathbb{Z}/2)^{n-1}$. These groups E_n and E_n^0 are non-abelian and fit into exact sequences of the form

$$1 \to \mathbb{Z}/2 \longrightarrow E_n \longrightarrow (\mathbb{Z}/2)^n \to 1, \quad 1 \to \mathbb{Z}/2 \longrightarrow E_n^0 \longrightarrow (\mathbb{Z}/2)^{n-1} \to 1,$$

in which each kernel $\mathbb{Z}/2$ is equal to the centre of the corresponding group E_n or E_n^0. This means they are examples of *extraspecial* 2*-groups*.

EXERCISES

5.1. In the Clifford algebra Cl_n, let $u, v \in \mathbb{R}^n \subseteq \mathrm{Cl}_n$.

a) If $|u| = 1$, by expressing v as a sum $v_1 + v_2$ with $v_1 = tu$ and $u \cdot v_2 = 0$, find a general formula for uvu.

b) Let $\{u_1, \ldots, u_n\}$ be an orthonormal basis for $\mathbb{R}^n$. Show that

$$u_j u_i = \begin{cases} -1 & \text{if } j = i, \\ -u_i u_j & \text{if } j \neq i. \end{cases}$$

Deduce that if i, j, k, ℓ are distinct numbers in the range 1 to n then

$$u_i u_j u_k u_\ell = u_k u_\ell u_i u_j.$$

c) Show that each element $A \in \mathrm{O}(n)$ induces an automorphism $A_*\colon \mathrm{Cl}_n \longrightarrow \mathrm{Cl}_n$ for which $A_* x = Ax$ if $x \in \mathbb{R}^n$. Determine $A_*(e_1 \cdots e_n)$.

5.2. Use the Universal Property of Theorem 5.4 in the following.

a) Show that the natural embedding

$$i_n\colon \mathbb{R}^n \longrightarrow \mathbb{R}^{n+1}; \quad \begin{bmatrix} x_1 \\ \vdots \\ x_n \end{bmatrix} \longmapsto \begin{bmatrix} x_1 \\ \vdots \\ x_n \\ 0 \end{bmatrix}$$

induces an $\mathbb{R}$-algebra homomorphism $i'_n\colon \mathrm{Cl}_n \longrightarrow \mathrm{Cl}_{n+1}$ for which $i'_n(x) = i_n(x)$ whenever $x \in \mathbb{R}^n$. Show that i'_n is injective and determine its image $i'_n\mathrm{Cl}_n \subseteq \mathrm{Cl}_{n+1}$.

b) Show that the $\mathbb{R}$-linear transformation

$$k_n\colon \mathbb{R}^n \longrightarrow \mathrm{Cl}_{n+1}; \quad k_n(x) = xe_{n+1},$$

induces an $\mathbb{R}$-algebra homomorphism $k'_n\colon \mathrm{Cl}_n \longrightarrow \mathrm{Cl}_{n+1}$ for which $k'_n(x) = k_n(x)$ whenever $x \in \mathbb{R}^n$. Show that k'_n is injective with image $k'_n\mathrm{Cl}_n = \mathrm{Cl}^+_{n+1}$.

5.3. Let $n \geqslant 1$ and $n \equiv 3 \bmod 4$. In the Clifford algebra Cl_n, consider the element $\omega_n = e_1 \cdots e_n$.

a) Show that ω_n is central in Cl_n.

b) Show that in Cl_n, the elements

$$\Omega_+ = \frac{1}{2}(1 + \omega_n), \quad \Omega_- = \frac{1}{2}(1 - \omega_n),$$

are central orthogonal idempotents for which $\Omega_+ + \Omega_- = 1$.

c) Decompose Cl_n as a product $\mathrm{Cl}_n = A_+ \times A_-$.

d) Use the decomposition of (c) to show that $\mathrm{Cl}_3 \cong \mathbb{H} \times \mathbb{H}$ and $\mathrm{Cl}_7 \cong \mathrm{M}_8(\mathbb{R}) \times \mathrm{M}_8(\mathbb{R})$.

5.4. For $n \geqslant 1$, define the *finite Clifford group* ClGp_n to be generated by $(n+1)$ elements $\varepsilon_1, \ldots, \varepsilon_n, \gamma$ with γ central and subject to the relations

$$\varepsilon_1^2 = \cdots = \varepsilon_n^2 = \gamma, \quad \varepsilon_i\varepsilon_j = \gamma\varepsilon_j\varepsilon_i \text{ if } i \neq j, \quad \gamma^2 = 1.$$

a) Show that the order of ClGp_n is 2^{n+1}, hence the real group algebra $\mathbb{R}[\mathrm{ClGp}_n]$ has dimension 2^{n+1}.

b) Show that ClGp_n is isomorphic to a subgroup of $\mathrm{Cl}_n^\times$.

c) Find an $\mathbb{R}$-algebra homomorphism $\varphi\colon \mathbb{R}[\mathrm{ClGp}_n] \longrightarrow \mathrm{Cl}_n$ which is surjective.

d) Show that the $\mathbb{R}$-algebra $\mathbb{R}[\mathrm{ClGp}_n]$ is the product algebra

$$\mathbb{R}[\mathrm{ClGp}_n] = A \times B,$$

where $B = \ker\varphi$ and $A \cong \mathrm{Cl}_n$.

5.5. ⚠ Let $S = [s_{ij}]$ be a $2m \times 2m$ real skew symmetric matrix. Working in the Clifford algebra Cl_{2m}, consider the element

$$s = \sum_{1 \leqslant i < j \leqslant 2m} s_{ij} e_i e_j = \frac{1}{2} \sum_{\substack{1 \leqslant i,j \leqslant m \\ i \neq j}} s_{ij} e_i e_j.$$

a) Expressing the element s^m in terms of the basis of monomials $e_{i_1} \cdots e_{i_r}$, show that the coefficient of $e_1 e_2 \cdots e_{2m}$ is the real number

$$c_S = m! \sum_{\sigma \in \Pi(2m)} \operatorname{sgn} \sigma \; s_{\sigma(1)\sigma(2)} \cdots s_{\sigma(2m-1)\sigma(2m)},$$

where $\Pi(2m)$ consists of all permutations σ of $\{1, 2, \ldots, 2m\}$ satisfying the conditions

- $\sigma(2r-1) < \sigma(2s-1)$ if $1 \leqslant r < s \leqslant m$;
- $\sigma(2t-1) < \sigma(2t)$ if $1 \leqslant t \leqslant m$.

Determine c_S for a few small values of m.

b) Determine c_S when $S = \begin{bmatrix} O_m & I_m \\ -I_m & O_m \end{bmatrix}$.

c) If $\Sigma_{2m} \subseteq \mathrm{M}_{2m}(\mathbb{R})$ is the subspace of all non-singular skew symmetric matrices, deduce that the function

$$\mathrm{pf} \colon \Sigma_{2m} \longrightarrow \mathbb{R}^{\times}; \quad \mathrm{pf}([s_{ij}]) = \frac{c_S}{m!}$$

is continuous. This function pf is known as a *Pfaffian*.

d) Show that pf takes both positive and negative values and in fact is surjective.

e) Let $P = [p_{ij}] \in \mathrm{GL}_{2m}(\mathbb{R})$. Using the well-known formula

$$\det P = \sum_{\sigma \in S_{2m}} \operatorname{sgn} \sigma \prod_{k=1}^{2m} p_{k\,\sigma(k)},$$

show that

$$\mathrm{pf}(P^T S P) = \mathrm{pf}(S) \det P.$$

5.6. a) Verify Proposition 5.31.

b) For $n \geqslant 2$, determine $\mathrm{Z}(\mathrm{O}(n))$.

6
Lorentz Groups

We met the Lorentz group Lor in Section 1.5. In this chapter, we will consider this kind of group in greater generality. We merely sketch some of the details, leaving the reader to fill in the more obvious gaps. The most important example is that for which $n = 3$ as this provides the geometric setting for Special Relativity. However, many of the main features can be seen in the cases $n = 1, 2$.

6.1 Lorentz Groups

For $n \geqslant 1$, consider the non-singular symmetric matrix

$$Q_{n,1} = \operatorname{diag}(\underbrace{1, \ldots, 1}_{n}, -1) \in \mathrm{M}_{n+1}(\mathbb{R}),$$

which defines an inner product on $\mathbb{R}^{n+1}$ given by

$$\langle \mathbf{x}, \mathbf{y} \rangle = \mathbf{x}^T Q_{n,1} \mathbf{y} = x_1 y_1 + \cdots + x_n y_n - x_{n+1} y_{n+1}. \tag{6.1}$$

This is the *Lorentz inner product* on $\mathbb{R}^{n+1}$ and it is standard to denote this inner product space by $\mathbb{R}^{n,1}$. Actually, physicists often adopt different sign conventions and use one of the following formulæ in place of Equation (6.1):

$$\langle \mathbf{x}, \mathbf{y} \rangle = \begin{cases} -x_1 y_1 + \cdots + x_n y_n + x_{n+1} y_{n+1}, \\ -x_1 y_1 + \cdots - x_n y_n + x_{n+1} y_{n+1}, \\ x_1 y_1 + \cdots - x_n y_n - x_{n+1} y_{n+1}. \end{cases}$$

This affects the signs in the following definitions. We will also use t in place of x_{n+1} since in Relativity, this coordinate is related to the time measurements while the others are related to spatial ones.

We will often write elements of $\mathbb{R}^{n,1}$ in the form

$$\mathbf{X} = \begin{bmatrix} \mathbf{x} \\ t \end{bmatrix} \quad (\mathbf{x} \in \mathbb{R}^n,\ t \in \mathbb{R}).$$

Then

$$\langle \mathbf{X}_1, \mathbf{X}_2 \rangle = \mathbf{x}_1 \cdot \mathbf{x}_2 - t_1 t_2.$$

Definition 6.1

A non-zero vector $\mathbf{X} \in \mathbb{R}^{n,1}$ is called

- *spacelike* if $\langle \mathbf{X}, \mathbf{X} \rangle > 0$ or equivalently $|\mathbf{x}|^2 > t^2$;
- *timelike* if $\langle \mathbf{X}, \mathbf{X} \rangle < 0$ or equivalently $|\mathbf{x}|^2 < t^2$;
- a *null vector* if $\langle \mathbf{X}, \mathbf{X} \rangle = 0$ or equivalently $|\mathbf{x}|^2 = t^2$.

We can view $\mathbb{R}^n$ as sitting inside $\mathbb{R}^{n,1}$ as the subset of all spacelike vectors of the form

$$\mathbf{X} = \begin{bmatrix} \mathbf{x} \\ 0 \end{bmatrix} = \begin{bmatrix} x_1 \\ \vdots \\ x_n \\ 0 \end{bmatrix}$$

and we usually identify $\mathbf{x} \in \mathbb{R}^n$ with the element $\mathbf{X} = \begin{bmatrix} \mathbf{x} \\ 0 \end{bmatrix} \in \mathbb{R}^{n,1}$. Of course we then have $\langle \mathbf{X}, \mathbf{X} \rangle = \mathbf{x} \cdot \mathbf{x}$, so this embedding of $\mathbb{R}^n$ into $\mathbb{R}^{n,1}$ is an *isometry*.

If we represent $\mathbb{R}^{n,1}$ by the (x,t)-plane with $\mathbb{R}^n$ as the x-axis, then the t-axis is timelike, while the null vectors correspond to the points lying on the line $x = t$. In fact, the null vectors in $\mathbb{R}^{n,1}$ lie in the set

$$\mathcal{H}_{n,1}(0) = \{\mathbf{X} \in \mathbb{R}^{n,1} : \mathbf{X} \neq \mathbf{0},\ \langle \mathbf{X}, \mathbf{X} \rangle = \mathbf{x} \cdot \mathbf{x} - t^2 = 0\}.$$

For $n > 1$, $\mathcal{H}_{n,1}(0)$ has two path connected components,

$$\mathcal{H}^{+}_{n,1}(0) = \{\mathbf{X} \in \mathcal{H}_{n,1}(0) : t > 0\}, \quad \mathcal{H}^{-}_{n,1}(0) = \{\mathbf{X} \in \mathcal{H}_{n,1}(0) : t < 0\}.$$

These are often referred to as the *positive* or *future pointing light cone* and the *negative* or *past pointing light cone* respectively since in Relativity these represent points moving at the speed of light.

For each positive real number r, the hyperboloid

$$\mathcal{H}_{n,1}(r) = \{\mathbf{X} \in \mathbb{R}^{n,1} : \langle \mathbf{X}, \mathbf{X} \rangle = \mathbf{x} \cdot \mathbf{x} - t^2 = -r\}$$

has two path connected components

$$\mathcal{H}^{+}_{n,1}(r) = \{\mathbf{X} \in \mathcal{H}_{n,1}(r) : t > 0\}, \quad \mathcal{H}^{-}_{n,1}(r) = \{\mathbf{X} \in \mathcal{H}_{n,1}(r) : t < 0\}.$$

For later use we extend the Lorentz inner product to an $\mathbb{R}$-bilinear inner product on $\mathbb{C}^{n+1}$ by setting

$$\langle \mathbf{U}, \mathbf{V} \rangle = \mathbf{U}^* Q_{n,1} \mathbf{V} = \mathbf{u}^* \mathbf{v} - \overline{u} v = \mathbf{u} \cdot \mathbf{v} - \overline{u} v$$

for each pair of vectors

$$\mathbf{U} = \begin{bmatrix} \mathbf{u} \\ u \end{bmatrix}, \ \mathbf{V} = \begin{bmatrix} \mathbf{v} \\ v \end{bmatrix} \in \mathbb{C}^{n,1}.$$

We will denote the $\mathbb{C}$-vector space $\mathbb{C}^{n+1}$ with this inner product by $\mathbb{C}^{n,1}$. Then $\langle \ , \ \rangle$ has all the properties of a hermitian inner product except that the real number $\langle \mathbf{U}, \mathbf{U} \rangle$ is not necessarily non-negative. In particular, for $z, w \in \mathbb{C}$,

$$\langle z\mathbf{U}, w\mathbf{V} \rangle = \overline{z} w \langle \mathbf{U}, \mathbf{V} \rangle . \tag{6.2}$$

We will refer to a non-zero $\mathbf{U} \in \mathbb{C}^{n,1}$ as spacelike, timelike or null according to whether $\langle \mathbf{U}, \mathbf{U} \rangle > 0$, $\langle \mathbf{U}, \mathbf{U} \rangle < 0$ or $\langle \mathbf{U}, \mathbf{U} \rangle = 0$ as we do for vectors in $\mathbb{R}^{n,1}$.

We will make repeated use of the following result.

Proposition 6.2

Let $\mathbf{U} = \begin{bmatrix} \mathbf{u} \\ u \end{bmatrix}$ and $\mathbf{X} = \begin{bmatrix} \mathbf{x} \\ t \end{bmatrix} \in \mathbb{C}^{n,1}$ be non-zero vectors with $\langle \mathbf{U}, \mathbf{X} \rangle = 0$.

i) If $\mathbf{U}$ is timelike then $\langle \mathbf{X}, \mathbf{X} \rangle > 0$.

ii) If $\mathbf{U}$ is null and $\mathbf{X}$ is not a complex multiple of $\mathbf{U}$, then $\langle \mathbf{X}, \mathbf{X} \rangle > 0$.

Proof

When (i) holds, then $|\mathbf{u}|^2 - |u|^2 < 0$, so $u \neq 0$; when (ii) holds then $\mathbf{U} \neq \mathbf{0}$ and $|\mathbf{u}|^2 = |u|^2$, also giving $u \neq 0$. Since

$$\mathbf{u} \cdot \mathbf{x} - \overline{u} t = 0, \tag{6.3}$$

in either case we have

$$\langle \mathbf{X}, \mathbf{X} \rangle = |\mathbf{x}|^2 - |t|^2 = |\mathbf{x}|^2 - \frac{|\mathbf{u} \cdot \mathbf{x}|^2}{|u|^2}.$$

In case (i), this gives

$$\langle \mathbf{X}, \mathbf{X} \rangle > |\mathbf{x}|^2 - \frac{|\mathbf{u} \cdot \mathbf{x}|^2}{|\mathbf{u}|^2} \geqslant 0,$$

since $|\mathbf{u}\cdot\mathbf{x}| \leqslant |\mathbf{u}||\mathbf{x}|$. In case (ii),

$$\langle \mathbf{X},\mathbf{X}\rangle = |\mathbf{x}|^2 - \frac{|\mathbf{u}\cdot\mathbf{x}|^2}{|\mathbf{u}|^2} \geqslant 0,$$

with equality if and only if $|\mathbf{u}\cdot\mathbf{x}| = |\mathbf{u}||\mathbf{x}|$, *i.e.*, $\mathbf{x} = z\mathbf{u}$ for some $z \in \mathbb{C}$ with $|z| = 1$. In the latter case, from Equation (6.3) we would have

$$t\bar{u} = z|u|^2 = zu\bar{u},$$

so $t = zu$, in turn giving $\mathbf{X} = z\mathbf{U}$ and so $\langle \mathbf{X},\mathbf{X}\rangle = 0$. □

We can consider the closed subgroup

$$\mathrm{O}(n,1) = \{A \in \mathrm{GL}_{n+1}(\mathbb{R}) : A^T Q_{n,1} A = Q_{n,1}\} \leqslant \mathrm{GL}_{n+1}(\mathbb{R}).$$

Note that $A \in \mathrm{O}(n,1)$ if and only if for all $\mathbf{x},\mathbf{y} \in \mathbb{R}^{n,1}$, $\langle A\mathbf{x}, A\mathbf{y}\rangle = \langle \mathbf{x},\mathbf{y}\rangle$, so $\mathrm{O}(n,1)$ consists of all the Lorentzian isometries of $\mathbb{R}^{n,1}$. Notice that for $A \in \mathrm{O}(n,1)$ we have $(\det A)^2 = 1$, so $\det A = \pm 1$. Also $A \in \mathrm{O}(n,1)$ preserves each of the sets of spacelike, timelike and null vectors; it also either preserves or interchanges the positive and negative light cones. It is standard to set

$$\mathrm{SO}(n,1) = \{A \in \mathrm{O}(n,1) : \det A = 1\} = \mathrm{O}(n,1) \cap \mathrm{SL}_{n+1}(\mathbb{R}) \leqslant \mathrm{O}(n,1).$$

However, $\mathrm{SO}(n,1)$ is not connected since its elements can interchange the sets $\mathcal{H}^{\pm}_{n,1}(r)$ for $r \geqslant 0$. We define the Lorentz group of $\mathbb{R}^{n,1}$ to be the closed subgroup of $\mathrm{SO}(n,1)$ preserving each of the connected sets $\mathcal{H}^{\pm}_{n,1}(1)$,

$$\mathrm{Lor}(n,1) = \{A \in \mathrm{SO}(n,1) : A\mathcal{H}^{\pm}_{n,1}(1) = \mathcal{H}^{\pm}_{n,1}(1)\} \leqslant \mathrm{SO}(n,1).$$

Proposition 6.3

For $r \geqslant 0$,

$$\mathrm{Lor}(n,1) = \{A \in \mathrm{SO}(n,1) : A\mathcal{H}^{\pm}_{n,1}(r) = \mathcal{H}^{\pm}_{n,1}(r)\} \leqslant \mathrm{SO}(n,1).$$

Proof

It is easy to verify this for $r > 0$. For $r = 0$, notice that every null vector $\mathbf{U} = \begin{bmatrix}\mathbf{u}\\ u\end{bmatrix}$ can be expressed as a limit

$$\mathbf{U} = \lim_{k\to\infty} \mathbf{V}_k$$

with each $\mathbf{V}_k$ timelike. We can even arrange that for every k, $\mathbf{V}_k = \begin{bmatrix} \mathbf{v}_k \\ v_k \end{bmatrix}$ with $uv_k > 0$ (or $uv_k < 0$). Since the action of each $A \in \mathrm{Lor}(n,1)$ on $\mathbb{R}^{n,1}$ is continuous,

$$A\mathbf{U} = \lim_{k\to\infty} A\mathbf{V}_k,$$

from which we see that A preserves each of the sets $\mathcal{H}^{\pm}_{n,1}(0)$.

On the other hand, suppose that $A \in \mathrm{SO}(n,1)$ preserves each of the connected sets $\mathcal{H}^{\pm}_{n,1}(0)$. If A fails to preserve one (and hence all) of the sets $\mathcal{H}^{+}_{n,1}(s)$ for $s > 0$, then as any $\mathbf{U} \in \mathcal{H}^{+}_{n,1}(0)$ is a limit

$$\mathbf{U} = \lim_{k\to\infty} \mathbf{V}_k$$

with $\mathbf{V}_k \in \mathcal{H}^{+}_{n,1}(s_k)$ for $s_k > 0$, we have

$$A\mathbf{U} = \lim_{k\to\infty} A\mathbf{V}_k$$

where $A\mathbf{V}_k \in \mathcal{H}^{-}_{n,1}(s'_k)$ for $s'_k > 0$. This implies that $A\mathbf{U} \in \mathcal{H}^{-}_{n,1}(0)$, contradicting the original assumption on A. □

Notice that $A^* = A^T$ if $A \in \mathrm{Lor}(n,1)$, so if $\mathbf{U}, \mathbf{V} \in \mathbb{C}^{n,1}$, we have

$$\langle A\mathbf{U}, A\mathbf{V}\rangle = \mathbf{U}^* A^* Q_{n,1} A\mathbf{V} = \mathbf{U}^* Q_{n,1}\mathbf{V} = \langle \mathbf{U}, \mathbf{V}\rangle. \tag{6.4}$$

When $n = 3$, $\mathrm{Lor}(3,1) = \mathrm{Lor}$ in the notation of Section 1.5.

Example 6.4

We have

$$\mathrm{Lor}(1,1) = \left\{ \begin{bmatrix} \cosh s & \sinh s \\ \sinh s & \cosh s \end{bmatrix} : s \in \mathbb{R} \right\},$$

$$\mathrm{SO}(1,1) = \mathrm{Lor}(1,1) \cup \begin{bmatrix} -1 & 0 \\ 0 & -1 \end{bmatrix} \mathrm{Lor}(1,1),$$

$$\mathrm{O}(1,1) = \mathrm{SO}(1,1) \cup \begin{bmatrix} 1 & 0 \\ 0 & -1 \end{bmatrix} \mathrm{SO}(1,1).$$

Proof

If $A \in \mathrm{O}(1,1)$ has $\det A = -1$ then

$$\begin{bmatrix} 1 & 0 \\ 0 & -1 \end{bmatrix} A \in \mathrm{SO}(1,1),$$

so it suffices to determine $\mathrm{SO}(1,1)$. Similarly, if $B \in \mathrm{SO}(1,1)$ with $B\mathcal{H}^{+}_{n,1}(0) = \mathcal{H}^{-}_{n,1}(0)$ then

$$\begin{bmatrix} -1 & 0 \\ 0 & -1 \end{bmatrix} B \in \mathrm{Lor}(1,1),$$

so it suffices to determine $\mathrm{Lor}(1,1)$.

Let $\begin{bmatrix} a & b \\ c & d \end{bmatrix} \in \mathrm{Lor}(1,1)$. Since

$$\begin{bmatrix} a & b \\ c & d \end{bmatrix} \begin{bmatrix} 0 \\ 1 \end{bmatrix} = \begin{bmatrix} b \\ d \end{bmatrix},$$

we must have $d > 0$. Also, the simultaneous equations

$$\begin{cases} a^2 - c^2 = 1, \\ d^2 - b^2 = 1, \\ cd - ab = 0, \\ ad - bc = 1, \end{cases}$$

are satisfied. Putting $d = \cosh s$ and $b = \sinh s$ for suitable $s \in \mathbb{R}$, we easily obtain $a = \cosh s$, $c = \sinh s$. Hence every element of $\mathrm{Lor}(1,1)$ has the form $\begin{bmatrix} \cosh s & \sinh s \\ \sinh s & \cosh s \end{bmatrix}$ for some $s \in \mathbb{R}$. □

Remark 6.5

The Lie algebra of $\mathrm{Lor}(1,1)$ is easily determined to be

$$\mathfrak{lor}(1,1) = \left\{ \begin{bmatrix} 0 & t \\ t & 0 \end{bmatrix} : t \in \mathbb{R} \right\}$$

with trivial Lie bracket. The exponential map $\exp\colon \mathfrak{lor}(1,1) \longrightarrow \mathrm{Lor}(1,1)$ is surjective since for $s \in \mathbb{R}$,

$$\exp\left(\begin{bmatrix} 0 & s \\ s & 0 \end{bmatrix} \right) = \begin{bmatrix} \cosh s & \sinh s \\ \sinh s & \cosh s \end{bmatrix}. \tag{6.5}$$

As we will eventually see, the exponential map for $\mathrm{Lor}(n,1)$ is actually surjective for every $n \geqslant 1$.

For future use we record the following explicit form for elements of $\mathrm{Lor}(n,1)$, whose proof is a direct consequence of the definition of the Lorentz group.

Proposition 6.6

A matrix

$$P = \begin{bmatrix}\mathbf{U}_1 & \cdots & \mathbf{U}_{n+1}\end{bmatrix} \in \mathrm{GL}_{n+1}(\mathbb{R})$$

is in $\mathrm{Lor}(n,1)$ if and only if its column vectors $\mathbf{U}_j = \begin{bmatrix}\mathbf{u}_j \\ u_j\end{bmatrix}$ satisfy the orthonormality equations

$$\langle \mathbf{U}_j, \mathbf{U}_k \rangle = \begin{cases} 0 & \text{if } j \neq k, \\ 1 & \text{if } j = k \leqslant n, \\ -1 & \text{if } j = k = n+1. \end{cases}$$

In particular $\mathbf{U}_1, \ldots, \mathbf{U}_n$ are spacelike and $\mathbf{U}_{n+1}$ is timelike with $u_{n+1} > 0$.

Theorem 6.7

For $n \geqslant 1$, $\mathrm{Lor}(n,1)$ is path connected.

Proof

We will show that $\mathrm{Lor}(n,1)$ has a connected subgroup H for which the homogeneous space $\mathrm{Lor}(n,1)/H$ is connected. Together with Proposition 9.10, this implies that $\mathrm{Lor}(n,1)$ is connected.

Consider the continuous action of $\mathrm{Lor}(n,1)$ on $\mathbb{R}^{n,1}$ by left multiplication. The stabiliser of the timelike vector $\mathbf{e}_{n+1}$ is

$$\mathrm{Stab}_{\mathrm{Lor}(n,1)}(\mathbf{e}_{n+1}) = \left\{ \begin{bmatrix} A & O_{n-1,1} \\ O_{1,n-1} & 1 \end{bmatrix} : A \in \mathrm{SO}(n) \right\} \leqslant \mathrm{Lor}(n,1)$$

which is a closed subgroup obviously isomorphic to $\mathrm{SO}(n)$, hence connected by Example 9.14. Next we need to identify the orbit of $\mathbf{e}_{n+1}$.

Suppose that $\mathbf{U} = \begin{bmatrix}\mathbf{u} \\ u\end{bmatrix} \in \mathcal{H}^{+}_{n,1}(1)$, so $|\mathbf{u}|^2 - u^2 = -1$ and $u > 0$. We will show that

$$\begin{bmatrix}\mathbf{u} \\ u\end{bmatrix} \in \mathrm{Orb}_{\mathrm{Lor}(n,1)}(\mathbf{e}_{n+1}).$$

Consider the equation

$$\mathbf{u} \cdot \mathbf{x} - ut = 0$$

for the pair $\mathbf{x} \in \mathbb{R}^n$ and $t \in \mathbb{R}$. The solution set is an n-dimensional vector subspace of $\mathbb{R}^{n,1}$; furthermore, by Proposition 6.2 its non-zero vectors are all spacelike. Choosing a basis of this spacelike subspace, then applying the Gram–Schmidt process, we can produce a basis consisting of vectors

$$\begin{bmatrix}\mathbf{w}_1\\ s_1\end{bmatrix},\ldots,\begin{bmatrix}\mathbf{w}_n\\ s_n\end{bmatrix}$$

for which

$$\begin{cases}\mathbf{w}_i\cdot\mathbf{w}_j - s_i s_j = \delta_{ij} & (i,j=1,\ldots,n),\\ \mathbf{u}\cdot\mathbf{w}_k - u s_k = 0 & (k=1,\ldots,n).\end{cases}$$

The resulting matrix

$$P=\begin{bmatrix}\mathbf{w}_1 & \cdots & \mathbf{w}_n & \mathbf{u}\\ s_1 & \cdots & s_n & u\end{bmatrix}$$

is in $\mathrm{Lor}(n,1)$ and satisfies

$$P\mathbf{e}_{n+1}=\begin{bmatrix}\mathbf{u}\\ u\end{bmatrix}.$$

So

$$\mathrm{Orb}_{\mathrm{Lor}(n,1)}(\mathbf{e}_{n+1})=\mathcal{H}^{+}_{n,1}(1).$$

This also shows that $\mathrm{Orb}_{\mathrm{Lor}(n,1)}(\mathbf{e}_{n+1})$ is path connected since we already know that $\mathcal{H}^{+}_{n,1}(1)$ is. Thus taking $H=\mathrm{Stab}_{\mathrm{Lor}(n,1)}(\mathbf{e}_{n+1})\cong\mathrm{SO}(n)$ we see that $\mathrm{Lor}(n,1)$ is path connected. □

This identification of the homogeneous space $\mathrm{Lor}(n,1)/\mathrm{SO}(n)$ can also be used to calculate the dimension of $\mathrm{Lor}(n,1)$.

Proposition 6.8

The dimension of the Lorentz group $\mathrm{Lor}(n,1)$ is

$$\dim\mathrm{Lor}(n,1)=\binom{n+1}{2}.$$

Proof

Using the Implicit Function Theorem 7.11, it is easy to see that

$$\dim\mathrm{Lor}(n,1)/\mathrm{SO}(n)=\dim\mathcal{H}^{+}_{n,1}(1)=n.$$

Also from Section 3.3 we know that

$$\dim\mathrm{SO}(n)=\binom{n}{2}.$$

Combining these we obtain the result. □

6.2 A Principal Axis Theorem for Lorentz Groups

We will make use of the Principal Axis Theorem for special orthogonal groups which will be discussed later as part of Theorem 10.13. This says that every matrix $A \in \mathrm{SO}(m)$ has the form

$$A = \begin{cases} PR_{2n}(\theta_1, \ldots, \theta_n)P^T & \text{if } m = 2n \text{ is even,} \\ PR_{2n+1}(\theta_1, \ldots, \theta_n)P^T & \text{if } m = 2n+1 \text{ is odd,} \end{cases}$$

where for $\theta_1, \ldots, \theta_n, \theta \in \mathbb{R}$,

$$R_{2n}(\theta_1, \ldots, \theta_n) = \begin{bmatrix} R(\theta_1) & O & \cdots & \cdots & \cdots & O \\ O & R(\theta_2) & O & \ddots & \ddots & \vdots \\ \vdots & \ddots & \ddots & \ddots & \ddots & \vdots \\ O & \cdots & \cdots & \cdots & O & R(\theta_n) \end{bmatrix},$$

$$R_{2n+1}(\theta_1, \ldots, \theta_n) = \begin{bmatrix} R(\theta_1) & O & \cdots & \cdots & \cdots & \cdots & O \\ O & R(\theta_2) & O & \ddots & \ddots & \ddots & \ddots \\ \vdots & \ddots & \ddots & \ddots & \ddots & \ddots & \vdots \\ \vdots & \ddots & \ddots & \ddots & O & R(\theta_n) & O \\ O & \cdots & \cdots & \cdots & \cdots & O & 1 \end{bmatrix},$$

$$R(\theta) = \begin{bmatrix} \cos\theta & -\sin\theta \\ \sin\theta & \cos\theta \end{bmatrix}.$$

The following result leads to the analogue for Lorentz groups.

Theorem 6.9

For $n \geqslant 1$, every Lorentz matrix $A \in \mathrm{Lor}(n, 1)$ has one of the forms

$$A = P \begin{bmatrix} B & O_{n-2,2} & \\ O_{2,n-2} & \begin{matrix} \cosh t & \sinh t \\ \sinh t & \cosh t \end{matrix} \end{bmatrix} P^{-1}, \tag{I}$$

$$A = P \begin{bmatrix} C & O_{n,1} \\ O_{1,n} & 1 \end{bmatrix} P^{-1}, \tag{II}$$

where $B \in \mathrm{SO}(n-1)$, $C \in \mathrm{SO}(n)$, $t \in \mathbb{R}$ and $P \in \mathrm{Lor}(n, 1)$.

Proof

We will prove this by induction on $n \geqslant 1$. The initial case $n = 1$ is dealt with in Example 6.4. So we suppose that the result holds for $\mathrm{Lor}(k, 1)$ whenever $1 \leqslant k \leqslant n-1$.

We begin by analysing the eigenvalues and eigenvectors of A. Suppose that $\lambda \in \mathbb{C}$ is an eigenvalue for $A \in \mathrm{Lor}(n, 1)$ with eigenvector $\mathbf{U} \in \mathbb{C}^{n,1}$. If $\mathbf{U}$ is not a null vector, we have

$$\begin{aligned}|\lambda|^2 \langle \mathbf{U}, \mathbf{U}\rangle &= \langle \lambda\mathbf{U}, \lambda\mathbf{U}\rangle \\ &= \langle A\mathbf{U}, A\mathbf{U}\rangle \\ &= \langle \mathbf{U}, \mathbf{U}\rangle,\end{aligned}$$

hence $|\lambda| = 1$. On the other hand, if $\mathbf{U}$ is a null vector then $|\lambda|$ is potentially unrestricted.

Next suppose that λ and μ are distinct eigenvalues for A with eigenvectors $\mathbf{U}$ and $\mathbf{V}$. Then

$$\begin{aligned}\overline{\lambda}\mu \langle \mathbf{U}, \mathbf{V}\rangle &= \langle \lambda\mathbf{U}, \mu\mathbf{V}\rangle \\ &= \langle A\mathbf{U}, A\mathbf{V}\rangle \\ &= \langle \mathbf{U}, \mathbf{V}\rangle,\end{aligned}$$

hence either $\langle \mathbf{U}, \mathbf{V}\rangle = 0$ or $\mu = \overline{\lambda}^{-1}$. In particular, if $|\lambda| = 1$ we must also have $\overline{\lambda}^{-1} = \lambda \neq \mu$, yielding $\langle \mathbf{U}, \mathbf{V}\rangle = 0$.

Next we will investigate in more detail the situation when A has a null eigenvector $\mathbf{U} = \begin{bmatrix}\mathbf{u}\\ u\end{bmatrix}$ for an eigenvalue λ. Since $|\mathbf{u}|^2 = |u|^2 \neq 0$, we can multiply by a suitable complex scalar to ensure that $u = 1$, so we might as well assume that $\mathbf{U} = \begin{bmatrix}\mathbf{u}\\ 1\end{bmatrix}$. Notice that

$$A\begin{bmatrix}\overline{\mathbf{u}}\\ 1\end{bmatrix} = \overline{\left(A\begin{bmatrix}\mathbf{u}\\ 1\end{bmatrix}\right)} = \overline{\lambda}\begin{bmatrix}\overline{\mathbf{u}}\\ 1\end{bmatrix},$$

showing that $\overline{\lambda}$ is also an eigenvalue with eigenvector $\overline{\mathbf{U}} = \begin{bmatrix}\overline{\mathbf{u}}\\ 1\end{bmatrix}$. Also,

$$\begin{aligned}\langle \overline{\mathbf{U}}, \mathbf{U}\rangle &= \langle A\overline{\mathbf{U}}, A\mathbf{U}\rangle \\ &= \langle \overline{\lambda\mathbf{U}}, \lambda\mathbf{U}\rangle \\ &= \lambda^2 \langle \overline{\mathbf{U}}, \mathbf{U}\rangle.\end{aligned}$$

As $\overline{\mathbf{U}}$ is a null vector, Proposition 6.2 implies that either $\overline{\mathbf{U}} = \mathbf{U}$ or $\lambda^2 = 1$. Each of these possibilities means that λ is real and in fact positive, since A

preserves the positive light cone. So for a null eigenvector, the corresponding eigenvalue has to be a positive real number.

Continuing with this situation, consider the solution set of the linear equation

$$\langle \mathbf{U}, \mathbf{X} \rangle = 0. \tag{6.6}$$

This forms a vector subspace of $\mathbb{R}^{n,1}$ of dimension n, containing the subspace of real multiples of $\mathbf{U}$ as well as the $(n-1)$-dimensional subspace of spacelike elements together with zero. In fact, the spacelike elements are all the vectors of the form

$$\begin{bmatrix} \mathbf{x} \\ 0 \end{bmatrix} \quad (\mathbf{x} \in \mathbb{R}^n,\ \mathbf{u} \cdot \mathbf{x} = 0).$$

Thus the general solution of Equation (6.6) is

$$\mathbf{X} = \begin{bmatrix} \mathbf{x} + z\mathbf{u} \\ z \end{bmatrix} \quad (\mathbf{x} \in \mathbb{R}^n,\ \mathbf{u} \cdot \mathbf{x} = 0,\ z \in \mathbb{R}).$$

Then there is a unique null vector $\widetilde{\mathbf{U}} \in \mathbb{R}^{n,1}$ possessing the two properties

- $\left\langle \widetilde{\mathbf{U}}, \mathbf{X} \right\rangle = 0$ for all spacelike solutions of Equation (6.6),
- $\left\langle \mathbf{U}, \widetilde{\mathbf{U}} \right\rangle = -2$.

This vector is easily seen to be $\widetilde{\mathbf{U}} = \begin{bmatrix} -\mathbf{u} \\ 1 \end{bmatrix}$. Clearly the spacelike solutions of Equation (6.6) are preserved by A, and indeed A acts as an isometry on this space so relative to a basis it is given by an orthogonal matrix. Also

$$\begin{aligned} \left\langle \mathbf{U}, A(\lambda \widetilde{\mathbf{U}}) \right\rangle &= \lambda \left\langle A^{-1}\mathbf{U}, \widetilde{\mathbf{U}} \right\rangle \\ &= \lambda \left\langle \lambda^{-1}\mathbf{U}, \widetilde{\mathbf{U}} \right\rangle \\ &= \lambda\lambda^{-1} \left\langle \mathbf{U}, \widetilde{\mathbf{U}} \right\rangle \\ &= -2. \end{aligned}$$

Using the above characterisation of $\widetilde{\mathbf{U}}$, we see that

$$A\widetilde{\mathbf{U}} = \lambda^{-1}\widetilde{\mathbf{U}},$$

i.e., $\widetilde{\mathbf{U}}$ is an eigenvector of A for the eigenvalue λ^{-1}.

The vectors

$$\mathbf{U}' = \frac{1}{2}\left(\mathbf{U} - \widetilde{\mathbf{U}}\right), \quad \mathbf{U}'' = \frac{1}{2}\left(\mathbf{U} + \widetilde{\mathbf{U}}\right),$$

satisfy

$$AU' = \frac{(\lambda + \lambda^{-1})}{2}U' + \frac{(\lambda - \lambda^{-1})}{2}U'',$$
$$AU'' = \frac{(\lambda - \lambda^{-1})}{2}U' + \frac{(\lambda + \lambda^{-1})}{2}U'',$$

and setting $s = \ln \lambda$, we have

$$AU' = \cosh s\, U' + \sinh s\, U'',$$
$$AU'' = \sinh s\, U' + \cosh s\, U''.$$

Also

$$\langle U', U'' \rangle = 0, \quad \langle U', U' \rangle = 1, \quad \langle U'', U'' \rangle = -1,$$

so U' is a unit spacelike vector, while U'' is timelike and these are orthogonal vectors with respect to the Lorentzian inner product.

Together with an orthonormal basis $\{U_1, \ldots, U_{n-1}\}$ of the spacelike solutions of Equation (6.6), we see that $\mathbb{R}^{n,1}$ has a basis

$$\{U_1, \ldots, U_{n-1}, U', U''\}$$

which provides the columns of a Lorentz matrix

$$P = \begin{bmatrix} U_1 & \cdots & U_{n-1} & U' & U'' \end{bmatrix}$$

for which

$$A = P \begin{bmatrix} B & O_{n-2,2} \\ O_{2,n-2} & \begin{matrix} \cosh t & \sinh t \\ \sinh t & \cosh t \end{matrix} \end{bmatrix} P^{-1}$$

with $B \in O(n-1)$; in fact $\det B = 1$ since $\det A = 1$ and also

$$\det \begin{bmatrix} \cosh t & \sinh t \\ \sinh t & \cosh t \end{bmatrix} = \cosh^2 t - \sinh^2 t = 1.$$

Hence A has the form (I) whenever it has a null vector for an eigenvector.

Now we consider the situation where A has a timelike eigenvector U for the eigenvalue λ. Using similar arguments to those for a null eigenvector, we find that $\lambda = 1$ and $U \in \mathbb{R}^{n,1}$. Moreover, A acts as an isometry on the solution set of

$$\langle U, X \rangle = 0,$$

which is an n-dimensional subspace of $\mathbb{R}^{n,1}$ with timelike non-zero vectors. Hence A has the form (II).

The remaining case to deal with is that where there are no null or timelike eigenvectors. So suppose that $U = \begin{bmatrix} \mathbf{u} \\ u \end{bmatrix}$ is a spacelike eigenvector of unit length with eigenvalue λ of unit modulus. By multiplying by a suitable complex number we can assume that $u \in \mathbb{R}$ and $u \geqslant 0$.

When $\mathbf{U} \in \mathbb{R}^{n,1}$, we have $\lambda = \pm 1$ and $\det A = 1$.

If $\lambda = 1$ then the solutions of the equation

$$\langle \mathbf{U}, \mathbf{X} \rangle = 0$$

form a subspace of $\mathbb{R}^{n,1}$ closed under the action of A and indeed, on choosing a basis for this subspace we find that it is isomorphic to $\mathbb{R}^{n-1,1}$ and the action of A is given by an element of $\mathrm{Lor}(n-1,1)$. Hence there is a $Q \in \mathrm{Lor}(n,1)$ and $B \in \mathrm{Lor}(n-1,1)$ for which

$$A = Q \begin{bmatrix} 1 & O_{1,n} \\ O_{n,1} & B \end{bmatrix} Q^{-1}.$$

By the induction hypothesis, B has one of the forms (I) or (II), from which we deduce that A does too.

If $\lambda = -1$, this is an eigenvalue of multiplicity at least 2 since non-real complex eigenvalues come in complex conjugate pairs. With the aid of the Gram–Schmidt process we may produce a second eigenvector $\mathbf{V}$ for this eigenvalue of unit length and orthogonal to $\mathbf{U}$. The solutions of the pair of simultaneous equations

$$\langle \mathbf{U}, \mathbf{X} \rangle = 0 = \langle \mathbf{V}, \mathbf{X} \rangle$$

form a subspace of $\mathbb{R}^{n,1}$ closed under the action of A. On choosing a basis for this subspace we find that it is isomorphic to $\mathbb{R}^{n-2,1}$ and the action of A is given by an element of $\mathrm{Lor}(n-2,1)$. Hence there is a $Q \in \mathrm{Lor}(n,1)$ and $B' \in \mathrm{Lor}(n-2,1)$ for which

$$A = Q \begin{bmatrix} -I_2 & O_{2,n-1} \\ O_{n-1,2} & B' \end{bmatrix} Q^{-1}.$$

Again the induction hypothesis implies that B' has one of the forms (I) or (II), and therefore A does too.

When $\mathbf{U} \notin \mathbb{R}^{n,1}$, $\overline{\mathbf{U}} = \begin{bmatrix} \overline{\mathbf{u}} \\ u \end{bmatrix}$ is also a spacelike eigenvector of unit length for the eigenvalue $\overline{\lambda}$. The vectors

$$\mathbf{U}' = \frac{1}{2}(\mathbf{U} + \overline{\mathbf{U}}), \quad \mathbf{U}'' = \frac{1}{2i}(\mathbf{U} - \overline{\mathbf{U}}),$$

are orthonormal and satisfy the equations

$$A\mathbf{U}' = \cos\theta\, \mathbf{U}' + \sin\theta\, \mathbf{U}'', \quad A\mathbf{U}'' = -\sin\theta\, \mathbf{U}' + \cos\theta\, \mathbf{U}'',$$

where $\lambda = e^{\theta i}$. As in the previous case, the set of solutions of the pair of simultaneous equations

$$\langle \mathbf{U}', \mathbf{X} \rangle = 0 = \langle \mathbf{U}'', \mathbf{X} \rangle$$

forms a Lorentzian subspace which can be identified with $\mathrm{Lor}(n-2,1)$ after choosing a basis, and A acts on it as an element of $\mathrm{Lor}(n-2,1)$. Again there is a $Q \in \mathrm{Lor}(n,1)$ and $B'' \in \mathrm{Lor}(n-2,1)$ for which

$$A = Q \begin{bmatrix} R(\theta) & O_{2,n-1} \\ O_{n-1,2} & B'' \end{bmatrix} Q^{-1}.$$

By the induction hypothesis, B'' has one of the forms (I) or (II), from which we deduce that A does too. □

Combining this result with the Principal Axis Theorem 10.13 for $\mathrm{SO}(k)$, we obtain the following.

Theorem 6.10 (Principal Axis Theorem for Lorentz groups)

For $n \geqslant 1$, each matrix $A \in \mathrm{Lor}(n,1)$ has one of the forms

$$A = Q \begin{bmatrix} R_{n-1}(\theta_1, \ldots, \theta_\ell) & & O_{n-2,2} \\ & \cosh t & \sinh t \\ O_{2,n-2} & \sinh t & \cosh t \end{bmatrix} Q^{-1}, \tag{I'}$$

$$A = Q \begin{bmatrix} R_n(\theta_1, \ldots, \theta_m) & O_{n,1} \\ O_{1,n} & 1 \end{bmatrix} Q^{-1}, \tag{II''}$$

where $t \in \mathbb{R}$, $Q \in \mathrm{Lor}(n,1)$ and in case (I′), $n = 2\ell + 1$ or $n = 2\ell + 2$, while in case (II″), $n = 2m$ or $n = 2m + 1$.

We can also deduce information on the exponential for Lorentz groups.

Theorem 6.11

For $n \geqslant 1$, the exponential map $\exp\colon \mathfrak{lor}(n,1) \longrightarrow \mathrm{Lor}(n,1)$ is surjective.

Proof

When $n = 1$, we know from Remark 6.5 that every element of $\mathrm{Lor}(1,1)$ has the form given in (I). Also, by Theorem 10.16, $\exp\colon \mathfrak{so}(k) \longrightarrow \mathrm{SO}(k)$ is surjective for $k \geqslant 1$. Let $A \in \mathrm{Lor}(n,1)$ for $n \geqslant 2$. When A has the form (I),

$$A = P \begin{bmatrix} \exp(S) & O_{n-2,2} \\ O_{2,n-2} & \exp\left(\begin{bmatrix} 0 & t \\ t & 0 \end{bmatrix}\right) \end{bmatrix} P^{-1} = \exp\left(P \begin{bmatrix} S & O_{n-2,2} \\ O_{2,n-2} & \begin{matrix} 0 & t \\ t & 0 \end{matrix} \end{bmatrix} P^{-1} \right)$$

for some $S \in \mathfrak{so}(n-1)$. On the other hand, when A has the form (II),

$$A = P \begin{bmatrix} \exp(S) & O_{n,1} \\ O_{1,n} & 1 \end{bmatrix} P^{-1} = \exp\left(P \begin{bmatrix} S & O_{n,1} \\ O_{1,n} & 0 \end{bmatrix} P^{-1} \right)$$

for some $S \in \mathfrak{so}(n)$. In either case, $A \in \exp \mathfrak{lor}(n,1)$. □

6.3 $\mathrm{SL}_2(\mathbb{C})$ and the Lorentz Group $\mathrm{Lor}(3,1)$

Let us now consider the Lie algebra of $\mathrm{SL}_2(\mathbb{C})$, $\mathfrak{sl}_2(\mathbb{C})$. By Equation (3.8),

$$\mathfrak{sl}_2(\mathbb{C}) = \ker \operatorname{tr} \subseteq \mathrm{M}_2(\mathbb{C})$$

and so $\dim_{\mathbb{C}} \mathfrak{sl}_2(\mathbb{C}) = 3$. The following matrices form a $\mathbb{C}$-basis for $\mathfrak{sl}_2(\mathbb{C})$:

$$H' = \begin{bmatrix} 1 & 0 \\ 0 & -1 \end{bmatrix}, \quad E' = \begin{bmatrix} 0 & 1 \\ 0 & 0 \end{bmatrix}, \quad F' = \begin{bmatrix} 0 & 0 \\ 1 & 0 \end{bmatrix}.$$

The elements $H', iH', E', iE', F', iF'$ form an $\mathbb{R}$-basis and $\dim \mathfrak{sl}_2(\mathbb{C}) = 6$. Notice also that $\mathfrak{su}(2) \subseteq \mathfrak{sl}_2(\mathbb{C})$ and the elements $H, E, F \in \mathfrak{su}(2)$ form a $\mathbb{C}$-basis of $\mathfrak{sl}_2(\mathbb{C})$, so H, iH, E, iE, F, iF form an $\mathbb{R}$-basis. The Lie brackets of H', E', F' are determined by

$$[H', E'] = 2E', \quad [H', F'] = -2F', \quad [E', F'] = H'.$$

Notice that the subspaces spanned by each of the pairs H', E' and H', F' are $\mathbb{C}$-Lie subalgebras. In fact, H', E' span the Lie algebra $\mathfrak{ut}_2^0(\mathbb{C})$ of the group of *upper triangular* complex matrices of determinant 1, while H', F' span the Lie algebra of the group of *lower triangular* complex matrices of determinant 1.

Given the existence of the double covering homomorphism $\overline{\mathrm{Ad}} \colon \mathrm{SU}(2) \longrightarrow \mathrm{SO}(3)$ of Section 3.5, it seems reasonable to ask if a similar homomorphism exists for $\mathrm{SL}_2(\mathbb{C})$. It does, but we need to use the Lorentz group $\mathrm{Lor}(3,1)$ and obtain an important double covering homomorphism $\mathrm{SL}_2(\mathbb{C}) \longrightarrow \mathrm{Lor}(3,1)$ which appears in Physics in connection with spinors and twistors.

Next we will describe the $\mathbb{R}$-Lie algebra $\mathfrak{lor}(3,1)$ of $\mathrm{Lor}(3,1) \leqslant \mathrm{SL}_4(\mathbb{R})$. Let $\alpha \colon (-\varepsilon, \varepsilon) \longrightarrow \mathrm{Lor}(3,1)$ be a differentiable curve with $\alpha(0) = I$. By definition, for $t \in (-\varepsilon, \varepsilon)$ we have

$$\alpha(t) Q \alpha(t)^T = Q,$$

where

$$Q = \begin{bmatrix} 1 & 0 & 0 & 0 \\ 0 & 1 & 0 & 0 \\ 0 & 0 & 1 & 0 \\ 0 & 0 & 0 & -1 \end{bmatrix}.$$

Differentiating and setting $t = 0$ we obtain

$$\alpha'(0)Q + Q\alpha'(0)^T = O,$$

giving

$$\begin{bmatrix} \alpha'(0)_{11} & \alpha'(0)_{12} & \alpha'(0)_{13} & -\alpha'(0)_{14} \\ \alpha'(0)_{21} & \alpha'(0)_{22} & \alpha'(0)_{23} & -\alpha'(0)_{24} \\ \alpha'(0)_{31} & \alpha'(0)_{32} & \alpha'(0)_{33} & -\alpha'(0)_{34} \\ \alpha'(0)_{41} & \alpha'(0)_{42} & \alpha'(0)_{43} & -\alpha'(0)_{44} \end{bmatrix} +$$

$$\begin{bmatrix} \alpha'(0)_{11} & \alpha'(0)_{21} & \alpha'(0)_{31} & \alpha'(0)_{41} \\ \alpha'(0)_{12} & \alpha'(0)_{22} & \alpha'(0)_{32} & \alpha'(0)_{42} \\ \alpha'(0)_{13} & \alpha'(0)_{23} & \alpha'(0)_{33} & \alpha'(0)_{43} \\ -\alpha'(0)_{14} & -\alpha'(0)_{42} & -\alpha'(0)_{34} & -\alpha'(0)_{44} \end{bmatrix} = O.$$

So we have

$$\alpha'(0) = \begin{bmatrix} 0 & \alpha'(0)_{12} & \alpha'(0)_{13} & \alpha'(0)_{14} \\ -\alpha'(0)_{12} & 0 & \alpha'(0)_{23} & \alpha'(0)_{24} \\ -\alpha'(0)_{13} & -\alpha'(0)_{23} & 0 & \alpha'(0)_{34} \\ \alpha'(0)_{14} & \alpha'(0)_{24} & \alpha'(0)_{34} & 0 \end{bmatrix}.$$

Notice that the trace of such a matrix is zero. In fact, every matrix of the form

$$A = \begin{bmatrix} 0 & a_{12} & a_{13} & a_{14} \\ -a_{12} & 0 & a_{23} & a_{24} \\ -a_{13} & -a_{23} & 0 & a_{34} \\ a_{14} & a_{24} & a_{34} & 0 \end{bmatrix}$$

or equivalently satisfying

$$AQ + QA^T = O \quad (\text{and hence } \operatorname{tr} A = 0)$$

is in $\mathfrak{lor}(3,1)$. To see this, consider the curve

$$\alpha\colon \mathbb{R} \longrightarrow \mathrm{GL}_4(\mathbb{R}); \quad \alpha(t) = \exp(tA),$$

for which $\alpha'(0) = A$. Since $QA^T = -AQ$, we have

$$\exp(tA)Q\exp(tA)^T = \exp(tA)Q\exp(tA^T) = \exp(tA)\exp(-tA)Q = Q,$$

By Lemma 3.23,

$$\det \exp(tA) = e^{\operatorname{tr}(tA)} = 1.$$

Finally, for each $t \in \mathbb{R}$, $\alpha(t)$ preserves each of the components $\mathcal{H}^{\pm}_{3,1}(1)$ of the hyperboloid

$$x_1^2 + x_2^2 + x_3^2 - x_4^2 = -1.$$

All of this shows that $\exp(tA) \in \mathrm{Lor}(3,1)$ so we might as well redefine

$$\alpha \colon \mathbb{R} \longrightarrow \mathrm{Lor}(3,1); \quad \alpha(t) = \exp(tA).$$

We also have $A \in \mathfrak{lor}(3,1)$. So we have

$$\begin{aligned} \mathfrak{lor}(3,1) &= \{A \in \mathrm{M}_4(\mathbb{R}) : AQ + QA^T = O\} \\ &= \left\{ A \in \mathrm{M}_4(\mathbb{R}) : A = \begin{bmatrix} 0 & a_{12} & a_{13} & a_{14} \\ -a_{12} & 0 & a_{23} & a_{24} \\ -a_{13} & -a_{23} & 0 & a_{34} \\ a_{14} & a_{24} & a_{34} & 0 \end{bmatrix} \right\} \end{aligned} \tag{6.7}$$

and therefore

$$\dim \mathrm{Lor}(3,1) = \dim \mathfrak{lor}(3,1) = 6. \tag{6.8}$$

An $\mathbb{R}$-basis for $\mathfrak{lor}(3,1)$ consists of the elements

$$P_{12} = \begin{bmatrix} 0 & -1 & 0 & 0 \\ 1 & 0 & 0 & 0 \\ 0 & 0 & 0 & 0 \\ 0 & 0 & 0 & 0 \end{bmatrix}, \quad P_{13} = \begin{bmatrix} 0 & 0 & -1 & 0 \\ 0 & 0 & 0 & 0 \\ 1 & 0 & 0 & 0 \\ 0 & 0 & 0 & 0 \end{bmatrix}, \quad P_{14} = \begin{bmatrix} 0 & 0 & 0 & 1 \\ 0 & 0 & 0 & 0 \\ 0 & 0 & 0 & 0 \\ 1 & 0 & 0 & 0 \end{bmatrix},$$

$$P_{23} = \begin{bmatrix} 0 & 0 & 0 & 0 \\ 0 & 0 & -1 & 0 \\ 0 & 1 & 0 & 0 \\ 0 & 0 & 0 & 0 \end{bmatrix}, \quad P_{24} = \begin{bmatrix} 0 & 0 & 0 & 0 \\ 0 & 0 & 0 & 1 \\ 0 & 0 & 0 & 0 \\ 0 & 1 & 0 & 0 \end{bmatrix}, \quad P_{34} = \begin{bmatrix} 0 & 0 & 0 & 0 \\ 0 & 0 & 0 & 0 \\ 0 & 0 & 0 & 1 \\ 0 & 0 & 1 & 0 \end{bmatrix}.$$

The non-trivial brackets for these are

$$\begin{array}{llll} [P_{12}, P_{13}] = P_{23}, & [P_{12}, P_{14}] = P_{24}, & [P_{12}, P_{23}] = -P_{13}, & [P_{12}, P_{24}] = -P_{14}, \\ [P_{12}, P_{34}] = 0, & [P_{13}, P_{14}] = P_{34}, & [P_{13}, P_{23}] = P_{12}, & [P_{13}, P_{24}] = 0, \\ [P_{13}, P_{34}] = 0, & [P_{14}, P_{23}] = 0, & [P_{14}, P_{24}] = -P_{12}, & [P_{14}, P_{34}] = -P_{13}, \\ [P_{23}, P_{24}] = P_{34}, & [P_{23}, P_{34}] = -P_{24}, & [P_{24}, P_{34}] = -P_{23}. & \end{array}$$

We will now define the homomorphism $\mathrm{SL}_2(\mathbb{C}) \longrightarrow \mathrm{Lor}(3,1)$. To do this we will identify the 2×2 skew hermitian matrices $\text{Sk-Herm}_2(\mathbb{C})$ with $\mathbb{R}^4$ using the correspondence

$$\begin{bmatrix} (t+x)i & y+zi \\ -y+zi & (t-x)i \end{bmatrix} \longleftrightarrow x\mathbf{e}_1 + y\mathbf{e}_2 + z\mathbf{e}_3 + t\mathbf{e}_4.$$

Define an $\mathbb{R}$-bilinear inner product on Sk-Herm$_2(\mathbb{C})$ by the formula

$$\langle S_1|S_2\rangle = \frac{1}{4}(\det(S_1+S_2) - \det(S_1-S_2)). \tag{6.9a}$$

When $S_1 = S_2 = S$ we have

$$\langle S|S\rangle = \frac{1}{4}(\det 2S - \det O) = \det S. \tag{6.9b}$$

It is easy to check that

$$\left\langle \begin{bmatrix} (t_1+x_1)i & y_1+z_1 i \\ -y_1+z_1 i & (t_1-x_1)i \end{bmatrix} \middle| \begin{bmatrix} (t_2+x_2)i & y_2+z_2 i \\ -y_2+z_2 i & (t_2-x_2)i \end{bmatrix} \right\rangle$$
$$= x_1x_2 + y_1y_2 + z_1z_2 - t_1t_2, \tag{6.10}$$

which is the Lorentzian inner product on $\mathbb{R}^{3,1} = \mathbb{R}^4$ given by

$$x_1x_2 + y_1y_2 + z_1z_2 - t_1t_2 = \begin{bmatrix} x_1 & y_1 & z_1 & t_1 \end{bmatrix} \begin{bmatrix} 1 & 0 & 0 & 0 \\ 0 & 1 & 0 & 0 \\ 0 & 0 & 1 & 0 \\ 0 & 0 & 0 & -1 \end{bmatrix} \begin{bmatrix} x_2 \\ y_2 \\ z_2 \\ t_2 \end{bmatrix}.$$

Now observe that for $A \in \mathrm{SL}_2(\mathbb{C})$ and $S \in$ Sk-Herm$_2(\mathbb{C})$,

$$(ASA^*)^* = AS^*A^* = -ASA^*,$$

so $ASA^* \in$ Sk-Herm$_2(\mathbb{C})$. By Equation (6.9a), for $S_1, S_2 \in$ Sk-Herm$_2(\mathbb{C})$, and the fact that $\det A = 1 = \det A^*$,

$$\begin{aligned} \langle AS_1A^*|AS_2A^*\rangle &= \frac{1}{4}(\det A(S_1+S_2)A^* - \det A(S_1-S_2)A^*) \\ &= \frac{1}{4}(\det A \det(S_1+S_2)\det A^* - \det A\det(S_1-S_2)\det A^*) \\ &= \frac{1}{4}(\det(S_1+S_2) - \det(S_1-S_2)) \\ &= \langle S_1|S_2\rangle . \end{aligned}$$

Hence the function

$$\text{Sk-Herm}_2(\mathbb{C}) \longrightarrow \text{Sk-Herm}_2(\mathbb{C}); \quad S \mapsto ASA^*,$$

is an $\mathbb{R}$-linear transformation preserving the inner product $\langle\ |\ \rangle$. We can identify this with an $\mathbb{R}$-linear transformation $\widetilde{\mathrm{Ad}}_A\colon \mathbb{R}^{3,1} \longrightarrow \mathbb{R}^{3,1}$ which preserves the Lorentzian inner product. In fact, $\det \widetilde{\mathrm{Ad}}_A = 1$ and $\widetilde{\mathrm{Ad}}_A$ preserves the components of the hyperboloid $x^2+y^2+z^2-t^2 = -1$. Let

$$\widetilde{\mathrm{Ad}}\colon \mathrm{SL}_2(\mathbb{C}) \longrightarrow \mathrm{Lor}(3,1); \quad \widetilde{\mathrm{Ad}}(A) = \widetilde{\mathrm{Ad}}_A.$$

Then $\widetilde{\mathrm{Ad}}$ is a homomorphism since

$$\widetilde{\mathrm{Ad}}_{AB}(S) = AB(S)(AB)^* = AB(S)B^*A^* = \widetilde{\mathrm{Ad}}_A(\widetilde{\mathrm{Ad}}_B(S)) = \widetilde{\mathrm{Ad}}_A\widetilde{\mathrm{Ad}}_B(S).$$

It is clearly continuous. Also, $A \in \ker\widetilde{\mathrm{Ad}}$ if and only if $ASA^* = S$ for all $S \in \text{Sk-Herm}_2(\mathbb{C})$, and it is easy to see that this occurs exactly when $A = \pm I$. Thus we have $\ker\widetilde{\mathrm{Ad}} = \{\pm I\}$.

Theorem 6.12

i) The derivative $\mathrm{d}\,\widetilde{\mathrm{Ad}}\colon \mathfrak{sl}_2(\mathbb{C}) \longrightarrow \mathfrak{lor}(3,1)$ is an isomorphism of $\mathbb{R}$-Lie algebras.
ii) $\widetilde{\mathrm{Ad}}\colon \mathrm{SL}_2(\mathbb{C}) \longrightarrow \mathrm{Lor}(3,1)$ is a continuous surjective Lie homomorphism for which $\ker\widetilde{\mathrm{Ad}} = \{\pm I\}$, hence

$$\mathrm{SL}_2(\mathbb{C})/\{\pm I\} \cong \mathrm{Lor}(3,1).$$

Proof

(i) As in the case of SU(2) and SO(3), we can determine the derivative $\mathrm{d}\,\widetilde{\mathrm{Ad}}$ by considering for each $C \in \mathfrak{sl}_2(\mathbb{C})$, the curve

$$\gamma\colon \mathbb{R} \longrightarrow \mathrm{SL}_2(\mathbb{C}); \quad \gamma(t) = \exp(tC),$$

which gives rise to the curve

$$\overline{\gamma}\colon \mathbb{R} \longrightarrow \mathrm{Lor}(3,1); \quad \overline{\gamma}(t) = \widetilde{\mathrm{Ad}}_{\gamma(t)}.$$

Using as an $\mathbb{R}$-basis for $\text{Sk-Herm}_2(\mathbb{C})$ the vectors

$$V_1 = \begin{bmatrix} i & 0 \\ 0 & -i \end{bmatrix}, \quad V_2 = \begin{bmatrix} 0 & 1 \\ -1 & 0 \end{bmatrix}, \quad V_3 = \begin{bmatrix} 0 & i \\ i & 0 \end{bmatrix}, \quad V_4 = \begin{bmatrix} i & 0 \\ 0 & i \end{bmatrix},$$

we can determine the action of $\widetilde{\mathrm{Ad}}_{\gamma(t)}$ on $\text{Sk-Herm}_2(\mathbb{C})$ and interpret it as an element of $\mathrm{Lor}(3,1)$. Differentiating at $t = 0$, we obtain the action of C as an element of $\mathfrak{lor}(3,1)$ and so $\mathrm{d}\,\widetilde{\mathrm{Ad}}(C)$. For $X \in \text{Sk-Herm}_2(\mathbb{C})$ we have

$$\widetilde{\mathrm{Ad}}_{\gamma(t)}(X) = \exp(tC)X\exp(tC)^* = \exp(tC)X\exp(tC^*),$$

hence

$$\frac{\mathrm{d}}{\mathrm{d}t}\widetilde{\mathrm{Ad}}_{\gamma(t)}(X)|t = 0 = CX + XC^*.$$

So for the $\mathbb{R}$-basis H, iH, E, iE, F, iF of $\mathfrak{sl}_2(\mathbb{C})$, we have

$$H(x_1V_1 + x_2V_2 + x_3V_3 + x_4V_4) + (x_1V_1 + x_2V_2 + x_3V_3 + x_4V_4)H^* = x_2V_3 - x_3V_2,$$

so

$$\mathrm{d}\,\widetilde{\mathrm{Ad}}(H) = \begin{bmatrix} 0 & 0 & 0 & 0 \\ 0 & 0 & -1 & 0 \\ 0 & 1 & 0 & 0 \\ 0 & 0 & 0 & 0 \end{bmatrix}.$$

Here is the complete list written in terms of the matrices P_{rs} which we know form an $\mathbb{R}$-basis of Lor(3, 1):

$$\mathrm{d}\,\widetilde{\mathrm{Ad}}(H) = \begin{bmatrix} 0 & 0 & 0 & 0 \\ 0 & 0 & -1 & 0 \\ 0 & 1 & 0 & 0 \\ 0 & 0 & 0 & 0 \end{bmatrix} = P_{23}, \quad \mathrm{d}\,\widetilde{\mathrm{Ad}}(iH) = \begin{bmatrix} 0 & 0 & 0 & -1 \\ 0 & 0 & 0 & 0 \\ 0 & 0 & 0 & 0 \\ -1 & 0 & 0 & 0 \end{bmatrix} = -P_{14},$$

$$\mathrm{d}\,\widetilde{\mathrm{Ad}}(E) = \begin{bmatrix} 0 & 0 & 1 & 0 \\ 0 & 0 & 0 & 0 \\ -1 & 0 & 0 & 0 \\ 0 & 0 & 0 & 0 \end{bmatrix} = -P_{13}, \quad \mathrm{d}\,\widetilde{\mathrm{Ad}}(iE) = \begin{bmatrix} 0 & 0 & 0 & 0 \\ 0 & 0 & 0 & -1 \\ 0 & 0 & 0 & 0 \\ 0 & -1 & 0 & 0 \end{bmatrix} = -P_{24},$$

$$\mathrm{d}\,\widetilde{\mathrm{Ad}}(F) = \begin{bmatrix} 0 & -1 & 0 & 0 \\ 1 & 0 & 0 & 0 \\ 0 & 0 & 0 & 0 \\ 0 & 0 & 0 & 0 \end{bmatrix} = P_{12}, \quad \mathrm{d}\,\widetilde{\mathrm{Ad}}(iF) = \begin{bmatrix} 0 & 0 & 0 & 0 \\ 0 & 0 & 0 & 0 \\ 0 & 0 & 0 & -1 \\ 0 & 0 & -1 & 0 \end{bmatrix} = -P_{34}.$$

This shows that $\mathrm{d}\,\widetilde{\mathrm{Ad}}(C)$ maps a basis for Sk-Herm$_2(\mathbb{C})$ to one for $\mathfrak{lor}(3,1)$, thus it is an isomorphism of Lie algebras.

(ii) The surjectivity of $\widetilde{\mathrm{Ad}}$ follows immediately from the surjectivity of the derivative in (i) together with Theorem 6.11. □

EXERCISES

6.1. a) For $t \in \mathbb{R}$, show that the matrix

$$A_t = \begin{bmatrix} \cosh t & 0 & 0 & -\sinh t \\ 0 & \cos t & -\sin t & 0 \\ 0 & \sin t & \cos t & 0 \\ -\sinh t & 0 & 0 & \cosh t \end{bmatrix}$$

is in Lor(3, 1).

b) Express A_t in one of the forms predicted by Theorem 6.10.

c) Express A_t in the form $\exp(tU)$ for some $U \in \mathfrak{lor}(3,1)$.

d) Find a differential equation satisfied by the function

$$\alpha\colon \mathbb{R} \longrightarrow \mathrm{Lor}(3,1); \quad \alpha(t) = A_t.$$

6.2. In this question, we freely use the notation of Section 6.3.
a) Consider the differential equation

$$\alpha'(t) = (P_{24} + P_{34})\alpha(t),$$

where $\alpha\colon \mathbb{R} \longrightarrow \mathrm{GL}_4(\mathbb{R})$. Solve this to obtain a one-parameter subgroup α in $\mathrm{GL}_4(\mathbb{R})$ and show that $\alpha(t) \in \mathrm{Lor}(3,1)$ for all $t \in \mathbb{R}$.
b) Consider the differential equation

$$\widetilde{\alpha}'(t) = -i(E + F)\widetilde{\alpha}(t),$$

where $\widetilde{\alpha}\colon \mathbb{R} \longrightarrow \mathrm{GL}_2(\mathbb{C})$. Solve this to obtain a one-parameter subgroup in $\mathrm{GL}_2(\mathbb{C})$ and show that $\widetilde{\alpha}(t) \in \mathrm{SL}_2(\mathbb{C})$ for all $t \in \mathbb{R}$.
c) Explain the relationship between these two curves.

6.3. ⚠⚠ For $n \geqslant 1$, let $\mathbb{R}^{n-1,1}$ be $\mathbb{R}^n$ equipped with the Lorentz inner product $\langle\ ,\ \rangle$. Define a *Lorentz Clifford algebra* $\mathrm{Cl}_{n-1,1}$ in a similar fashion to the way Cl_n was defined in Section 5.1, except that the algebra generators $e_1, \ldots, e_{n-1}, e_n$ satisfy the relations

$$\begin{cases} e_s e_r = -e_s e_r & \text{if } s \neq r, \\ e_t^2 = -1 & \text{if } 1 \leqslant t \leqslant n-1, \\ e_n^2 = 1. \end{cases}$$

Let $j\colon \mathbb{R}^{n-1,1} \longrightarrow \mathrm{Cl}_{n-1,1}$ be the $\mathbb{R}$-linear transformation embedding $\mathbb{R}^{n-1,1}$ into $\mathrm{Cl}_{n-1,1}$.
a) Show that there is an analogue for $\mathrm{Cl}_{n-1,1}$ of the Universal Property of Theorem 5.4.
b) For $n \geqslant 2$, define $f_{n-1,1}\colon \mathbb{R}^{n-1,1} \longrightarrow \mathrm{M}_2(\mathrm{Cl}_{n-2})$ by

$$f_{n-1,1}\Big(\sum_{1\leqslant r\leqslant n} a_r e_r\Big) = \begin{bmatrix} \displaystyle\sum_{1\leqslant r\leqslant n-2} a_r e_r + a_n & -a_{n-1} \\ a_{n-1} & \displaystyle -\sum_{1\leqslant r\leqslant n-2} a_r e_r - a_n \end{bmatrix}.$$

Using the Universal Property of (a), show that $f_{n-1,1}$ induces an isomorphism of $\mathbb{R}$-algebras

$$F_{n-1,1}\colon \mathrm{Cl}_{n-1,1} \longrightarrow \mathrm{M}_2(\mathrm{Cl}_{n-2})$$

for which $F_{n-1,1} \circ j_{n-1,1} = f_{n-1,1}$.
c) Discuss the analogue $\mathrm{Spin}(n-1,1) \leqslant \mathrm{Cl}_{n-1,1}^{\times}$ of the spinor group $\mathrm{Spin}(n) \leqslant \mathrm{Cl}_n^{\times}$. In particular, deduce that there is a connected subgroup $\mathrm{Spin}^0(n-1,1) \leqslant \mathrm{Spin}(n-1,1)$ and a double covering homomorphism

$$\mathrm{Spin}^0(n-1,1) \longrightarrow \mathrm{Lor}(n-1,1).$$

6.4. ⚠⚠ a) Using the previous question in the case where $n = 4$, deduce that there is an isomorphism of $\mathbb{R}$-algebras

$$\mathrm{Cl}_{3,1} \longrightarrow \mathrm{M}_2(\mathbb{H}).$$

Using the regular representation $\mathrm{M}_2(\mathbb{H}) \longrightarrow \mathrm{M}_4(\mathbb{C})$, this provides an action of $\mathrm{Cl}_{3,1}$ by $\mathbb{C}$-linear transformations on the $\mathbb{C}$-vector space $\mathbb{C}^4$. Identifying $\mathrm{SL}_2(\mathbb{C})$ with $\mathrm{Spin}^0(3,1)$, this gives a continuous $\mathbb{C}$-linear action of $\mathrm{SL}_2(\mathbb{C})$ on $\mathbb{C}^4$.
b) Let $\varphi\colon \mathbb{R}^{3,1} \longrightarrow \mathbb{C}^4$ be a smooth function with

$$\varphi = (\varphi_1, \varphi_2, \varphi_3, \varphi_4).$$

Show that the *D'Alembertian operator* $\Box^2$ given by

$$\Box^2\varphi = \frac{\partial^2\varphi_1}{\partial {x_1}^2} + \frac{\partial^2\varphi_2}{\partial {x_2}^2} + \frac{\partial^2\varphi_3}{\partial {x_3}^2} - \frac{\partial^2\varphi_4}{\partial t^2}$$

can be expressed as

$$\Box^2 = \left(e_1\frac{\partial}{\partial x_1} + e_2\frac{\partial}{\partial x_2} + e_3\frac{\partial}{\partial x_3} + e_4\frac{\partial}{\partial t}\right)^2,$$

where the terms inside the brackets are to be thought of as differential operators with coefficients in $\mathrm{Cl}_{3,1}$ which act on $\mathbb{C}^4$ via the matrix representation defined in (a).
[This was originally observed by the physicist Dirac, who used it to factorise a relativistic version of the Schrödinger equation in order to study the electron. The connection with Clifford algebras seems to have been observed by Atiyah, Bott & Shapiro in their work on Index Theory [3].]

Part II

Matrix Groups as Lie Groups

7
Lie Groups

In this chapter we will introduce some of the basic ideas of *smooth manifolds* and then define *Lie groups*. Full details can be found in books on Differential Geometry such as [6, 8, 29] while [7, 23] contain briefer introductions. One of our main aims is to prove that every matrix subgroup of $\mathrm{GL}_n(\mathbb{R})$ is a *Lie subgroup* and we follow the proof of this result described in Howe [12]. We will also show that not every Lie group is a matrix group by exhibiting the simplest counterexample. In fact every *compact* Lie group is a matrix group, but the proof of this requires the theory of *Haar measure* on a Lie group; we will discuss compact Lie groups in Chapter 10.

7.1 Smooth Manifolds

The concept of a *smooth manifold* is fundamental in much of Mathematics and Physics and provides a natural setting for many geometric concepts, suitable for calculus and analysis involving global topological aspects.

Definition 7.1

A continuous map $g\colon V_1 \longrightarrow V_2$ where each $V_k \subseteq \mathbb{R}^{m_k}$ is open is *smooth* if it is infinitely differentiable. A smooth bijection g is a *diffeomorphism* if its inverse $g^{-1}\colon V_2 \longrightarrow V_1$ is also smooth.

We will require the following topological notion.

Definition 7.2

A topological space X is *separable* if it has a countable basis, *i.e.*, a basis of the form $\{U_j\}_{j=1}^{\infty} = \{U_1, U_2, U_3, \ldots\}$.

Every compact topological space is separable, as is any subspace of $\mathbb{R}^n$ or $\mathbb{C}^n$ for $n \geqslant 1$, so all matrix groups are separable. A *quotient space* (in the sense to be introduced in Section 8.1) of a separable space is also separable.

From now on, let M be a separable Hausdorff topological space.

Definition 7.3

If $U \subseteq M$ and $V \subseteq \mathbb{R}^n$ are open subsets, a homeomorphism $f\colon U \longrightarrow V$ is called an *n-chart* for U.

If $\mathfrak{U} = \{U_\alpha : \alpha \in A\}$ is an open covering of M and $\mathfrak{F} = \{f_\alpha\colon U_\alpha \longrightarrow V_\alpha\}$ is a collection of n-charts, then $\mathfrak{F}$ is called an *atlas* for M if, whenever $U_\alpha \cap U_\beta \neq \varnothing$,

$$f_\beta \circ f_\alpha^{-1}\colon f_\alpha(U_\alpha \cap U_\beta) \longrightarrow f_\beta(U_\alpha \cap U_\beta)$$

is a diffeomorphism.

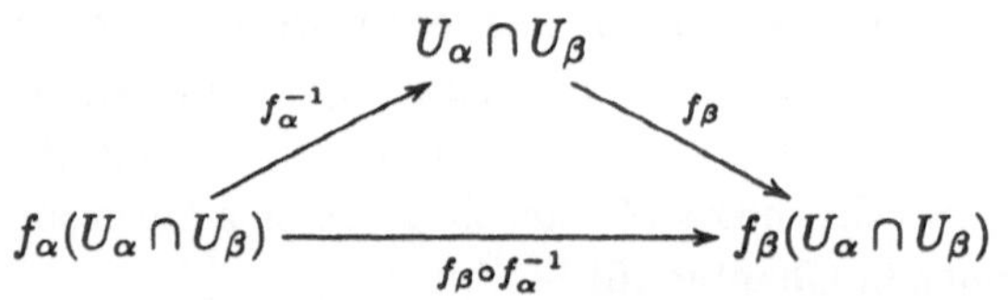

(7.1)

Sometimes we will denote an atlas by $(M, \mathfrak{U}, \mathfrak{F})$ and refer to it as a *smooth manifold of dimension n* or *smooth n-manifold.*

Definition 7.4

Let $(M, \mathfrak{U}, \mathfrak{F})$ and $(M', \mathfrak{U}', \mathfrak{F}')$ be atlases on topological spaces M and M'. A *smooth map* $h\colon (M, \mathfrak{U}, \mathfrak{F}) \longrightarrow (M', \mathfrak{U}', \mathfrak{F}')$ is a continuous map $h\colon M \longrightarrow M'$ such that for each pair α, α' with $h(U_\alpha) \cap U'_{\alpha'} \neq \varnothing$, the composite

$$f'_{\alpha'} \circ h \circ f_\alpha^{-1}\colon f_\alpha(h^{-1}U'_{\alpha'}) \longrightarrow V'_{\alpha'}$$

is smooth.

$$\begin{array}{ccc} f_\alpha(h^{-1}U'_{\alpha'}) & \xrightarrow{f'_{\alpha'} o h o f_\alpha^{-1}} & V'_{\alpha'} \\ {\scriptstyle f_\alpha^{-1}}\big\downarrow & & \big\downarrow{\scriptstyle f'_{\alpha'}{}^{-1}} \\ h^{-1}U'_{\alpha'} & \xrightarrow[h]{} & h(U_\alpha) \cap U'_{\alpha'} \end{array} \tag{7.2}$$

7.2 Tangent Spaces and Derivatives

Let $(M, \mathfrak{U}, \mathfrak{F})$ be a smooth n-manifold and $p \in M$. Let $\gamma\colon (a,b) \longrightarrow M$ be a continuous curve with $a < 0 < b$.

Definition 7.5

γ is *differentiable* at $t \in (a,b)$ if for every chart $f\colon U \longrightarrow V$ with $\gamma(t) \in U$, the curve $f \circ \gamma\colon (a,b) \longrightarrow V$ is differentiable at $t \in (a,b)$, *i.e.*, $(f \circ \gamma)'(t)$ exists. γ is *smooth* at $t \in (a,b)$ if all the derivatives of $f \circ \gamma$ exist at t.

The curve γ is *differentiable* if it is differentiable at all points in (a,b). Similarly, γ is *smooth* if it is smooth at all points in (a,b).

From now on we will often write $fg = f \circ g$ for the composition of functions f and g when no confusion seems likely to result.

Lemma 7.6

Let $f_0\colon U_0 \longrightarrow V_0$ be a chart with $\gamma(t) \in U_0$ and suppose that

$$f_0 \circ \gamma\colon (a,b) \cap \gamma^{-1} f_0^{-1} V_0 \longrightarrow V_0$$

is differentiable (respectively smooth) at t. Then for any chart $f\colon U \longrightarrow V$ with $\gamma(t) \in U$,

$$f \circ \gamma\colon (a,b) \cap \gamma^{-1} f^{-1} V \longrightarrow V$$

is differentiable (respectively smooth) at t.

Proof

This follows using the ideas of Definition 7.3. The composition $f \circ \gamma$ is defined on a subinterval of (a,b) containing t and is smooth. There is the usual *chain rule* or *function of a function rule* for the derivative

$$(f\gamma)'(t) = \mathrm{Jac}_{ff_0^{-1}}(f_0\gamma(t))(f_0\gamma)'(t). \tag{7.3}$$

Here, for a differentiable function

$$h\colon W_1 \longrightarrow W_2; \quad h(\mathbf{x}) = \begin{bmatrix} h_1(\mathbf{x}) \\ \vdots \\ h_{m_2}(\mathbf{x}) \end{bmatrix}$$

where $W_1 \subseteq \mathbb{R}^{m_1}$ and $W_2 \subseteq \mathbb{R}^{m_2}$ are open subsets, $\mathbf{x} \in W_1$ and

$$\mathrm{Jac}_h(\mathbf{x}) = \left[\frac{\partial h_i}{\partial x_j}(\mathbf{x})\right] \in \mathrm{M}_{m_2,m_1}(\mathbb{R})$$

is the *Jacobian matrix* of h at $\mathbf{x}$. □

If $\gamma(0) = p$ and γ is differentiable at 0, then for any (and hence every) chart $f_0\colon U_0 \longrightarrow V_0$ with $\gamma(0) \in U_0$, there is a derivative vector $\mathbf{v}_0 = (f_0\gamma)'(0) \in \mathbb{R}^n$. In passing to another chart $f\colon U \longrightarrow V$ with $\gamma(0) \in U$ by Equation (7.3) we have

$$(f\gamma)'(0) = \mathrm{Jac}_{ff_0^{-1}}(f_0\gamma(0))(f_0\gamma)'(0).$$

In order to define the notion of the *tangent space* $\mathrm{T}_p M$ to the manifold M at p, we consider all pairs of the form

$$((f\gamma)'(0), f\colon U \longrightarrow V)$$

where $\gamma(0) = p \in U$, and then impose an equivalence relation $\sim$ under which

$$((f_1\gamma)'(0), f_1\colon U_1 \longrightarrow V_1) \sim ((f_2\gamma)'(0), f_2\colon U_2 \longrightarrow V_2).$$

Since

$$(f_2\gamma)'(0) = \mathrm{Jac}_{f_2f_1^{-1}}(f_1\gamma(0))(f_1\gamma)'(0),$$

we can also write this as

$$(\mathbf{v}, f_1\colon U_1 \longrightarrow V_1) \sim (\mathrm{Jac}_{f_2f_1^{-1}}(f_1(p))\mathbf{v}, f_2\colon U_2 \longrightarrow V_2),$$

whenever there is a curve α in M for which

$$\gamma(0) = p, \quad (f_1\gamma)'(0) = \mathbf{v}.$$

The set of equivalence classes is $\mathrm{T}_p M$ and we will sometimes denote the equivalence class of $(\mathbf{v}, f\colon U \longrightarrow V)$ by $[\mathbf{v}, f\colon U \longrightarrow V]$.

Proposition 7.7

For $p \in M$, $\mathrm{T}_p M$ is an $\mathbb{R}$-vector space of dimension n.

Proof

For any chart $f\colon U \longrightarrow V$ with $p \in U$, we can identify the elements of $\mathrm{T}_p M$ with objects of the form $(\mathbf{v}, f\colon U \longrightarrow V)$. Every vector $\mathbf{v} \in \mathbb{R}^n$ arises as the derivative of a curve $\overline{\gamma}\colon (-\varepsilon, \varepsilon) \longrightarrow V$ for which $\overline{\gamma}(0) = f(p)$. For example, for small enough ε, we could take

$$\overline{\gamma}(t) = f(p) + t\mathbf{v}.$$

There is an associated curve in M,

$$\gamma\colon (-\varepsilon, \varepsilon) \longrightarrow M; \quad \gamma(t) = f^{-1}\overline{\gamma}(t),$$

for which $\gamma(0) = p$. So using such a chart we can identify $\mathrm{T}_p M$ with $\mathbb{R}^n$ by

$$[\mathbf{v}, f\colon U \longrightarrow V] \longleftrightarrow \mathbf{v}.$$

The same argument as used to prove Proposition 3.16 shows that $\mathrm{T}_p M$ is a vector space and that the above correspondence is a linear isomorphism. □

Let $h\colon (M, \mathcal{U}, \mathcal{F}) \longrightarrow (M', \mathcal{U}', \mathcal{F}')$ be a smooth map between manifolds of dimensions n, n'. We will use the notation of Definition 7.4. For $p \in M$, consider a pair of charts as in Diagram (7.2) with $p \in U_\alpha$ and $h(p) \in U'_{\alpha'}$. Since $h_{\alpha',\alpha} = f'_{\alpha'} \circ h \circ f_\alpha^{-1}$ is differentiable, the Jacobian matrix $\mathrm{Jac}_{h_{\alpha',\alpha}}(f_\alpha(p))$ has an associated $\mathbb{R}$-linear transformation

$$\mathrm{d}\, h_{\alpha',\alpha}\colon \mathbb{R}^n \longrightarrow \mathbb{R}^{n'}; \quad \mathrm{d}\, h_{\alpha',\alpha}(\mathbf{x}) = \mathrm{Jac}_{h_{\alpha',\alpha}}(f_\alpha(p))\mathbf{x}.$$

It can be verified that this passes to equivalence classes to give a well-defined $\mathbb{R}$-linear transformation

$$\mathrm{d}\, h_p\colon \mathrm{T}_p M \longrightarrow \mathrm{T}_{h(p)} M'.$$

The following result summarises the properties of the derivative and should be compared with Proposition 3.20.

Proposition 7.8

Let $h\colon (M, \mathcal{U}, \mathcal{F}) \longrightarrow (M', \mathcal{U}', \mathcal{F}')$ and $g\colon (M', \mathcal{U}', \mathcal{F}') \longrightarrow (M'', \mathcal{U}'', \mathcal{F}'')$ be smooth maps between manifolds M, M', M'' of dimensions n, n', n''. Then for each $p \in M$ we have

i) $\mathrm{d}\, g_{h(p)} \circ \mathrm{d}\, h_p = \mathrm{d}(g \circ h)_p$;

ii) $\mathrm{d}\,\mathrm{Id}_p = \mathrm{Id}_{\mathrm{T}_p M}$, where $\mathrm{Id}\colon M \longrightarrow M$ is the identity map.

Definition 7.9

Let $(M, \mathcal{U}, \mathcal{F})$ be a manifold of dimension n. A subset $N \subseteq M$ is a *submanifold of dimension* k if for every $p \in N$ there is an open neighbourhood $U \subseteq M$ of p and an n-chart $f: U \longrightarrow V$ such that

$$p \in f^{-1}(V \cap \mathbb{R}^k) = N \cap U.$$

For such an N we can form k-charts of the form

$$f_0: N \cap U \longrightarrow V \cap \mathbb{R}^k; \quad f_0(x) = f(x).$$

We will denote this manifold by $(N, \mathcal{U}_N, \mathcal{F}_N)$. The following result is immediate.

Proposition 7.10

For a submanifold $N \subseteq M$ of dimension k, the inclusion function $\mathrm{incl}: N \longrightarrow M$ is smooth and for every $p \in N$, $\mathrm{d}\,\mathrm{incl}_p: \mathrm{T}_p N \longrightarrow \mathrm{T}_p M$ is an injection.

The next result allows us to recognise submanifolds as inverse images of points under smooth mappings.

Theorem 7.11 (Implicit Function Theorem for manifolds)

Let $h: (M, \mathcal{U}, \mathcal{F}) \longrightarrow (M', \mathcal{U}', \mathcal{F}')$ be a smooth map between manifolds of dimensions n, n'. Suppose that for some $q \in M'$, $\mathrm{d}\,h_p: \mathrm{T}_p M \longrightarrow \mathrm{T}_{h(p)} M'$ is surjective for every $p \in N = h^{-1}q$. Then $N \subseteq M$ is submanifold of dimension $n - n'$ and the tangent space at $p \in N$ is given by $\mathrm{T}_p N = \ker \mathrm{d}\,h_p$.

Proof

This follows from the Implicit Function Theorem of Calculus. □

Another important application of the Implicit Function Theorem is to the following version of the Inverse Function Theorem.

Theorem 7.12 (Inverse Function Theorem for manifolds)

Let $h: (M, \mathcal{U}, \mathcal{F}) \longrightarrow (M', \mathcal{U}', \mathcal{F}')$ be a smooth map between manifolds of dimensions n, n'. Suppose that for some $p \in M$, $\mathrm{d}\,h_p: \mathrm{T}_p M \longrightarrow \mathrm{T}_{h(p)} M'$ is an isomorphism. Then there is an open neighbourhood $U \subseteq M$ of p and an open neighbourhood $V \subseteq M'$ of $h(p)$ such that $hU = V$ and the restriction of h to the map $h_1: U \longrightarrow V$ is a diffeomorphism. In particular, $n = n'$.

When this occurs we say that h is *locally a diffeomorphism at* p.

Example 7.13

Consider the exponential function $\exp\colon \mathrm{M}_n(\mathbb{R}) \longrightarrow \mathrm{GL}_n(\mathbb{R})$. Then by Proposition 2.5, $\mathrm{d}\exp_O(X) = X$. Hence exp is locally a diffeomorphism at O.

7.3 Lie Groups

The following should be compared with Definition 1.15.

Definition 7.14

Let G be a smooth manifold which is also a topological group with multiplication map $\mathrm{mult}\colon G \times G \longrightarrow G$ and inverse map $\mathrm{inv}\colon G \longrightarrow G$ and view $G \times G$ as the product manifold. Then G is a *Lie group* if mult, inv are smooth maps.

Definition 7.15

Let G be a Lie group. A closed subgroup $H \leqslant G$ that is also a submanifold is called a *Lie subgroup* of G. It is then automatic that the restrictions to H of the multiplication and inverse maps on G are smooth, hence H is also a Lie group.

For a Lie group G, and an element $g \in G$ there is a tangent space $\mathrm{T}_g G$ and when G is a matrix group this agrees with the tangent space defined in Chapter 3. We will use the notation $\mathfrak{g} = \mathrm{T}_1 G$ for the tangent space at the identity of G. A smooth homomorphism of Lie groups $G \longrightarrow H$ has the properties of a Lie homomorphism in the sense of Definition 3.19.

For a Lie group G, let $g \in G$. The following three functions are of particular importance:

$$\begin{aligned}
&\mathrm{L}_g\colon G \longrightarrow G; \quad \mathrm{L}_g(x) = gx, && \text{(Left multiplication)}\\
&\mathrm{R}_g\colon G \longrightarrow G; \quad \mathrm{R}_g(x) = xg, && \text{(Right multiplication)}\\
&\chi_g\colon G \longrightarrow G; \quad \chi_g(x) = gxg^{-1}. && \text{(Conjugation)}
\end{aligned}$$

Proposition 7.16

For each $g \in G$, the maps L_g, R_g, χ_g are diffeomorphisms with inverses

$$\mathrm{L}_g^{-1} = \mathrm{L}_{g^{-1}}, \quad \mathrm{R}_g^{-1} = \mathrm{R}_{g^{-1}}, \quad \chi_g^{-1} = \chi_{g^{-1}}.$$

Proof

Charts for $G \times G$ have the form

$$\varphi_1 \times \varphi_2 \colon U_1 \times U_2 \longrightarrow V_1 \times V_2,$$

where $\varphi_k \colon U_k \longrightarrow V_k$ are charts for G. Now suppose that $\mu U_1 \times U_2 \subseteq W \subseteq G$ where there is a chart $\theta \colon W \longrightarrow Z$. By assumption, the composition

$$\theta \circ \mu \circ (\varphi_1 \times \varphi_2)^{-1} = \theta \circ \mu \circ (\varphi_1^{-1} \times \varphi_2^{-1}) \colon V_1 \times V_2 \longrightarrow Z$$

is smooth. Then $\mathrm{L}_g(x) = \mu(g, x)$, so if $g \in U_1$ and $x \in U_2$, we have

$$\mathrm{L}_g(x) = \theta^{-1} \circ (\theta \circ \mathrm{L}_g \circ \varphi_2^{-1}) \circ \varphi_2(x).$$

But then it is clear that

$$\theta \circ \varphi_2^{-1} \colon V_2 \longrightarrow Z$$

is smooth since it is obtained from $\theta \circ \mu \circ (\varphi_1 \times \varphi_2)^{-1}$, but treating the first variable as a constant. A similar argument deals with R_g.

For χ_g, notice that

$$\chi_g = \mathrm{L}_g \circ \mathrm{R}_g = \mathrm{R}_g \circ \mathrm{L}_g,$$

and a composite of smooth maps is smooth. □

The derivatives of these maps at the identity $1 \in G$ are worth studying. Since L_g and R_g are diffeomorphisms with inverses $\mathrm{L}_{g^{-1}}$ and $\mathrm{R}_{g^{-1}}$, the derivatives

$$\mathrm{d}(\mathrm{L}_g)_1,\ \mathrm{d}(\mathrm{R}_g)_1 \colon \mathfrak{g} = \mathrm{T}_1 G \longrightarrow \mathrm{T}_g G$$

are $\mathbb{R}$-linear isomorphisms. We can use this to identify every tangent space of G with $\mathfrak{g}$. The conjugation map χ_g fixes 1, so it induces an $\mathbb{R}$-linear isomorphism

$$\mathrm{Ad}_g = \mathrm{d}(\chi_g)_1 \colon \mathfrak{g} \longrightarrow \mathfrak{g}.$$

This is the *adjoint action* of $g \in G$ on $\mathfrak{g}$. For G a matrix group this agrees with the adjoint action defined in Chapter 3. There is also a natural Lie bracket $[\ ,\ ]$ defined on $\mathfrak{g}$, making it into an $\mathbb{R}$-Lie algebra. The construction follows that for matrix groups. The following Lie group analogue of Theorem 3.21 is true.

Theorem 7.17

Let G, H be Lie groups and $\varphi\colon G \longrightarrow H$ be a Lie homomorphism. Then the derivative is a homomorphism of Lie algebras. In particular, if $G \leqslant H$ is a Lie subgroup, the inclusion map $\mathrm{incl}\colon G \longrightarrow H$ induces an injection of Lie algebras $\mathrm{d}\,\mathrm{incl}\colon \mathfrak{g} \longrightarrow \mathfrak{h}$.

7.4 Some Examples of Lie Groups

Example 7.18

When $\Bbbk = \mathbb{R}$ or $\mathbb{C}$, $\mathrm{GL}_n(\Bbbk)$ is a Lie group.

Proof

This follows from Proposition 1.14(i) which implies that $\mathrm{GL}_n(\Bbbk) \subseteq \mathrm{M}_n(\Bbbk)$ is an open subset, where as usual we identify $\mathrm{M}_n(\Bbbk)$ with $\Bbbk^{n^2}$. For charts we take the open sets $U \subseteq \mathrm{GL}_n(\Bbbk)$ and the identity function $\mathrm{Id}\colon U \longrightarrow U$. The tangent space at each point $A \in \mathrm{GL}_n(\Bbbk)$ is just $\mathrm{M}_n(\Bbbk)$. So the notions of tangent space and dimension of Sections 7.1, 7.2 and Chapter 3 agree here. The multiplication and inverse maps are obviously smooth as they are defined by polynomial and rational functions between open subsets of $\mathrm{M}_n(\Bbbk)$. □

Example 7.19

For $\Bbbk = \mathbb{R}$ or $\mathbb{C}$, $\mathrm{SL}_n(\Bbbk) \leqslant \mathrm{GL}_n(\Bbbk)$ is a Lie subgroup.

Proof

Following Proposition 1.14(ii), we have

$$\mathrm{SL}_n(\Bbbk) = \det^{-1} 1 \leqslant \mathrm{GL}_n(\Bbbk),$$

where $\det\colon \mathrm{GL}_n(\Bbbk) \longrightarrow \Bbbk$ is smooth. $\Bbbk$ is a smooth manifold of dimension $\dim_{\mathbb{R}} \Bbbk$ with tangent space the $\Bbbk$-vector space $\mathrm{T}_r \Bbbk = \Bbbk$ at each $r \in \Bbbk$. In order to apply Theorem 7.11, first we must show that the derivative $\mathrm{d}\det_A\colon \mathrm{M}_n(\Bbbk) \longrightarrow \Bbbk$ is surjective for every $A \in \mathrm{GL}_n(\Bbbk)$. To do this, consider a smooth curve $\alpha\colon (-\varepsilon, \varepsilon) \longrightarrow \mathrm{GL}_n(\Bbbk)$ with $\alpha(0) = A$. The derivative $\mathrm{d}\det_A$ applied to $\alpha'(0)$ can be found using the formula

$$\mathrm{d}\det_A(\alpha'(0)) = \frac{\mathrm{d}\det\alpha(t)}{\mathrm{d}t}\Big|_{t=0}.$$

The modified curve

$$\alpha_0 \colon (-\varepsilon, \varepsilon) \longrightarrow \mathrm{GL}_n(\Bbbk); \quad \alpha_0(t) = A^{-1}\alpha(t)$$

satisfies $\alpha_0(0) = I$ and Lemma 3.22 implies that

$$\mathrm{d}\det{}_I(\alpha_0'(0)) = \frac{\mathrm{d}\det \alpha_0(t)}{\mathrm{d}t}\Big|_{t=0} = \mathrm{tr}\,\alpha_0'(0).$$

Hence

$$\mathrm{d}\det{}_A(\alpha'(0)) = \frac{\mathrm{d}\det(A\alpha_0(t))}{\mathrm{d}t}\Big|_{t=0} = \det A \frac{\mathrm{d}\det(\alpha_0(t))}{\mathrm{d}t}\Big|_{t=0} = \det A\, \mathrm{tr}\,\alpha_0'(0).$$

So $\mathrm{d}\det_A$ is the $\Bbbk$-linear transformation

$$\mathrm{d}\det{}_A \colon \mathrm{M}_n(\Bbbk) \longrightarrow \Bbbk; \quad \mathrm{d}\det{}_A(X) = (\det A)\,\mathrm{tr}(A^{-1}X).$$

The kernel of $\mathrm{d}\det_A$ is $\ker \mathrm{d}\det_A = A\mathfrak{sl}_n(\Bbbk)$ and it is also surjective since tr is. In particular this is true for $A \in \mathrm{SL}_n(\Bbbk)$. By Theorem 7.11, $\mathrm{SL}_n(\Bbbk) \longrightarrow \mathrm{GL}_n(\Bbbk)$ is a submanifold and so is a Lie subgroup. Again the two notions of tangent space and dimension agree. □

There are some useful general principles behind this last proof. Although we state the following two results for matrix groups, it is worth noting that they also apply when $\mathrm{GL}_n(\mathbb{R})$ is replaced by an arbitrary Lie group.

Proposition 7.20 (Left Translation Trick)

Let $F \colon \mathrm{GL}_n(\mathbb{R}) \longrightarrow M$ be a smooth function into a smooth manifold M and suppose that $B \in \mathrm{GL}_n(\mathbb{R})$ satisfies $F(BC) = F(C)$ for all $C \in \mathrm{GL}_n(\mathbb{R})$. Let $A \in \mathrm{GL}_n(\mathbb{R})$ with $\mathrm{d}F_A$ surjective. Then $\mathrm{d}F_{BA}$ is surjective.

Proof

Left multiplication by $B \in G$, $\mathrm{L}_B \colon \mathrm{GL}_n(R) \longrightarrow \mathrm{GL}_n(R)$, is a diffeomorphism and its derivative at $A \in \mathrm{GL}_n(R)$ is

$$\mathrm{d}(\mathrm{L}_B) \colon \mathrm{M}_n(R) \longrightarrow \mathrm{M}_n(R); \quad \mathrm{d}\,\mathrm{L}_B(X) = BX.$$

By assumption, $F \circ \mathrm{L}_B = F$ as a function on $\mathrm{GL}_n(R)$. Then

$$\begin{aligned}
\mathrm{d}F_{BA}(X) &= \mathrm{d}F_{BA}(B(B^{-1}X)) \\
&= \mathrm{d}F_{BA} \circ \mathrm{d}(\mathrm{L}_B)_A(B^{-1}X) \\
&= \mathrm{d}(F \circ \mathrm{L}_B)_A(B^{-1}X) \\
&= \mathrm{d}F_A(B^{-1}X).
\end{aligned}$$

Since left multiplication by B^{-1} on $\mathrm{M}_n(\mathbb{R})$ is surjective, the result follows. □

Proposition 7.21 (Identity Check Trick)

Let $G \leqslant \mathrm{GL}_n(\mathbb{R})$ be a matrix subgroup and M be a smooth manifold. Let $F\colon \mathrm{GL}_n(\mathbb{R}) \longrightarrow M$ be a smooth function with $F^{-1}q = G$ for some $q \in M$. Suppose that for every $B \in G$, $F(BC) = F(C)$ for all $C \in \mathrm{GL}_n(\mathbb{R})$. If $\mathrm{d}\,F_I$ is surjective then $\mathrm{d}\,F_A$ is surjective for all $A \in G$ and $\ker \mathrm{d}\,F_A = A\mathfrak{g}$.

Example 7.22

$\mathrm{O}(n)$ is a Lie subgroup of $\mathrm{GL}_n(\mathbb{R})$.

Proof

Recall from Chapter 3 that we can specify $\mathrm{O}(n) \subseteq \mathrm{GL}_n(\mathbb{R})$ as the solution set of a family of polynomial equations in n^2 variables arising from the matrix equation $A^T A = I$. In fact, the following equations in the entries of the matrix $A = [a_{ij}]$ are sufficient:

$$\sum_{k=1}^{n} a_{kr}^2 - 1 = 0 \quad (1 \leqslant r \leqslant n), \qquad \sum_{k=1}^{n} a_{kr}a_{ks} = 0 \quad (1 \leqslant r < s \leqslant n).$$

Notice that there are

$$n + \binom{n}{2} = \binom{n+1}{2}$$

of these equations. We can combine the left-hand sides of these in some order to give a function $F\colon \mathrm{GL}_n(\mathbb{R}) \longrightarrow \mathbb{R}^{\binom{n+1}{2}}$, for example

$$F([a_{ij}]) = \begin{bmatrix} \sum_{k=1}^{n} a_{k1}^2 - 1 \\ \vdots \\ \sum_{k=1}^{n} a_{kn}^2 - 1 \\ \sum_{k=1}^{n} a_{k1}a_{k2} \\ \vdots \\ \sum_{k=1}^{n} a_{k1}a_{kn} \\ \vdots \\ \sum_{k=1}^{n} a_{k(n-1)}a_{kn} \end{bmatrix}.$$

We need to investigate the derivative $\mathrm{d}\,F_A\colon \mathrm{M}_n(\mathbb{R}) \longrightarrow \mathbb{R}^{\binom{n+1}{2}}$.

By the Identity Check Trick 7.21, to show that $\mathrm{d}\,F_A$ is surjective for all $A \in \mathrm{O}(n)$, it is sufficient to check the case $A = I$. The Jacobian matrix of F at

$A = [a_{ij}] = I$ is the $\binom{n+1}{2} \times n^2$ matrix

$$\mathrm{d}\,F_I = \begin{bmatrix} 2 & 0 & 0 & 0 & \cdots & 0 & 0 \\ \vdots & & & \ddots & & \vdots & \vdots \\ 0 & 0 & 0 & 0 & \cdots & 0 & 2 \\ 0 & 1 & 1 & 0 & \cdots & 0 & 0 \\ \vdots & & & \ddots & & \vdots & \vdots \\ 0 & 1 & 0 & \cdots & 0 & 1 & 0 \\ \vdots & & & \ddots & & \vdots & \vdots \\ 0 & 0 & 0 & \cdots & 1 & 1 & 0 \end{bmatrix}$$

where in the top block of n rows, the rth row has a 2 corresponding to the variable a_{rr} and in the bottom block, each row has a 1 in each column corresponding to one of the pair a_{rs}, a_{sr} with $r < s$. The rank of this matrix is $n + \binom{n}{2} = \binom{n+1}{2}$, so $\mathrm{d}\,F_I$ is surjective. It is also true that

$$\ker \mathrm{d}\,F_I = \text{Sk-Sym}_n(\mathbb{R}) = \mathfrak{o}(n).$$

Hence $\mathrm{O}(n) \leqslant \mathrm{GL}_n(\mathbb{R})$ is a Lie subgroup and at each element, the tangent space and dimension agree with those obtained using the definitions of Chapter 3. □

This example is typical of what happens for any matrix group that is a Lie subgroup of $\mathrm{GL}_n(\mathbb{R})$. We summarise the situation in the following theorem, whose proof involves a careful comparison between the ideas introduced in Chapter 3 and the definitions involving manifolds.

Theorem 7.23

Let $G \leqslant \mathrm{GL}_n(\mathbb{R})$ be a matrix group which is also a submanifold, hence a Lie subgroup. Then the tangent space to G at I agrees with the Lie algebra $\mathfrak{g}$ and the dimension of the smooth manifold G is $\dim G$; more generally, $\mathrm{T}_A\, G = A\mathfrak{g}$.

In the following sections, our goal will be to prove an important result.

Theorem 7.24

Let $G \leqslant \mathrm{GL}_n(\mathbb{R})$ be a matrix subgroup. Then G is a Lie subgroup of $\mathrm{GL}_n(\mathbb{R})$.

In fact a more general result also holds but we will not give a proof.

Theorem 7.25

Let $G \leqslant H$ be a closed subgroup of a Lie group H. Then G is a Lie subgroup of H.

7.5 Some Useful Formulæ in Matrix Groups

Let $G \leqslant \mathrm{GL}_n(\mathbb{R})$ be a closed matrix subgroup. Applying Proposition 2.4, we may choose $r \in \mathbb{R}$ so that $0 < r \leqslant 1/2$ and if $A, B \in \mathrm{N}_{\mathrm{M}_n(\mathbb{R})}(O; r)$ then $\exp(A)\exp(B) \in \exp(\mathrm{N}_{\mathrm{M}_n(\mathbb{R})}(O; 1/2))$. Since exp is injective on $\mathrm{N}_{\mathrm{M}_n(\mathbb{R})}(O; r)$, there is a unique $C \in \mathrm{M}_n(\mathbb{R})$ for which

$$\exp(A)\exp(B) = \exp(C). \tag{7.4}$$

We also set

$$S = C - A - B - \frac{1}{2}[A, B] \in \mathrm{M}_n(\mathbb{R}). \tag{7.5}$$

For $X \in \mathrm{M}_n(\mathbb{R})$ we have

$$\exp(X) = I + X + R_1(X),$$

where the remainder term $R_1(X)$ is given by

$$R_1(X) = \sum_{2 \leqslant k} \frac{1}{k!} X^k.$$

Hence,

$$\|R_1(X)\| \leqslant \|X\|^2 \sum_{2 \leqslant k} \frac{1}{k!} \|X\|^{k-2},$$

and therefore if $\|X\| \leqslant 1$,

$$\|R_1(X)\| \leqslant \|X\|^2 \left(\sum_{2 \leqslant k} \frac{1}{k!} \right) = \|X\|^2 (e - 2) < \|X\|^2.$$

Since $\|C\| < \dfrac{1}{2}$ we obtain

$$\|R_1(C)\| < \|C\|^2. \tag{7.6}$$

Similar considerations lead to

$$\exp(C) = \exp(A)\exp(B) = I + A + B + R_1(A, B),$$

where

$$R_1(A,B) = \sum_{k\geqslant 2} \frac{1}{k!}\left(\sum_{r=0}^{k}\binom{k}{r}A^r B^{k-r}\right).$$

This gives

$$\begin{aligned}\|R_1(A,B)\| &\leqslant \sum_{k\geqslant 2}\frac{1}{k!}\left(\sum_{r=0}^{k}\binom{k}{r}\|A\|^r\|B\|^{k-r}\right)\\ &= \sum_{k\geqslant 2}\frac{(\|A\|+\|B\|)^k}{k!}\\ &= (\|A\|+\|B\|)^2\sum_{k\geqslant 2}\frac{(\|A\|+\|B\|)^{k-2}}{k!}\\ &\leqslant (\|A\|+\|B\|)^2\end{aligned}$$

since $\|A\|+\|B\|<1$.

Combining the two ways of writing $\exp(C)$ from above, we have

$$C = A + B + R_1(A,B) - R_1(C) \tag{7.7}$$

and so

$$\begin{aligned}\|C\| &\leqslant \|A\|+\|B\|+\|R_1(A,B)\|+\|R_1(C)\|\\ &< \|A\|+\|B\|+(\|A\|+\|B\|)^2+\|C\|^2\\ &\leqslant 2(\|A\|+\|B\|)+\frac{1}{2}\|C\|,\end{aligned}$$

since $\|A\|, \|B\|, \|C\| \leqslant \frac{1}{2}$. Finally this gives

$$\|C\| \leqslant 4(\|A\|+\|B\|).$$

Equation (7.7) also gives

$$\begin{aligned}\|C-A-B\| &\leqslant \|R_1(A,B)\|+\|R_1(C)\|\\ &\leqslant (\|A\|+\|B\|)^2+(4(\|A\|+\|B\|))^2,\end{aligned}$$

giving

$$\|C-A-B\| \leqslant 17(\|A\|+\|B\|)^2. \tag{7.8}$$

Now we will refine these estimates further. Write

$$\exp(C) = I + C + \frac{1}{2}C^2 + R_2(C)$$

where

$$R_2(C) = \sum_{k\geqslant 3} \frac{1}{k!} C^k$$

which satisfies the estimate

$$\|R_2(C)\| \leqslant \frac{1}{3}\|C\|^3$$

since $\|C\| \leqslant 1$. With the aid of Equation (7.5) we obtain

$$\begin{aligned} \exp(C) &= I + A + B + \frac{1}{2}[A,B] + S + \frac{1}{2}C^2 + R_2(C) \\ &= I + A + B + \frac{1}{2}[A,B] + \frac{1}{2}(A+B)^2 + T \\ &= I + A + B + \frac{1}{2}(A^2 + 2AB + B^2) + T, \end{aligned} \tag{7.9}$$

where

$$T = S + \frac{1}{2}(C^2 - (A+B)^2) + R_2(C). \tag{7.10}$$

Also,

$$\exp(A)\exp(B) = I + A + B + \frac{1}{2}(A^2 + 2AB + B^2) + R_2(A,B) \tag{7.11}$$

where

$$R_2(A,B) = \sum_{k\geqslant 3} \frac{1}{k!}\left(\sum_{r=0}^{k}\binom{k}{r} A^r B^{k-r}\right),$$

which satisfies

$$\|R_2(A,B)\| \leqslant \frac{1}{3}(\|A\| + \|B\|)^3$$

since $\|A\| + \|B\| \leqslant 1$.

Comparing Equations (7.9) and (7.11) and using (7.4) we see that

$$S = R_2(A,B) + \frac{1}{2}((A+B)^2 - C^2) - R_2(C).$$

Taking norms we have

$$\begin{aligned} \|S\| &\leqslant \|R_2(A,B)\| + \frac{1}{2}\|(A+B)(A+B-C) - (A+B-C)C\| + \|R_2(C)\| \\ &\leqslant \frac{1}{3}(\|A\| + \|B\|)^3 + \frac{1}{2}(\|A\| + \|B\| + \|C\|)\|A+B-C\| + \frac{1}{3}\|C\|^3 \\ &\leqslant \frac{1}{3}(\|A\| + \|B\|)^3 + \frac{5}{2}(\|A\| + \|B\|)\cdot 17(\|A\| + \|B\|)^2 + \frac{1}{3}(4\|A\| + \|B\|)^3 \\ &\leqslant 65(\|A\| + \|B\|)^3, \end{aligned}$$

which yields the estimate

$$\|S\| \leqslant 65(\|A\| + \|B\|)^3. \tag{7.12}$$

Theorem 7.26

For $U, V \in \mathrm{M}_n(\mathbb{R})$ we have the following formulæ.

Trotter Product Formula:

$$\exp(U+V) = \lim_{r\to\infty} \left(\exp((1/r)U)\exp((1/r)V)\right)^r .$$

Commutator Formula:

$$\exp([U,V]) = \lim_{r\to\infty} \left(\exp((1/r)U)\exp((1/r)V)\exp(-(1/r)U)\exp(-(1/r)V)\right)^{r^2} .$$

Proof

For large r we may take $A = (1/r)U$ and $B = (1/r)V$ and apply Equation (7.5) to give

$$\exp((1/r)U)\exp((1/r)V) = \exp(C_r)$$

with

$$\|C_r - (1/r)(U+V)\| \leqslant \frac{17(\|U\| + \|V\|)^2}{r^2}.$$

As $r \to \infty$,

$$\|rC_r - (U+V)\| = \frac{17(\|U\| + \|V\|)^2}{r} \to 0,$$

hence $rC_r \to (U+V)$. Since $\exp(rC_r) = \exp(C_r)^r$, the Trotter formula follows by continuity of exp.

We also have

$$C_r = \frac{1}{r}(U+V) + \frac{1}{2r^2}[U,V] + S_r$$

where

$$\|S_r\| \leqslant 65\frac{(\|U\| + \|V\|)^3}{r^3}.$$

Similarly, replacing U, V with $-U, -V$ we obtain

$$\exp((-1/r)U)\exp((-1/r)V) = \exp(C_r'),$$

where

$$C_r' = \frac{1}{r}(U+V) + \frac{1}{2r^2}[U,V] + S_r'$$

and

$$\|S_r'\| \leqslant 65\frac{(\|U\| + \|V\|)^3}{r^3}.$$

Combining these we obtain

$$\exp((1/r)U)\exp((1/r)V)\exp((-1/r)U)\exp((-1/r)V)$$
$$= \exp(C_r)\exp(C'_r) = \exp(E_r),$$

where

$$\begin{aligned} E_r &= C_r + C'_r + \frac{1}{2}[C_r, C'_r] + T_r \\ &= \frac{1}{r^2} + \frac{1}{2}[C_r, C'_r] + S_r + S'_r + T_r. \end{aligned} \tag{7.13}$$

Here T_r is defined from Equation (7.5) by setting $C_r = A$, $C'_r = B$ and $T_r = S$.

Another tedious computation shows that

$$\begin{aligned} [C_r, C'_r] &= \left[\frac{1}{r}(U+V) + \frac{1}{2r^2}[U,V] + S_r, \frac{-1}{r}(U+V) + \frac{1}{2r^2}[U,V] + S'_r\right] \\ &= \frac{1}{r^3}[U+V,[U,V]] + \frac{1}{r}[U+V, S_r + S'_r] \\ &\qquad + \frac{1}{2r^2}[[U,V], S'_r - S_r] + [S_r, S'_r]. \end{aligned}$$

By the estimate of (7.12), all four of these terms has norm bounded by an expression of the form (constant)$/r^3$, so the same is true of $[C_r, C'_r]$. Estimate (7.12) also implies that S_r, S'_r, T_r have similarly bounded norms. Setting

$$Q_r = r^2 E_r - [U,V],$$

we obtain

$$\|Q_r\| = \left\|E_r - \frac{1}{r^2}[U,V]\right\| \leqslant \frac{\text{(constant)}}{r^3} \to 0$$

as $r \to \infty$, so

$$\exp(E_r)^{r^2} = \exp([U,V] + Q_r) \to \exp([U,V]).$$

The commutator formula now follows using continuity of exp. □

Proof (Another proof of Lemma 3.23)

As an application of the Trotter formula, we will reprove the formula of Lemma 3.23:

$$\det\exp(A) = \exp(\operatorname{tr} A).$$

The case $n = 1$ is immediate, so assume that $n > 1$.

If $U, V \in \mathrm{M}_n(\mathbb{C})$ then by the Trotter formula together with the fact that det is continuous and multiplicative,

$$\begin{aligned}
\det\exp(U+V) &= \det\left(\lim_{r\to\infty}(\exp((1/r)U)\exp((1/r)V))^r\right)\\
&= \lim_{r\to\infty}\det(\exp((1/r)U)\exp((1/r)V))^r\\
&= \lim_{r\to\infty}\det\exp((1/r)U)^r\,\det\exp((1/r)V)^r\\
&= \lim_{r\to\infty}\det(\exp((1/r)U)^r)\,\det(\exp((1/r)V)^r)\\
&= \lim_{r\to\infty}\det\exp(U)\,\det\exp(V)\\
&= \det\exp(U)\,\det\exp(V).
\end{aligned}$$

More generally, given $U_1, \ldots, U_k \in \mathrm{M}_n(\mathbb{C})$ we have

$$\det\exp(U_1 + \cdots + U_k) = \det\exp(U_1)\cdots\det\exp(U_k). \tag{7.14}$$

So if $A = A_1 + \cdots + A_k$ where the A_j satisfy

$$\det\exp(A_j) = \exp(\operatorname{tr} A_j) \quad (j = 1, \ldots, k),$$

we have

$$\begin{aligned}
\det\exp(A) &= \det\exp(A_1 + \cdots + A_k)\\
&= \exp(\operatorname{tr} A_1)\cdots\exp(\operatorname{tr} A_k)\\
&= \exp(\operatorname{tr} A_1 + \cdots + \operatorname{tr} A_k)\\
&= \exp(\operatorname{tr} A).
\end{aligned}$$

So it suffices to show that every matrix A has this form.

Recall that $A = [a_{ij}]$ can be expressed as

$$A = \sum_{\substack{1\leqslant r\leqslant n\\ 1\leqslant s\leqslant n}} a_{rs}E^{rs},$$

where E^{rs} is the matrix having 1 in the (r, s) place and 0 everywhere else, *i.e.*,

$$E^{rs}{}_{ij} = \delta_{ir}\delta_{js}.$$

For $z \in \mathbb{C}$,

$$\begin{aligned}
\det\exp(zE^{rs}) &= \det\left(\sum_{k\geqslant 0}\frac{1}{k!}(z^k(E^{rs})^k\right)\\
&= \begin{cases}\det((e^z-1)E^{rr}+I_n) & \text{if } r = s,\\ \det I_n & \text{if } r \neq s\end{cases}\\
&= \begin{cases}e^z & \text{if } r = s,\\ 1 & \text{if } r \neq s.\end{cases}
\end{aligned}$$

On the other hand,

$$\operatorname{tr} zE^{rs} = \begin{cases} z & \text{if } r = s, \\ 0 & \text{if } r \neq s. \end{cases}$$

Thus

$$\exp(\operatorname{tr} zE^{rs}) = \begin{cases} e^{z} & \text{if } r = s, \\ 1 & \text{if } r \neq s, \end{cases}$$

so we obtain

$$\det \exp(zE^{rs}) = \exp(\operatorname{tr} zE^{rs}),$$

which is the desired equation. □

7.6 Matrix Groups are Lie Groups

Our aim in this section is to prove Theorem 7.24. Let $G \leqslant \mathrm{GL}_n(\mathbb{R})$ be a matrix subgroup. Recall that the Lie algebra $\mathfrak{g} = \mathrm{T}_I G$ is an $\mathbb{R}$-Lie subalgebra of $\mathfrak{gl}_n(\mathbb{R}) = \mathrm{M}_n(\mathbb{R})$. Let

$$\widetilde{\mathfrak{g}} = \{A \in \mathrm{M}_n(\mathbb{R}) : \forall t \in \mathbb{R},\ \exp(tA) \in G\}.$$

Theorem 7.27

$\widetilde{\mathfrak{g}}$ is an $\mathbb{R}$-Lie subalgebra of $\mathrm{M}_n(\mathbb{R})$.

Proof

By definition $\widetilde{\mathfrak{g}}$ is closed under multiplication by real scalars. If $U, V \in \widetilde{\mathfrak{g}}$ and $r \geqslant 1$, then the following are in G:

$$(\exp((1/r)U)\exp((1/r)V)),\ (\exp((1/r)U)\exp((1/r)V))^{r},$$
$$(\exp((1/r)U)\exp((1/r)V)\exp(-(1/r)U)\exp(-(1/r)V))^{r^2},$$
$$(\exp((1/r)U)\exp((1/r)V)\exp(-(1/r)U)\exp(-(1/r)V))^{r^2}.$$

By Theorem 7.26, for $t \in \mathbb{R}$, we have

$$\exp(tU + tV) = \lim_{r\to\infty} (\exp((1/r)tU)\exp((1/r)tV))^{r},$$

and

$$\begin{aligned}\exp(t[U, V]) &= \exp([tU, V]) \\ &= \lim_{r\to\infty} (\exp((1/r)tU)\exp((1/r)V)\exp(-(1/r)tU)\exp(-(1/r)V))^{r^2}.\end{aligned}$$

As these are both limits of elements of the closed subgroup $G \leqslant \mathrm{GL}_n(\mathbb{R})$ they are also in G. This shows that $\widetilde{\mathfrak{g}}$ is a Lie subalgebra of $\mathfrak{gl}_n(\mathbb{R}) = \mathrm{M}_n(\mathbb{R})$. □

Proposition 7.28

For a matrix subgroup $G \leqslant \mathrm{GL}_n(\mathbb{R})$, $\widetilde{\mathfrak{g}}$ is an $\mathbb{R}$-Lie subalgebra of $\mathfrak{g}$.

Proof

Let $U \in \widetilde{\mathfrak{g}}$. Then the curve

$$\gamma\colon \mathbb{R} \longrightarrow G; \quad \gamma(t) = \exp(tU),$$

has $\gamma(0) = I$ and $\gamma'(0) = U$, hence $U \in \mathfrak{g}$. □

Remark 7.29

Eventually we will see that $\widetilde{\mathfrak{g}} = \mathfrak{g}$.

We will require a technical result.

Lemma 7.30

Let $\{A_n \in \exp^{-1} G\}_{n\geqslant 1}$ and $\{s_n \in \mathbb{R}\}_{n\geqslant 1}$ be sequences for which $\|A_n\| \to 0$ and $s_n A_n \to A \in \mathrm{M}_n(\mathbb{R})$ as $n \to \infty$. Then $A \in \widetilde{\mathfrak{g}}$.

Proof

Let $t \in \mathbb{R}$. For each n, choose an integer $m_n \in \mathbb{Z}$ so that $|ts_n - m_n| \leqslant 1$. Then

$$\begin{aligned}\|m_n A_n - tA\| &\leqslant \|(m_n - ts_n)A_n\| + \|ts_n A_n - tA\| \\ &= |m_n - ts_n|\,\|A_n\| + \|ts_n A_n - tA\| \\ &\leqslant \|A_n\| + \|ts_n A_n - tA\| \to 0\end{aligned}$$

as $n \to \infty$, showing that $m_n A_n \to tA$. Since

$$\exp(m_n A_n) = \exp(A_n)^{m_n} \in G,$$

and G is closed in $\mathrm{GL}_n(\mathbb{R})$, we have

$$\exp(tA) = \lim_{n\to\infty} \exp(m_n A_n) \in G.$$

Thus every real scalar multiple tA is in $\exp^{-1} G$, showing that $A \in \widetilde{\mathfrak{g}}$. □

Choose a complementary $\mathbb{R}$-subspace $\mathfrak{w}$ to $\widetilde{\mathfrak{g}}$ in $\mathfrak{gl}_n(\mathbb{R}) = \mathrm{M}_n(\mathbb{R})$, *i.e.*, any vector subspace such that

$$\widetilde{\mathfrak{g}} + \mathfrak{w} = \mathrm{M}_n(\mathbb{R}),$$
$$\dim_{\mathbb{R}} \widetilde{\mathfrak{g}} + \dim_{\mathbb{R}} \mathfrak{w} = \dim_{\mathbb{R}} \mathrm{M}_n(\mathbb{R}) = n^2.$$

The second of these conditions is equivalent to

$$\widetilde{\mathfrak{g}} \cap \mathfrak{w} = 0.$$

This gives a direct sum decomposition of $\mathrm{M}_n(\mathbb{R})$, so every element $X \in \mathrm{M}_n(\mathbb{R})$ has a unique expression of the form

$$X = U + V \quad (U \in \widetilde{\mathfrak{g}},\ V \in \mathfrak{w}).$$

Consider the map

$$\Phi\colon \mathrm{M}_n(\mathbb{R}) \longrightarrow \mathrm{GL}_n(\mathbb{R}); \qquad \Phi(U+V) = \exp(U)\exp(V) \quad (U \in \widetilde{\mathfrak{g}},\ V \in \mathfrak{w}).$$

Φ is a smooth function which maps 0 to I. Notice that the factor $\exp(U)$ is in G. Consider the derivative at O,

$$\mathrm{d}\,\Phi_O\colon \mathrm{M}_n(\mathbb{R}) \longrightarrow \mathfrak{gl}_n(\mathbb{R}) = \mathrm{M}_n(\mathbb{R}).$$

To determine $\mathrm{d}\,\Phi_O(A+B)$, where $A \in \widetilde{\mathfrak{g}}$ and $B \in \mathfrak{w}$, we differentiate the curve $t \mapsto \Phi(t(A+B))$ at $t = 0$. Assuming that A, B are small enough and using the notation of Equations (7.4) and (7.5) for small $t \in \mathbb{R}$, there is a unique $C(t)$ depending on t for which

$$\Phi(t(A+B)) = \exp(C(t)).$$

Estimate (7.12) gives

$$\|(C(t) - tA - tB) - \frac{t^2}{2}[A,B]\| \leqslant 65|t|^3(\|A\| + \|B\|)^3.$$

From this we obtain

$$\begin{aligned}\|(C(t) - tA - tB\| &\leqslant \frac{t^2}{2}\|[A,B]\| + 65|t|^3(\|A\| + \|B\|)^3 \\ &= \frac{t^2}{2}\left(\|[A,B]\| + 130|t|(\|A\| + \|B\|)^3\right)\end{aligned}$$

and so

$$\frac{\mathrm{d}}{\mathrm{d}t}\Phi(t(A+B))_{|t=0} = \frac{\mathrm{d}}{\mathrm{d}t}\exp(C(t))_{|t=0} = A + B.$$

By linearity of the derivative, for small A, B,

$$\mathrm{d}\,\Phi_O(A+B) = A + B,$$

so $d\Phi_O$ is the identity function on $M_n(\mathbb{R})$. By the Inverse Function Theorem 7.12, Φ is a diffeomorphism onto its image when restricted to a small open neighbourhood of O, and we might as well take this to be an open disc $N_{M_n(\mathbb{k})}(O;\delta)$ for some $\delta > 0$. Thus the restriction of Φ to

$$\Phi_1 \colon N_{M_n(\mathbb{k})}(O;\delta) \longrightarrow \Phi N_{M_n(\mathbb{k})}(O;\delta)$$

is a diffeomorphism.

Now we must show that Φ maps some open subset (which we could assume to be an open disc) of $N_{M_n(\mathbb{k})}(O;\delta) \cap \widetilde{\mathfrak{g}}$ containing O onto an open neighbourhood of I in G. Suppose not; then there is a sequence of elements $U_n \in G$ with $U_n \to I$ as $n \to \infty$ but $U_n \notin \Phi\widetilde{\mathfrak{g}}$. For large enough n, $U_n \in \Phi N_{M_n(\mathbb{k})}(O;\delta)$, hence there are unique elements $A_n \in \widetilde{\mathfrak{g}}$ and $B_n \in \mathfrak{w}$ with $\Phi(A_n + B_n) = U_n$. Notice that $B_n \neq O$ since otherwise $U_n \in \Phi\widetilde{\mathfrak{g}}$. As Φ_1 is a diffeomorphism, $A_n + B_n \to O$ and this implies that $A_n \to O$ and $B_n \to O$. By definition of Φ,

$$\exp(B_n) = \exp(A_n)^{-1} U_n \in G,$$

hence $B_n \in \exp^{-1} G$. Consider the elements $\overline{B}_n = (1/\|B_n\|)B_n$ of unit norm. Each $\overline{B}_n$ is in the unit sphere in $M_n(\mathbb{R})$, which is compact hence there is a convergent subsequence of $\{\overline{B}_n\}$. By renumbering this subsequence, we can assume that $\overline{B}_n \to B$ as $n \to \infty$, where $\|B\| = 1$. Applying Lemma 7.30 to the sequences $\{B_n\}$ and $\{1/\|B_n\|\}$ we find that $B \in \widetilde{\mathfrak{g}}$. But each B_n (and hence $\overline{B}_n$) is in $\mathfrak{w}$, so B must be too. Thus $B \in \widetilde{\mathfrak{g}} \cap \mathfrak{w}$, which contradicts the fact that $B \neq O$.

So there must be an open disc

$$N_{\widetilde{\mathfrak{g}}}(O;\delta_1) = N_{M_n(\mathbb{R})}(O;\delta_1) \cap \widetilde{\mathfrak{g}}$$

which is mapped by Φ onto an open neighbourhood of I in G. So the restriction of Φ to this open disc is a local diffeomorphism at O. The inverse map gives a chart for $GL_n(\mathbb{R})$ at I and moreover $N_{\widetilde{\mathfrak{g}}}(O;\delta_1)$ is then a submanifold of $N_{M_n(\mathbb{R})}(O;\delta_1)$.

We can use left translation to move this chart to a new chart at any other point $U \in G$, by considering $L_U \circ \Phi$. The details are left as an exercise.

So we have shown that $G \leqslant GL_n(\mathbb{R})$ is a Lie subgroup, proving Theorem 7.24. This is a fundamental result that can be usefully reformulated as follows. The proof of the second part is similar to our proof of the first with minor adjustments required for the general case.

Theorem 7.31

A subgroup of $GL_n(\mathbb{R})$ is a Lie subgroup if and only if it is a matrix subgroup, *i.e.*, a closed subgroup. More generally, a subgroup of an arbitrary Lie group G is a Lie subgroup if and only if it is a closed subgroup.

Notice that the dimension of a matrix group G as a manifold is $\dim_{\mathbb{R}} \widetilde{\mathfrak{g}}$. By Proposition 7.28, $\widetilde{\mathfrak{g}} \subseteq \mathfrak{g}$ and $\dim_{\mathbb{R}} \widetilde{\mathfrak{g}} \leqslant \dim_{\mathbb{R}} \mathfrak{g}$. By Theorem 7.23, these dimensions are in fact equal, giving

$$\widetilde{\mathfrak{g}} = \mathfrak{g}. \tag{7.15}$$

Combined with Proposition 2.4, this gives the following theorem.

Theorem 7.32

For a matrix group $G \leqslant \mathrm{GL}_n(\mathbb{R})$, the exponential map $\exp\colon \mathfrak{g} \longrightarrow \mathrm{M}_n(\mathbb{R})$ has image in G, $\operatorname{im}\exp \leqslant G$. Moreover, exp is a local diffeomorphism at O, mapping some open neighbourhood of O onto an open neighbourhood of I in G.

7.7 Not All Lie Groups are Matrix Groups

For completeness we describe the simplest example of a Lie group which is not a matrix group. In fact there are infinitely many related examples of such *Heisenberg groups* Heis_n and the example Heis_3 that we will discuss in detail is particularly important in Quantum Physics.

For $n \geqslant 3$, the Heisenberg group Heis_n is defined as follows. Recall the group of $n \times n$ real unipotent matrices $\mathrm{SUT}_n = \mathrm{SUT}_n(\mathbb{R})$, whose elements have the form

$$\begin{bmatrix} 1 & a_{12} & \cdots & \cdots & \cdots & a_{1n} \\ 0 & 1 & a_{23} & \ddots & \ddots & a_{2n} \\ 0 & 0 & \ddots & \ddots & \ddots & \vdots \\ \vdots & \vdots & \ddots & 1 & a_{n-2\,n-1} & \vdots \\ \vdots & \vdots & \ddots & 0 & 1 & a_{n-1\,n} \\ 0 & 0 & \cdots & 0 & 0 & 1 \end{bmatrix},$$

with $a_{ij} \in \mathbb{R}$. Then SUT_n is a matrix subgroup of $\mathrm{GL}_n(\mathbb{R})$ whose Lie algebra $\mathfrak{sut}_n = \mathfrak{sut}_n(\mathbb{R})$ consists of the matrices of the form

$$\begin{bmatrix} 0 & t_{12} & \cdots & \cdots & \cdots & t_{1n} \\ 0 & 0 & t_{23} & \ddots & \ddots & t_{2n} \\ 0 & 0 & \ddots & \ddots & \ddots & \vdots \\ \vdots & \vdots & \ddots & 0 & t_{n-2\,n-1} & \vdots \\ \vdots & \vdots & \ddots & 0 & 0 & t_{n-1\,n} \\ 0 & 0 & \cdots & 0 & 0 & 0 \end{bmatrix}$$

with $t_{ij} \in \mathbb{R}$, hence $\dim \mathrm{SUT}_n = \binom{n}{2}$. It is a nice exercise to show that the following hold in general.

Proposition 7.33

For $n \geqslant 3$, the centre $Z(\mathrm{SUT}_n)$ of SUT_n consists of all the matrices $[a_{ij}] \in \mathrm{Heis}_n$ with $a_{ij} = 0$ except when $i = 1$ and $j = n$. Furthermore, $Z(\mathrm{SUT}_n)$ is contained in the commutator subgroup of SUT_n.

Notice that there is an isomorphism of Lie groups $\mathbb{R} \cong Z(\mathrm{SUT}_n)$ under which the subgroup of integers $\mathbb{Z} \subseteq \mathbb{R}$ corresponds to the matrices with $a_{1\,n} \in \mathbb{Z}$ and these form a discrete normal (in fact central) subgroup $Z_n \triangleleft \mathrm{SUT}_n$. We can form the quotient group

$$\mathrm{Heis}_n = \mathrm{SUT}_n / Z_n.$$

Heis_n has the quotient space topology and since Z_n is a discrete subgroup, the quotient map $q\colon \mathrm{SUT}_n \longrightarrow \mathrm{Heis}_n$ is a local homeomorphism. This can be used to show that Heis_n is also a Lie group since charts for SUT_n defined on small open sets will give rise to charts for Heis_n. The Lie algebra of Heis_n is the same as that of SUT_n, $\mathfrak{heis}_n = \mathfrak{sut}_n$.

Proposition 7.34

For $n \geqslant 3$, the centre $Z(\mathrm{Heis}_n)$ of Heis_n consists of the image under q of $Z(\mathrm{SUT}_n)$. Furthermore, $Z(\mathrm{Heis}_n)$ is contained in the commutator subgroup of Heis_n.

We note that $Z(\mathrm{Heis}_n) = Z(\mathrm{SUT}_n)/Z_n$ is isomorphic to the circle group

$$\mathbb{T} = \{z \in \mathbb{C} : |z| = 1\},$$

with the correspondence coming from the map

$$\mathbb{R} \longrightarrow \mathbb{T}; \quad t \longmapsto e^{2\pi i t}.$$

When $n = 3$, there is a surjective Lie homomorphism

$$p\colon \mathrm{SUT}_3 \longrightarrow \mathbb{R}^2; \quad \begin{bmatrix} 1 & x & t \\ 0 & 1 & y \\ 0 & 0 & 1 \end{bmatrix} \mapsto \begin{bmatrix} x \\ y \end{bmatrix}$$

whose kernel is $\ker p = Z(\mathrm{SUT}_3)$. Since $Z_3 \leqslant \ker p$, there is an induced surjective Lie homomorphism $\overline{p}\colon \mathrm{Heis}_3 \longrightarrow \mathbb{R}^2$ satisfying $\overline{p} \circ q = p$. In this case the isomorphism $Z(\mathrm{Heis}_n) \cong \mathbb{T}$ is given by

$$\begin{bmatrix} 1 & 0 & t \\ 0 & 1 & 0 \\ 0 & 0 & 1 \end{bmatrix} Z_3 \longleftrightarrow e^{2\pi i t}.$$

From now on we will denote the coset

$$\begin{bmatrix} 1 & x & t \\ 0 & 1 & y \\ 0 & 0 & 1 \end{bmatrix} Z_3 \in \mathrm{Heis}_3$$

by $[x, y, e^{2\pi i t}]$. Thus a general element of Heis_3 has the form $[x, y, z]$ for $x, y \in \mathbb{R}$ and $z \in \mathbb{T}$; the identity element is $1 = [0, 0, 1]$. Similarly, the element

$$\begin{bmatrix} 1 & x & t \\ 0 & 1 & y \\ 0 & 0 & 1 \end{bmatrix} \in \mathfrak{heis}_3$$

will be denoted (x, y, t).

Proposition 7.35

Multiplication, inverses and commutators in Heis_3 are given by

$$\begin{aligned} [x_1, y_1, z_1][x_2, y_2, z_2] &= [x_1 + x_2, y_1 + y_2, z_1 z_2 e^{2\pi i x_1 y_2}], \\ [x, y, z]^{-1} &= [-x, -y, z^{-1} e^{2\pi i x y}] \\ [x_1, y_1, z_1][x_2, y_2, z_2][x_1, y_1, z_1]^{-1}[x_2, y_2, z_2]^{-1} &= [0, 0, e^{2\pi i (x_1 y_2 - y_1 x_2)}]. \end{aligned}$$

The Lie bracket in $\mathfrak{heis}_3$ is given by

$$[(x_1, y_1, t_1), (x_2, y_2, t_2)] = (0, 0, x_1 y_2 - y_1 x_2).$$

The Lie algebra $\mathfrak{heis}_3$ is often called a *Heisenberg (Lie) algebra* and occurs throughout Quantum Physics. It is essentially the same as the Lie algebra of operators on differentiable functions $f\colon \mathbb{R} \longrightarrow \mathbb{R}$ spanned by the three operators $\mathbf{1}, \mathbf{p}, \mathbf{q}$ defined by

$$\mathbf{1} f(x) = f(x), \quad \mathbf{p} f(x) = \frac{\mathrm{d} f(x)}{\mathrm{d} x}, \quad \mathbf{q} f(x) = x f(x).$$

The non-trivial commutator involving these three operators is given by the *canonical commutation relation*

$$[\mathbf{p}, \mathbf{q}] = \mathbf{pq} - \mathbf{qp} = \mathbf{1}.$$

In $\mathfrak{heis}_3$ the elements $(1, 0, 0), (0, 1, 0), (0, 0, 1)$ form a basis with the only non-trivial commutator

$$[(1, 0, 0), (0, 1, 0)] = (0, 0, 1).$$

Theorem 7.36

There are no continuous homomorphisms $\varphi\colon \mathrm{Heis}_3 \longrightarrow \mathrm{GL}_n(\mathbb{C})$ with trivial kernel $\ker\varphi = 1$.

Proof

Suppose that $\varphi\colon \mathrm{Heis}_3 \longrightarrow \mathrm{GL}_n(\mathbb{C})$ is a continuous homomorphism with trivial kernel and suppose that n is minimal with this property. For each $g \in \mathrm{Heis}_3$, the matrix $\varphi(g)$ acts on vectors in $\mathbb{C}^n$.

We will identify $Z(\mathrm{Heis}_3)$ with the circle $\mathbb{T}$ as above. Then $\mathbb{T}$ has a *topological generator* z_0; this is an element whose powers form a cyclic subgroup $\langle z_0\rangle \leqslant \mathbb{T}$ which has closure $\mathbb{T}$. Proposition 10.7 will provide a more general version of this phenomenon. For now we point out that for any irrational number $r \in \mathbb{R}$, the following is true: for any real number $s \in \mathbb{R}$ and any $\varepsilon > 0$, there are integers $p, q \in \mathbb{Z}$ such that

$$|s - pr - q| < \varepsilon.$$

This implies that $e^{2\pi i r}$ is a topological generator of $\mathbb{T}$ since its powers are dense.

Let λ be an eigenvalue for the matrix $\varphi(z_0)$, with eigenvector $\mathbf{v}$. If necessary replacing z_0 with z_0^{-1}, we may assume that $|\lambda| \geqslant 1$. If $|\lambda| > 1$, then

$$\varphi(z_0^k)\mathbf{v} = \varphi(z_0)^k\mathbf{v} = \lambda^k\mathbf{v}$$

and so

$$\|\varphi(z_0^k)\| \geqslant |\lambda|^k.$$

Thus $\|\varphi(z_0^k)\| \to \infty$ as $k \to \infty$, which implies that $\varphi\mathbb{T}$ is unbounded. But φ is continuous and $\mathbb{T}$ is compact hence $\varphi\mathbb{T}$ is bounded. So in fact $|\lambda| = 1$.

Since φ is a homomorphism and $z_0 \in Z(\mathrm{Heis}_3)$, for any $g \in \mathrm{Heis}_3$ we have

$$\varphi(z_0)\varphi(g)\mathbf{v} = \varphi(z_0 g)\mathbf{v} = \varphi(g z_0)\mathbf{v} = \varphi(g)\varphi(z_0)\mathbf{v} = \lambda\varphi(g)\mathbf{v},$$

which shows that $\varphi(g)\mathbf{v}$ is another eigenvector of $\varphi(z_0)$ for the eigenvalue λ. If we set

$$V_\lambda = \{\mathbf{v} \in \mathbb{C}^n : \exists k \geqslant 1 \text{ s.t. } (\varphi(z_0) - \lambda I_n)^k\mathbf{v} = \mathbf{0}\},$$

then $V_\lambda \subseteq \mathbb{C}^n$ is a vector subspace which is also closed under the actions of all the matrices $\varphi(g)$ with $g \in \mathrm{Heis}_3$. Choose $k_0 \geqslant 1$ to be the largest number for which there is a vector $\mathbf{v}_0 \in V_\lambda$ satisfying

$$(\varphi(z_0) - \lambda I_n)^{k_0}\mathbf{v}_0 = \mathbf{0}, \quad (\varphi(z_0) - \lambda I_n)^{k_0-1}\mathbf{v}_0 \neq \mathbf{0}.$$

If $k_0 > 1$, there are non-zero vectors $\mathbf{u}, \mathbf{v} \in V_\lambda$ for which

$$\varphi(z_0)\mathbf{u} = \lambda\mathbf{u} + \mathbf{v}, \quad \varphi(z_0)\mathbf{v} = \lambda\mathbf{v}.$$

Then

$$\varphi(z_0^k)\mathbf{u} = \varphi(z_0)^k\mathbf{u} = \lambda^k\mathbf{u} + k\lambda^{k-1}\mathbf{v}$$

and since $\mathbf{v} \neq \mathbf{0}$,

$$\|\varphi(z_0^k)\mathbf{u}\| = \|\varphi(z_0)^k\mathbf{u}\| \geqslant |\lambda\mathbf{u} + k\mathbf{v}| \to \infty$$

as $k \to \infty$. This also contradicts the fact that $\varphi\mathbb{T}$ is bounded. So $k_0 = 1$ and V_λ is just the eigenspace for the eigenvalue λ. This argument actually proves the following important general result, which in particular applies to finite groups viewed as zero-dimensional compact Lie groups.

Proposition 7.37

Let G be a compact Lie group and $\rho\colon G \longrightarrow \mathrm{GL}_n(\mathbb{C})$ be a continuous homomorphism. Then for any $g \in G$, $\rho(g)$ is diagonalisable.

Having chosen a basis for V_λ, we obtain a continuous homomorphism $\theta\colon \mathrm{Heis}_3 \longrightarrow \mathrm{GL}_d(\mathbb{C})$ for which $\theta(z_0) = \lambda I_d$. By continuity, every element of $\mathbb{T}$ also has the form (scalar)I_d. The minimality of n implies that we must have $d = n$ and so we can assume that $\varphi(z_0) = \lambda I_n$.

By the equation for commutators given in Proposition 7.35, each element $z \in \mathbb{T} \leqslant \mathrm{Heis}_3$ is a commutator $z = ghg^{-1}h^{-1}$ in Heis_3. Since det and φ are homomorphisms,

$$\det\varphi(z) = \varphi(ghg^{-1}h^{-1}) = 1.$$

Thus there is a continuous function $\mu\colon \mathbb{T} \longrightarrow \mathbb{C}^\times$ such that for every $z \in \mathbb{T}$,

$$\varphi(z) = \mu(z)I_d, \quad \mu(z)^d = 1.$$

As $\mathbb{T}$ is path connected, $\mu(z) = 1$ for every $z \in \mathbb{T}$. Therefore for each $z \in \mathbb{T}$, the only eigenvalue of $\varphi(z)$ is 1. This shows that $\mathbb{T} \leqslant \ker\varphi$, contradicting the assumption that $\ker\varphi$ is trivial. □

The argument used in this proof can be modified to show that none of the Heisenberg groups Heis_n with $n \geqslant 3$ can be realised as a matrix group.

EXERCISES

7.1. a) Show that the subset

$$M = \{(A, b) \in \mathrm{M}_n(\mathbb{R}) \times \mathbb{R} : b\det A = 1\} \subseteq \mathrm{M}_n(\mathbb{R}) \times \mathbb{R}$$

is a closed submanifold of $M_n(\mathbb{R}) \times \mathbb{R}$ and determine $T_{(A,b)} M$ for $(A, b) \in M$.
b) Show that M has the structure of a Lie group with multiplication μ given by

$$\mu((A_1, b_1), (A_2, b_2)) = (A_1 A_2, b_1 b_2).$$

To which standard matrix group is M isomorphic?
c) Repeat this with $\mathbb{R}$ replaced by $\mathbb{C}$.

7.2. Work through the details of the calculation in Example 7.22 for the cases $n = 2, 3$.

7.3. a) Modify the details of Example 7.22 to show that $U(n) \leqslant GL_n(\mathbb{C})$ is a Lie subgroup. It might be helpful to work through the cases $n = 1, 2, 3$ first.
b) Show that $SU(n) \leqslant U(n)$ is a Lie subgroup by using the determinant function

$$\det\colon U(n) \longrightarrow \mathbb{T} = \{z \in \mathbb{C} : |z| = 1\}$$

together with the Identity Check Trick 7.21.

7.4. Let G be a matrix group. Use Theorem 7.24 to show that each of the following subgroups of G is a Lie subgroup. In each case, try to find a proof that works when G is an arbitrary Lie group.
a) For $g \in G$, the centraliser of g, $Z_G(g) = \{x \in G : xgx^{-1} = g\}$.
b) The centre of G, $Z(G) = \bigcap_{g \in G} Z_G(g)$.
c) For a closed subgroup $H \leqslant G$, the normaliser of H,

$$N_G(H) = \{x \in G : xHx^{-1} = H\}.$$

d) The kernel of φ, $\ker \varphi$, where $\varphi\colon G \longrightarrow H$ is a continuous homomorphism into a matrix group H.

7.5. Let G be a matrix group and M be a smooth manifold. Suppose that $\mu\colon G \times M \longrightarrow M$ is a continuous group action as defined in Section 1.9 and investigated in the exercises for Chapter 1. Also suppose that μ is smooth, *i.e.*, is a *smooth group action*.
a) Show that for each $x \in M$, $\mathrm{Stab}_G(x) \leqslant G$ is a Lie subgroup.
b) If $X \subseteq M$ is a closed subset, show that its stabiliser

$$\mathrm{Stab}_G(X) = \{g \in G : gX = X\} \leqslant G$$

is a Lie subgroup.

7.6. For a Lie group G and a closed subgroup $H \leqslant G$, show that the cosets gH and Hg and the conjugate gHg^{-1} are submanifolds of G. In each case, identify the tangent space at a point in terms of a suitable tangent space to H.

7.7. Let G and H be Lie groups and $\varphi\colon G \longrightarrow H$ be a Lie homomorphism. Show that $\ker\varphi \leqslant G$ is a Lie subgroup and identify the tangent space $\mathrm{T}_g \ker\varphi$ at $g \in \ker\varphi$.

7.8. For $n \geqslant 1$, consider the subset

$$\mathcal{C}_{2n} = \{J \in \mathrm{GL}_{2n}(\mathbb{R}) : J^2 = -I_{2n}\} \subseteq \mathrm{GL}_{2n}(\mathbb{R}).$$

a) Show that $\mathcal{C}_{2n} \subseteq \mathrm{GL}_{2n}(\mathbb{R})$ is a closed subset.
b) Show that there is a continuous action of $\mathrm{GL}_{2n}(\mathbb{R})$ on $\mathcal{C}_{2n}$ given by

$$A \cdot J = AJA^{-1} \quad (A \in \mathrm{GL}_{2n}(\mathbb{R})).$$

By considering the eigenvalues of elements of $\mathcal{C}_{2n}$, show that this action has only one orbit.
c) Find the stabilisers of the matrices J_{2n} and J'_{2n} of Section 1.6.
d) ⚠⚠ Show that $\mathcal{C}_{2n} \subseteq \mathrm{GL}_{2n}(\mathbb{R})$ is a submanifold and find its tangent space at a point $J \in \mathcal{C}_{2n}$.
[An element $J \in \mathrm{GL}_{2n}(\mathbb{R})$ satisfying $J^2 = -I_{2n}$ is often called a *complex structure* on $\mathbb{R}^{2n}$ and $\mathcal{C}_{2n}$ can be interpreted as the space of all complex structures on $\mathbb{R}^{2n}$.]

7.9. Modify the previous exercise by replacing $\mathrm{GL}_{2n}(\mathbb{R})$ with $\mathrm{O}(2n)$ and $\mathcal{C}_{2n}$ with

$$\mathcal{C}'_{2n} = \{J \in \mathrm{O}(2n) : J^2 = -I_{2n}\} \subseteq \mathrm{O}(2n).$$

In particular show that $\mathcal{C}'_{2n} \subseteq \mathrm{O}(2n)$ is a submanifold and find its dimension. Also show that $\mathcal{C}'_{2n} \subseteq \mathrm{SO}(2n)$.

7.10. Determine the Lie bracket [,] of the Lie algebra $\mathfrak{heis}_4$ of the Heisenberg group Heis_4.

8
Homogeneous Spaces

Homogeneous spaces are as important in connection with Lie groups and their applications as sets of cosets are in ordinary group theory. Indeed, in the Kleinian view, a geometry consists of a homogeneous space with the group acting as its symmetry group. For detailed discussions of homogeneous spaces and the related notion of symmetric space see [6, 11]. We describe the basic ideas although we omit the proof that a homogeneous space is actually a smooth manifold since that would require somewhat more differential geometry than we have developed. Instead we focus on homogeneous spaces arising as orbits for smooth group actions and these can be studied as submanifolds.

We give a number of examples based on standard constructions in linear algebra, such as the Gram–Schmidt process and the transformation theory of real quadratic forms. In Chapter 9 we will use homogeneous spaces to investigate connectivity of various families of matrix groups.

8.1 Homogeneous Spaces as Manifolds

Let G be a Lie group of dimension $\dim G = n$ and $H \leqslant G$ be a closed subgroup, which is therefore a Lie subgroup of dimension $\dim H = k$. The set of left cosets

$$G/H = \{gH : g \in G\}$$

has an associated quotient map

$$\pi \colon G \longrightarrow G/H; \quad \pi(g) = gH.$$

We give G/H a topology by requiring that a subset $W \subseteq G/H$ is open if and only if $\pi^{-1}W \subseteq G$ is open; this is called the *quotient topology* on G/H.

Remark 8.1

This idea can be generalised to the case of an arbitrary surjection $q\colon X \longrightarrow Y$ where X is a topological space and Y is a set. The resulting topology on Y is also called the *quotient topology with respect to* q. A particular case of this occurs when Y is the set of equivalence classes of an equivalence relation $\sim$ on X and $q(x)$ is the equivalence class of $x \in X$.

Lemma 8.2

The projection map $\pi\colon G \longrightarrow G/H$ is an open mapping and G/H is a topological space which is separable and Hausdorff.

Proof

For $U \subseteq G$,

$$\pi^{-1}(\pi U) = \bigcup_{h \in H} Uh,$$

where

$$Uh = \{uh \in G : u \in U\} \subseteq G.$$

If $U \subseteq G$ is open, then each Uh ($h \in H$) is open, implying that $\pi U \subseteq G/H$ is also open.

G/H is separable since a countable basis of G is mapped by π to a countable collection of open subsets of G/H that is also a basis.

To see that G/H is Hausdorff, consider the continuous map

$$\theta\colon G \times G \longrightarrow G; \quad \theta(x, y) = x^{-1}y.$$

Then

$$\theta^{-1}H = \{(x, y) \in G \times G : xH = yH\},$$

and this is a closed subset since $H \subseteq G$ is closed. Hence,

$$\{(x, y) \in G \times G : xH = yH\} \subseteq G \times G$$

is open. By definition of the product topology, this means that whenever $x, y \in G$ satisfy $xH \neq yH$, there are open subsets $U, V \subseteq G$ with $x \in U$, $y \in V$, $U \neq V$ and $\pi U \cap \pi V = \varnothing$. Since $\pi U, \pi V \subseteq G/H$ are open, this shows that G/H is Hausdorff. □

The quotient map $\pi\colon G \longrightarrow G/H$ is characterised by a universal property which is similar in spirit to that of Theorem 5.4.

Proposition 8.3 (Universal property of the quotient topology)

Let X be a topological space and $f\colon G \longrightarrow X$ be a continuous function for which $f(gh) = f(g)$ whenever $g \in G$ and $h \in H$. Then there is a unique continuous function $\overline{f}\colon G/H \longrightarrow X$ for which $\overline{f} \circ \pi = f$.

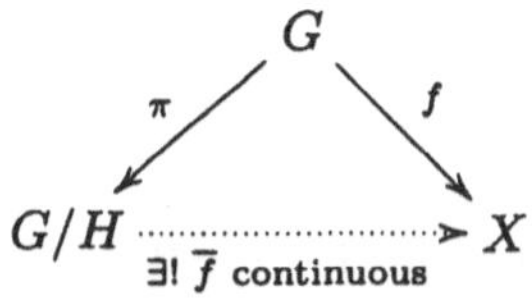

The following consequence characterises the topological space G/H up to homeomorphism and is proved in a similar fashion to Corollary 5.5.

Corollary 8.4

Let Q be a topological space and $q\colon G \longrightarrow Q$ be a continuous function for which $q(gh) = q(g)$ whenever $g \in G$ and $h \in H$, and suppose that it has the following universal property:

Let X be a topological space and $f\colon G \longrightarrow X$ be a continuous function for which $f(gh) = f(g)$ whenever $g \in G$ and $h \in H$. Then there is a unique continuous function $\widetilde{f}\colon Q \longrightarrow X$ for which $\widetilde{f} \circ q = f$.

Then there is a unique homeomorphism $\overline{q}\colon G/H \longrightarrow Q$ satisfying $\overline{q} \circ \pi = q$.

We would like to make G/H into a smooth manifold so that $\pi\colon G \longrightarrow G/H$ is smooth. The explicit construction of an atlas is rather complicated so we merely state a result on this and then consider some examples where the smooth structure comes from an existing manifold which is diffeomorphic to a quotient. The reader is referred to [6, 29] for further details.

Theorem 8.5

G/H can be given the unique structure of a smooth manifold of dimension

$$\dim G/H = \dim G - \dim H$$

so that the projection map $\pi\colon G \longrightarrow G/H$ is smooth and at each $g \in G$,

$$\ker(\mathrm{d}\,\pi\colon\ \mathrm{T}_g\, G \longrightarrow \mathrm{T}_{gH}\, G/H) = \mathrm{d}\,\mathrm{L}_g \mathfrak{h},$$

and there is an atlas for G/H consisting of charts of the form $\theta\colon W \longrightarrow \theta W \subseteq$ $\mathbb{R}^{n-k}$ for which there is a diffeomorphism $\Theta\colon W \times H \longrightarrow \pi^{-1}W$ satisfying the conditions

$$\Theta(w, h_1 h_2) = \Theta(w, h_1)h_2, \quad \pi(\Theta(w, h)) = w \qquad (w \in W,\ h, h_1, h_2 \in H).$$

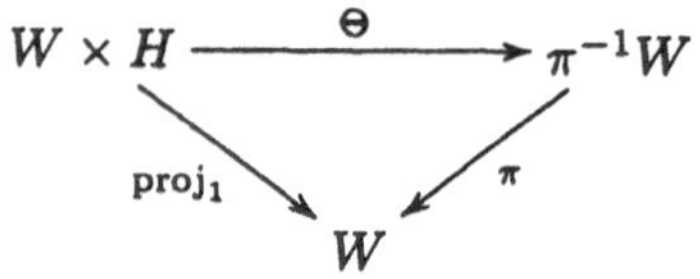

The projection π looks like $\mathrm{proj}_1\colon \pi^{-1}W \longrightarrow W$, the projection onto W, when restricted to $\pi^{-1}W$. For such a chart, the map Θ is said to provide a *local trivialisation of π over W*. An atlas consisting of such charts and local trivialisations $(\theta\colon W \longrightarrow \theta W, \Theta)$ provides a *local trivialisation of π*. This is related to the important notion of a *principal H-bundle* over G/H.

Notice that given such an atlas, an atlas for G can be obtained by taking each pair $(\theta\colon W \longrightarrow \theta W, \Theta)$ and combining the map θ with a chart $\psi\colon U \longrightarrow$ $\psi U \subseteq \mathbb{R}^k$ for H to obtain a chart

$$(\theta \times \psi) \circ \Theta^{-1}\colon \Theta(W \times U) \longrightarrow \theta W \times \psi U \subseteq \mathbb{R}^{n-k} \times \mathbb{R}^k = \mathbb{R}^n.$$

Such a manifold G/H is called a *homogeneous space* since each left translation map L_g on G gives rise to a diffeomorphism

$$\overline{\mathrm{L}}_g\colon G/H \longrightarrow G/H; \quad \overline{\mathrm{L}}_g(xH) = gxH,$$

for which $\pi \circ \mathrm{L}_g = \overline{\mathrm{L}}_g \circ \pi$.

$$
\begin{array}{ccc}
G & \xrightarrow{\ \mathrm{L}_g\ } & G \\[2pt]
{\scriptstyle \pi}\downarrow & & \downarrow{\scriptstyle \pi} \\[2pt]
G/H & \xrightarrow{\ \overline{\mathrm{L}}_g\ } & G/H
\end{array}
$$

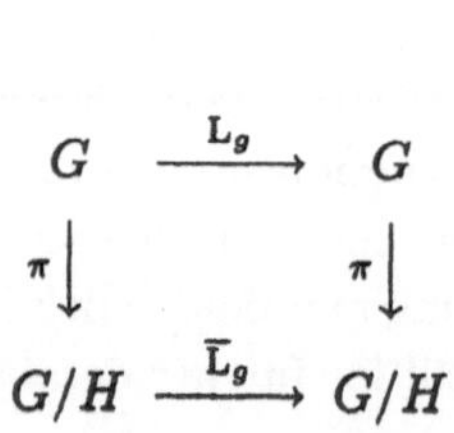

So each point gH has a neighbourhood diffeomorphic under $\overline{\mathrm{L}}_g^{-1}$ to a neighbourhood of $1H$; so locally G/H is unchanged as gH is varied. This is the basic insight in Felix Klein's view of a Geometry which is characterised as a homogeneous space G/H for some group of transformations G and subgroup $H \leqslant G$.

8.2 Homogeneous Spaces as Orbits

Just as in ordinary group theory, group actions have orbits equivalent to sets of cosets G/H, so homogeneous spaces also arise as orbits associated to smooth groups actions of G on a manifolds. From [6] we have

Theorem 8.6

Suppose that a Lie group G acts smoothly on a manifold M. If the element $x \in M$ has stabiliser $\mathrm{Stab}_G(x) \leqslant G$ and the orbit $\mathrm{Orb}_G(x) \subseteq M$ is a closed submanifold, then the function

$$f\colon G/\mathrm{Stab}_G(x) \longrightarrow \mathrm{Orb}_G(x); \quad f(g\,\mathrm{Stab}_G(x)) = gx$$

is a diffeomorphism.

Example 8.7

For $n \geqslant 1$, $\mathrm{O}(n)$ acts smoothly on $\mathbb{R}^n$ by matrix multiplication. For any non-zero vector $\mathbf{v} \in \mathbb{R}^n$, the orbit $\mathrm{Orb}_{\mathrm{O}(n)}(\mathbf{v}) \subseteq \mathbb{R}^n$ is diffeomorphic to $\mathrm{O}(n)/\mathrm{O}(n-1)$.

Proof

First observe that when $\mathbf{v}$ is the standard basis vector $\mathbf{e}_n$, for $A \in \mathrm{O}(n)$, $A\mathbf{e}_n = \mathbf{e}_n$ if and only if $\mathbf{e}_n$ is the last column of A, while all the other columns of A are orthogonal to $\mathbf{e}_n$. Since the columns of A must be an orthonormal set of vectors, this means that each of the first $(n-1)$ columns of A has the form

$$\begin{bmatrix} a_{1k} \\ a_{2k} \\ \vdots \\ a_{nk} \\ 0 \end{bmatrix}$$

where the matrix

$$\begin{bmatrix} a_{11} & a_{12} & \cdots & a_{1\,n-1} \\ a_{21} & a_{22} & \cdots & a_{2\,n-1} \\ \vdots & \ddots & \ddots & \vdots \\ a_{n-1\,1} & a_{n-1\,2} & \cdots & a_{n-1\,n-1} \end{bmatrix}$$

is orthogonal and hence in $O(n-1)$. We identify $O(n-1)$ with the subset of $O(n)$ consisting of matrices of the form

$$\begin{bmatrix} a_{11} & a_{12} & \cdots & a_{1\,n-1} & 0 \\ a_{21} & a_{22} & \cdots & a_{2\,n-1} & 0 \\ \vdots & \ddots & \ddots & \ddots & \vdots \\ a_{n-1\,1} & a_{n-1\,2} & \cdots & a_{n-1\,n-1} & 0 \\ 0 & 0 & \cdots & 0 & 1 \end{bmatrix}$$

and then have $\mathrm{Stab}_{O(n)}(\mathbf{e}_n) = O(n-1)$. The orbit of $\mathbf{e}_n$ is the whole unit sphere $\mathbb{S}^{n-1} \subseteq \mathbb{R}^n$ since given a unit vector $\mathbf{u}$ we can extend it to an orthonormal basis $\mathbf{u}_1, \ldots, \mathbf{u}_{n-1}, \mathbf{u}_n = \mathbf{u}$ whose vectors form the columns of an orthogonal matrix $U \in O(n)$ for which $U\mathbf{e}_n = \mathbf{u}$. Then there is a diffeomorphism

$$O(n)/\,\mathrm{Stab}_{O(n)}(\mathbf{e}_n) = O(n)/\,O(n-1) \longrightarrow \mathrm{Orb}_{O(n)}(\mathbf{e}_n) = \mathbb{S}^{n-1}.$$

Now for a general non-zero vector $\mathbf{v}$ notice that $\mathrm{Stab}_{O(n)}(\mathbf{v}) = \mathrm{Stab}_{O(n)}(\hat{\mathbf{v}})$ where $\hat{\mathbf{v}} = (1/|\mathbf{v}|)\mathbf{v}$ and

$$\mathrm{Orb}_{O(n)}(\mathbf{v}) = \mathbb{S}^{n-1}(|\mathbf{v}|),$$

the sphere of radius $|\mathbf{v}|$. If we choose any $P \in O(n)$ with $\hat{\mathbf{v}} = P\mathbf{e}_n$, we have

$$\mathrm{Stab}_{O(n)}(\mathbf{v}) = P\,\mathrm{Stab}_{O(n)}(\mathbf{e}_n)P^{-1}$$

and so the map

$$\mathrm{Orb}_{O(n)}(\mathbf{v}) \longrightarrow O(n)/P\,O(n-1)P^{-1} \xrightarrow{\chi_{P^{-1}}} O(n)/\,O(n-1)$$

is a diffeomorphism. □

A similar result holds for $SO(n)/\,SO(n-1)$ as a homogeneous space of $SO(n)$. We can also obtain the homogeneous spaces $U(n)/\,U(n-1)$ and $SU(n)/\,SU(n-1)$ of the unitary and special unitary groups as orbits of non-zero vectors in $\mathbb{C}^n$ on which these groups act by matrix multiplication; these are diffeomorphic to $\mathbb{S}^{2n-1}$. The action of the quaternionic symplectic group $Sp(n)$ on $\mathbb{H}^n$ leads to orbits of non-zero vectors diffeomorphic to $Sp(n)/\,Sp(n-1)$ and $\mathbb{S}^{4n-1}$.

8.3 Projective Spaces

More exotic orbit spaces are obtained as follows. Let $\mathbb{k} = \mathbb{R}$, $\mathbb{C}$ or $\mathbb{H}$ and $d = \dim_{\mathbb{R}} \mathbb{k}$. Consider $\mathbb{k}^{n+1}$ as a right $\mathbb{k}$-vector space. There is an action of the group of units $\mathbb{k}^{\times}$ on the subset of non-zero vectors $\mathbb{k}_0^{n+1} = \mathbb{k}^{n+1} - \{\mathbf{0}\}$, given by

$$z \cdot \mathbf{x} = \mathbf{x}z^{-1}.$$

The set of orbits is denoted $\mathbb{k}\mathrm{P}^n$ and is called *n-dimensional $\mathbb{k}$-projective space.* An element of $\mathbb{k}\mathrm{P}^n$ is a subset of $\mathbb{k}_0^{n+1}$ of the form

$$[\mathbf{x}] = \{\mathbf{x}z^{-1} : z \in \mathbb{k}^{\times}\}.$$

Clearly $[\mathbf{x}] = [\mathbf{y}]$ if and only if there is a $z \in \mathbb{k}^{\times}$ for which $\mathbf{y} = \mathbf{x}z^{-1}$.

Remark 8.8

Because of this we can identify elements $\mathbb{k}\mathrm{P}^n$ with $\mathbb{k}$-*lines* in $\mathbb{k}^{n+1}$, *i.e.*, 1-dimensional $\mathbb{k}$-vector subspaces. $\mathbb{k}\mathrm{P}^n$ is often taken to be the set of all such lines, particularly in the subject of Projective Geometry.

Associated to the map

$$q_n : \mathbb{k}_0^{n+1} \longrightarrow \mathbb{k}\mathrm{P}^n; \quad q_n(\mathbf{x}) = [\mathbf{x}],$$

is a quotient topology on $\mathbb{k}\mathrm{P}^n$ (see Remark 8.1) which is Hausdorff and separable.

Proposition 8.9

$\mathbb{k}\mathrm{P}^n$ is a smooth manifold of dimension $\dim \mathbb{k}\mathrm{P}^n = n \dim_{\mathbb{R}} \mathbb{k}$. Moreover, the quotient map $q_n : \mathbb{k}_0^{n+1} \longrightarrow \mathbb{k}\mathrm{P}^n$ is smooth with surjective derivative at every point in $\mathbb{k}_0^{n+1}$.

Proof

As usual, for $\mathbf{x} \in \mathbb{k}^{n+1}$ write $\mathbf{x} = \begin{bmatrix} x_1 \\ \vdots \\ x_{n+1} \end{bmatrix}$. Now for each $r = 1, 2, \ldots, n$, set

$$\mathbb{k}\mathrm{P}_r^n = \{[\mathbf{x}] : x_r \neq 0\} \subseteq \mathbb{k}\mathrm{P}^n.$$

Then $\Bbbk P^n_r \subseteq \Bbbk P^n$ is open and there is a function

$$\sigma_r \colon \Bbbk P^n_r \longrightarrow \Bbbk^n; \quad \sigma_r([\mathbf{x}]) = \begin{bmatrix} x_1 x_r^{-1} \\ \vdots \\ x_{r-1} x_r^{-1} \\ x_{r+1} x_r^{-1} \\ \vdots \\ x_{n+1} x_r^{-1} \end{bmatrix}$$

which is a continuous bijection that is actually a homeomorphism. Whenever $r \neq s$, the induced map

$$\sigma_s \circ \sigma_r^{-1} \colon \sigma_r \Bbbk P^n_r \cap \Bbbk P^n_s \longrightarrow \sigma_s \Bbbk P^n_r \cap \Bbbk P^n_s$$

is given by

$$\sigma_s \circ \sigma_r^{-1}(\mathbf{x}) = \begin{bmatrix} y_1 \\ \vdots \\ y_{s-1} \\ y_{s+1} \\ \vdots \\ y_{n+1} \end{bmatrix}$$

where

$$y_j = \begin{cases} x_j x_s^{-1} & \text{if } s \neq j < r, \\ x_{j-1} x_s^{-1} & \text{if } s \neq j > r. \end{cases}$$

These $(n+1)$ charts form the standard atlas for n-dimensional projective space over $\Bbbk$. □

An alternative description of $\Bbbk P^n$ is obtained by considering the action of the subgroup

$$\Bbbk_1^\times = \{z \in \Bbbk^\times : |z| = 1\} \leqslant \Bbbk^\times$$

on the unit sphere $S^{(n+1)d-1} \subseteq \Bbbk_0^{n+1}$. Notice that every element $[\mathbf{x}] \in \Bbbk P^n$ contains elements of $S^{(n+1)d-1}$. Also, if $\mathbf{x}, \mathbf{y} \in \Bbbk_0^{n+1}$ have unit length $|\mathbf{x}| = |\mathbf{y}| = 1$, then $[\mathbf{x}] = [\mathbf{y}]$ if and only if $\mathbf{y} = \mathbf{x}z^{-1}$ for some $z \in \Bbbk_1^\times$. This means we can also view $\Bbbk P^n$ as the orbit space of this action of $\Bbbk_1^\times$ on $S^{(n+1)d-1}$, and we also write the quotient map as $q_n \colon S^{(n+1)d-1} \longrightarrow \Bbbk P^n$; this map is also smooth.

Proposition 8.10

The quotient space associated to the map $q_n \colon S^{(n+1)d-1} \longrightarrow \Bbbk P^n$ is compact.

Proof

This follows from the standard fact that the image of a compact space under a continuous mapping is compact. Of course $\Bbbk P^n$ is also Hausdorff. □

Consider the action of $O(n+1)$ on the unit sphere $\mathbb{S}^n \subseteq \mathbb{R}^{n+1}$. Then for $A \in O(n+1)$, $z = \pm 1$ and $\mathbf{x} \in \mathbb{S}^n$, we have

$$A(\mathbf{x}z^{-1}) = (A\mathbf{x})z^{-1}.$$

Hence there is an induced action of $O(n+1)$ on $\mathbb{R}P^n$ given by

$$A \cdot [\mathbf{x}] = [A\mathbf{x}].$$

This action is transitive and also the matrices $\pm I_{n+1}$ fix every point of $\mathbb{R}P^n$. There is also an action of $SO(n+1)$ on $\mathbb{R}P^n$; notice that $-I_{n+1} \in SO(n+1)$ only if n is odd.

Similarly, $U(n+1)$ and $SU(n+1)$ act on $\mathbb{C}P^n$ with scalar matrices wI_{n+1} ($w \in \mathbb{C}_1^\times$) fixing every element. Notice that if $wI_{n+1} \in SU(n+1)$ then $w^{n+1} = 1$, so there are exactly $(n+1)$ such values.

Finally, $Sp(n+1)$ acts on $\mathbb{H}P^n$ and the matrices $\pm I_{n+1}$ fix every element.

There are some important new quotient Lie groups associated to these actions, the *projective unitary*, *special unitary* and *quaternionic symplectic groups*

$$\begin{aligned} PU(n+1) &= U(n+1)/\{wI_{n+1} : w \in \mathbb{C}_1^\times\}, \\ PSU(n+1) &= SU(n+1)/\{wI_{n+1} : w^{n+1} = 1\}, \\ PSp(n+1) &= Sp(n+1)/\{\pm I_{n+1}\}. \end{aligned}$$

Each of these is a subgroup of one of the three *projective linear groups* $PGL_n(\mathbb{R})$, $PGL_n(\mathbb{C})$ and $PGL_n(\mathbb{H})$ where the first two were mentioned in Section 4.6, while the last is defined by

$$PGL_n(\mathbb{H}) = PGL_n(\mathbb{H})/\{tI_n : t \in \mathbb{R}^\times\}.$$

Projective spaces are themselves homogeneous spaces. Consider the subgroup of $O(n+1)$ consisting of elements of the form

$$\begin{bmatrix} a_{11} & a_{12} & \cdots & a_{1\,n-1} & 0 \\ a_{21} & \ddots & \ddots & \ddots & 0 \\ \vdots & \ddots & \ddots & \ddots & \vdots \\ a_{n-1\,1} & \ddots & \ddots & \ddots & 0 \\ 0 & 0 & \cdots & 0 & \pm 1 \end{bmatrix}$$

We denote this subgroup of $\mathrm{O}(n+1)$ by $\mathrm{O}(n) \times \mathrm{O}(1)$. There is a subgroup $\widetilde{\mathrm{O}(n)} \leqslant \mathrm{SO}(n+1)$ whose elements have the form

$$\begin{bmatrix} a_{11} & a_{12} & \cdots & a_{1\,n-1} & 0 \\ a_{21} & \ddots & \ddots & \ddots & 0 \\ \vdots & \ddots & \ddots & \ddots & \vdots \\ a_{n-1\,1} & \ddots & \ddots & \ddots & 0 \\ 0 & 0 & \cdots & 0 & w \end{bmatrix}$$

where

$$\begin{bmatrix} a_{11} & a_{12} & \cdots & a_{1\,n-1} \\ a_{21} & \ddots & \ddots & \ddots \\ \vdots & \ddots & \ddots & \ddots \\ a_{n-1\,1} & \ddots & \ddots & a_{n-1\,n-1} \end{bmatrix} \in \mathrm{O}(n),$$

$$\det \begin{bmatrix} a_{11} & a_{12} & \cdots & a_{1\,n-1} \\ a_{21} & \ddots & \ddots & \ddots \\ \vdots & \ddots & \ddots & \ddots \\ a_{n-1\,1} & \ddots & \ddots & a_{n-1\,n-1} \end{bmatrix} = w.$$

Similarly, there is a subgroup $\mathrm{U}(n) \times \mathrm{U}(1) \leqslant \mathrm{U}(n+1)$ whose elements have the form

$$\begin{bmatrix} a_{11} & a_{12} & \cdots & a_{1\,n-1} & 0 \\ a_{21} & \ddots & \ddots & \ddots & 0 \\ \vdots & \ddots & \ddots & \ddots & \vdots \\ a_{n-1\,1} & \ddots & \ddots & a_{n-1\,n-1} & 0 \\ 0 & 0 & \cdots & 0 & w \end{bmatrix}$$

and $\widetilde{\mathrm{U}}(n) \leqslant \mathrm{SU}(n+1)$ with elements

$$\begin{bmatrix} a_{11} & a_{12} & \cdots & a_{1\,n-1} & 0 \\ a_{21} & \ddots & \ddots & \ddots & 0 \\ \vdots & \ddots & \ddots & \ddots & \vdots \\ a_{n-1\,1} & \ddots & \ddots & a_{n-1\,n-1} & 0 \\ 0 & 0 & \cdots & 0 & w \end{bmatrix}$$

where

$$\begin{bmatrix} a_{11} & a_{12} & \cdots & a_{1\,n-1} \\ a_{21} & \ddots & \ddots & \ddots \\ \vdots & \ddots & \ddots & \ddots \\ a_{n-1\,1} & \ddots & \ddots & a_{n-1\,n-1} \end{bmatrix} \in \mathrm{U}(n),$$

$$\det \begin{bmatrix} a_{11} & a_{12} & \cdots & a_{1\,n-1} \\ a_{21} & \ddots & \ddots & \ddots \\ \vdots & \ddots & \ddots & \ddots \\ a_{n-1\,1} & \ddots & \ddots & a_{n-1\,n-1} \end{bmatrix}^{-1} = w.$$

Finally we have $\mathrm{Sp}(n) \times \mathrm{Sp}(1) \in \mathrm{Sp}(n+1)$ consisting of matrices of the form

$$\begin{bmatrix} a_{11} & a_{12} & \cdots & a_{1\,n-1} & 0 \\ a_{21} & \ddots & \ddots & \ddots & 0 \\ \vdots & \ddots & \ddots & \ddots & \vdots \\ a_{n-1\,1} & \ddots & \ddots & a_{n-1\,n-1} & 0 \\ 0 & 0 & \cdots & 0 & w \end{bmatrix}.$$

Proposition 8.11

There are diffeomorphisms

$$\mathbb{R}\mathrm{P}^n \longrightarrow \mathrm{O}(n+1)/\,\mathrm{O}(n) \times \mathrm{O}(1), \qquad \mathbb{R}\mathrm{P}^n \longrightarrow \mathrm{SO}(n+1)/\widetilde{\mathrm{O}(n)};$$
$$\mathbb{C}\mathrm{P}^n \longrightarrow \mathrm{U}(n+1)/\,\mathrm{U}(n) \times \mathrm{U}(1), \qquad \mathbb{C}\mathrm{P}^n \longrightarrow \mathrm{SU}(n+1)/\widetilde{\mathrm{U}(n)};$$
$$\mathbb{H}\mathrm{P}^n \longrightarrow \mathrm{Sp}(n+1)/\,\mathrm{Sp}(n) \times \mathrm{Sp}(1).$$

There are similar homogeneous spaces of the general and special linear groups also diffeomorphic to these projective spaces. We illustrate this with one example.

The special linear group $\mathrm{SL}_2(\mathbb{C})$ contains the matrix subgroup P consisting of its lower triangular matrices

$$\begin{bmatrix} u & 0 \\ w & v \end{bmatrix} \in \mathrm{SL}_2(\mathbb{C}).$$

P is often referred to as a *parabolic subgroup* of $\mathrm{SL}_2(\mathbb{C})$.

Proposition 8.12

$\mathrm{SL}_2(\mathbb{C})/\mathrm{P}$ is diffeomorphic to $\mathbb{C}\mathrm{P}^1$.

Proof

There is smooth map

$$\psi\colon \mathrm{SL}_2(\mathbb{C}) \longrightarrow \mathbb{C}\mathrm{P}^1; \quad \psi(A) = [A\mathbf{e}_2].$$

Notice that for $B = \begin{bmatrix} u & 0 \\ w & v \end{bmatrix} \in \mathrm{P}$,

$$\begin{bmatrix} u & 0 \\ w & v \end{bmatrix} \begin{bmatrix} 0 \\ 1 \end{bmatrix} = \begin{bmatrix} 0 \\ v \end{bmatrix},$$

so $[(AB)\mathbf{e}_2] = [A\mathbf{e}_2]$ for any $A \in \mathrm{SL}_2(\mathbb{C})$. This means that $\psi(A)$ only depends on the coset $A\mathrm{P} \in \mathrm{SL}_2(\mathbb{C})/\mathrm{P}$. It is easy to see that is onto and that the induced map $\mathrm{SL}_2(\mathbb{C})/\mathrm{P} \longrightarrow \mathbb{C}\mathrm{P}^1$ is injective. □

8.4 Grassmannians

There are some important families of homogeneous spaces directly generalising projective spaces. These are the real, complex and quaternionic *Grassmannians*, which we now define.

Let $\mathrm{O}(k) \times \mathrm{O}(n-k) \leqslant \mathrm{O}(n)$ be the closed subgroup whose elements have the form

$$\begin{bmatrix} A & O_{k,n-k} \\ O_{n-k,k} & B \end{bmatrix} \quad (A \in \mathrm{O}(k),\ B \in \mathrm{O}(n-k)).$$

Similarly there are closed subgroups $\mathrm{U}(k) \times \mathrm{U}(n-k) \leqslant \mathrm{U}(n)$ and $\mathrm{Sp}(k) \times \mathrm{Sp}(n-k) \leqslant \mathrm{Sp}(n)$ with elements

$$\mathrm{U}(k) \times \mathrm{U}(n-k): \begin{bmatrix} A & O_{k,n-k} \\ O_{n-k,k} & B \end{bmatrix} \quad (A \in \mathrm{U}(k),\ B \in \mathrm{U}(n-k));$$

$$\mathrm{Sp}(k) \times \mathrm{Sp}(n-k): \begin{bmatrix} A & O_{k,n-k} \\ O_{n-k,k} & B \end{bmatrix} \quad (A \in \mathrm{Sp}(k),\ B \in \mathrm{Sp}(n-k)).$$

The associated homogeneous spaces are the Grassmannians

$$\mathrm{Gr}_{k,n}(\mathbb{R}) = \mathrm{O}(n)/\,\mathrm{O}(k) \times \mathrm{O}(n-k);$$
$$\mathrm{Gr}_{k,n}(\mathbb{C}) = \mathrm{U}(n)/\,\mathrm{U}(k) \times \mathrm{U}(n-k);$$
$$\mathrm{Gr}_{k,n}(\mathbb{H}) = \mathrm{Sp}(n)/\,\mathrm{Sp}(k) \times \mathrm{Sp}(n-k).$$

Proposition 8.13

For $\mathbb{k} = \mathbb{R}, \mathbb{C}, \mathbb{H}$, the Grassmannian $\mathrm{Gr}_{k,n}(\mathbb{k})$ can be viewed as the set of all k-dimensional $\mathbb{k}$-vector subspaces in $\mathbb{k}^n$.

Proof

We describe the real case $\mathbb{k} = \mathbb{R}$, the others being analogous.

Associated to element $W \in \mathrm{O}(n)$ is the subspace spanned by the first k columns of W, say $\mathbf{w}_1, \ldots, \mathbf{w}_k$; we will denote this subspace by $\langle \mathbf{w}_1, \ldots, \mathbf{w}_k \rangle$. As the columns of W are an orthonormal set, they are linearly independent, hence $\dim_{\mathbb{R}} \langle \mathbf{w}_1, \ldots, \mathbf{w}_k \rangle = k$. Notice that the remaining $(n-k)$ columns give rise to another subspace $\langle \mathbf{w}_{k+1}, \ldots, \mathbf{w}_n \rangle$ of dimension $\dim_{\mathbb{R}} \langle \mathbf{w}_{k+1}, \ldots, \mathbf{w}_n \rangle = n - k$. In fact these are mutually orthogonal in the sense that

$$\begin{aligned}
\langle \mathbf{w}_{k+1}, \ldots, \mathbf{w}_n \rangle &= \langle \mathbf{w}_1, \ldots, \mathbf{w}_k \rangle^{\perp} \\
&= \{\mathbf{x} \in \mathbb{R}^n : \mathbf{x} \cdot \mathbf{w}_r = 0, r = 1, \ldots, k\}, \\
\langle \mathbf{w}_1, \ldots, \mathbf{w}_k \rangle &= \langle \mathbf{w}_{k+1}, \ldots, \mathbf{w}_n \rangle^{\perp} \\
&= \{\mathbf{x} \in \mathbb{R}^n : \mathbf{x} \cdot \mathbf{w}_r = 0, r = k+1, \ldots, n\}.
\end{aligned}$$

For a matrix

$$\begin{bmatrix} A & O_{k,n-k} \\ O_{n-k,k} & B \end{bmatrix} \in \mathrm{O}(k) \times \mathrm{O}(n-k),$$

the columns in the product

$$W' = W \begin{bmatrix} A & O_{k,n-k} \\ O_{n-k,k} & B \end{bmatrix}$$

span subspaces $\langle \mathbf{w}'_1, \ldots, \mathbf{w}'_k \rangle$ and $\langle \mathbf{w}'_{k+1}, \ldots, \mathbf{w}'_n \rangle$. Note that $\mathbf{w}'_1, \ldots, \mathbf{w}'_k$ are orthonormal and also linear combinations of $\mathbf{w}_1, \ldots, \mathbf{w}_k$; similarly, $\mathbf{w}'_{k+1}, \ldots, \mathbf{w}'_n$ are linear combinations of $\mathbf{w}_{k+1}, \ldots, \mathbf{w}_n$. Hence

$$\langle \mathbf{w}'_1, \ldots, \mathbf{w}'_k \rangle = \langle \mathbf{w}_1, \ldots, \mathbf{w}_k \rangle, \quad \langle \mathbf{w}'_{k+1}, \ldots, \mathbf{w}'_n \rangle = \langle \mathbf{w}_{k+1}, \ldots, \mathbf{w}_n \rangle.$$

So there is a well-defined function

$$\mathrm{O}(n)/\mathrm{O}(k) \times \mathrm{O}(n-k) \longrightarrow k\text{-dimensional vector subpaces of } \mathbb{R}^n$$

which sends the coset of W to the subspace $\langle \mathbf{w}_1, \ldots, \mathbf{w}_k \rangle$. This is actually a bijection.

Notice also that there is another bijection

$$\mathrm{O}(n)/\mathrm{O}(k) \times \mathrm{O}(n-k) \longrightarrow (n-k)\text{-dimensional vector subpaces of } \mathbb{R}^n$$

which sends the coset of W to the subspace $\langle \mathbf{w}_{k+1}, \ldots, \mathbf{w}_n \rangle$. This corresponds to a diffeomorphism $\mathrm{Gr}_{k,n}(\mathbb{R}) \longrightarrow \mathrm{Gr}_{n-k,n}(\mathbb{R})$ which in turn corresponds to the obvious isomorphism $\mathrm{O}(k) \times \mathrm{O}(n-k) \longrightarrow \mathrm{O}(n-k) \times \mathrm{O}(k)$ induced by conjugation by a suitable element $P \in \mathrm{O}(n)$. □

8.5 The Gram–Schmidt Process

The Gram–Schmidt process provides a useful algorithm which allows an arbitrary basis of $\mathbb{R}^n$ to be replaced by an orthonormal basis. First we recall this and then explain how it gives rise to a homogeneous space of $\mathrm{GL}_n(\mathbb{R})$.

Let $\{\mathbf{u}_1, \ldots, \mathbf{u}_n\}$ be a basis of $\mathbb{R}^n$. Writing $\mathbf{u}_i^{(0)} = \mathbf{u}_i$, we construct a new sequence of vectors $\mathbf{u}_1^{(1)}, \ldots, \mathbf{u}_n^{(1)}$ by

$$\mathbf{u}_1^{(1)} = \frac{1}{|\mathbf{u}_1^{(0)}|}\mathbf{u}_1^{(0)}, \quad \mathbf{u}_i^{(1)} = \mathbf{u}_i^{(0)} - (\mathbf{u}_1^{(1)} \cdot \mathbf{u}_i^{(0)})\mathbf{u}_1^{(0)} \quad (i = 2, \ldots, n).$$

It is clear that these form a basis of $\mathbb{R}^n$ for which

$$|\mathbf{u}_1^{(1)}| = 1, \quad \mathbf{u}_1^{(1)} \cdot \mathbf{u}_i^{(1)} = 0 \quad (i = 2, \ldots, n).$$

If $U^{(0)}$ is the matrix whose columns are the $\mathbf{u}_i^{(0)}$, then the matrix $U^{(1)}$ whose columns are the $\mathbf{u}_i^{(1)}$ is obtained as

$$\begin{aligned} U^{(1)} &= U^{(0)} \operatorname{diag}\left(\frac{1}{|\mathbf{u}_1^{(0)}|}, 1, \ldots, 1\right) \begin{bmatrix} 1 & -\mathbf{u}_1^{(1)} \cdot \mathbf{u}_2^{(0)} & \cdots & \cdots & -\mathbf{u}_1^{(1)} \cdot \mathbf{u}_n^{(0)} \\ 0 & 1 & 0 & \cdots & 0 \\ \vdots & 0 & 1 & 0 & 0 \\ \vdots & \ddots & \ddots & \ddots & \vdots \\ 0 & \cdots & \cdots & 0 & 1 \end{bmatrix} \\ &= U^{(0)} T^{(1)}, \end{aligned}$$

where $T^{(1)}$ is upper triangular and has positive diagonal entries.

This construction can be iterated by defining for each $k = 1, \ldots, n$, a sequence $\mathbf{u}_1^{(k)}, \ldots, \mathbf{u}_n^{(k)}$ with

$$\mathbf{u}_i^{(k)} = \begin{cases} \mathbf{u}_i^{(k-1)} & \text{if } i = 1, \ldots, k-1, \\ \dfrac{1}{|\mathbf{u}_k^{(k-1)}|}\mathbf{u}_k^{(k-1)} & \text{if } i = k, \\ \mathbf{u}_i^{(k-1)} - \displaystyle\sum_{r=1}^{k} (\mathbf{u}_r^{(k)} \cdot \mathbf{u}_i^{(k-1)})\mathbf{u}_r^{(k)} & \text{if } i = k+1, \ldots, n. \end{cases}$$

It is straightforward to check that these vectors form a basis satisfying

$$|\mathbf{u}_i^{(k)}| = 1 \quad (i = 1, \ldots, k),$$
$$\mathbf{u}_i^{(k)} \cdot \mathbf{u}_j^{(k)} = 0 \quad (i = 1, \ldots, k,\ j = 1, \ldots, n).$$

In particular, the basis $\{\mathbf{u}_1^{(n)}, \ldots, \mathbf{u}_n^{(n)}\}$ is orthonormal,

$$|\mathbf{u}_i^{(n)}| = 1 \quad (i = 1, \ldots, n),$$
$$\mathbf{u}_i^{(n)} \cdot \mathbf{u}_j^{(n)} = 0 \quad (i, j = 1, \ldots, n).$$

At each stage of the iteration the matrix $U^{(k)}$ whose ith column is the vector $\mathbf{u}_i^{(k)}$ given by

$$U^{(k)} = U^{(k-1)} T^{(k)},$$

where $T^{(k)}$ is upper triangular with positive diagonal entries.

From all of this we can deduce that

$$U^{(n)} = U^{(0)} T$$

where $T = T^{(1)} \cdots T^{(n)}$ is upper triangular. Since $U^{(n)}$ is orthogonal and T^{-1} is upper triangular with positive diagonal entries, we obtain

$$U^{(0)} = U^{(n)} T^{-1} \in \mathrm{O}(n)\,\mathrm{UT}_n(\mathbb{R}).$$

If $\mathrm{PDUT}_n(\mathbb{R}) \leqslant \mathrm{UT}_n(\mathbb{R})$ is the matrix subgroup consisting of all upper triangular matrices with positive diagonal entries, then

$$\mathrm{PDUT}_n(\mathbb{R}) \cap \mathrm{O}(n) = \{I\},$$

hence we have shown the following.

Proposition 8.14

Every matrix $A \in$ has a unique factorisation $A = UT$ with $U \in \mathrm{O}(n)$ and $T \in \mathrm{PDUT}_n(\mathbb{R})$.

Corollary 8.15

There is a diffeomorphism $\mathrm{GL}_n(\mathbb{R}) / \mathrm{PDUT}_n(\mathbb{R}) \longrightarrow \mathrm{O}(n)$.

By conjugating and considering the matrix subgroup $\mathrm{PDLT}_n(\mathbb{R}) \leqslant \mathrm{GL}_n(\mathbb{R})$ of lower triangular matrices with positive diagonal entries, we also obtain the following.

Corollary 8.16

There is a diffeomorphism $\mathrm{GL}_n(\mathbb{R})/\mathrm{O}(n) \longrightarrow \mathrm{PDLT}_n(\mathbb{R})$.

Here the space $\mathrm{PDLT}_n(\mathbb{R})$ is contractible, hence so is the homogeneous space $\mathrm{GL}_n(\mathbb{R})/\mathrm{O}(n)$.

A similar construction works for a $\mathbb{C}$-basis of $\mathbb{C}^n$, using the standard hermitian inner product.

8.6 Reduced Echelon Form

The basic theory of systems of linear equations makes use of *Gaussian elimination* to produce the *reduced echelon form* of a matrix. For example, starting with an invertible matrix $A \in \mathrm{M}_n(\mathbb{k})$ where $\mathbb{k} = \mathbb{R}$ or $\mathbb{C}$, a sequence of elementary column operations will convert A into a lower triangular matrix. If A is invertible, $A \in \mathrm{GL}_n(\mathbb{k})$, this lower triangular matrix is the identity matrix I_n. If we only use elementary column operations which involve adding a multiple of column j to column i where $j < i$, then the best result will be a lower triangular matrix with non-zero diagonal terms; but even this may not be possible. To ensure that we obtain a lower triangular matrix we may also need to permute columns. This leads to the *LPU-decomposition* of A, see [28].

Proposition 8.17

Let $A \in \mathrm{GL}_n(\mathbb{k})$. Then there are matrices $L, P, U \in \mathrm{GL}_n(\mathbb{k})$ where L is lower triangular, P is a permutation matrix, U is upper triangular and $A = LPU$. Moreover, P is unique and if we require that the diagonal entries of L are all 1, then so are L and U. Therefore the homogeneous space $\mathrm{GL}_n(\mathbb{k})/\mathrm{UT}_n(\mathbb{k})$ is diffeomorphic to the manifold

$$\mathbb{k}^{\binom{n}{2}} \times \mathrm{Sym}_n, \mathrm{S}_n,$$

where S_n is the symmetric group on n elements.

This is a closely related to the important concept of a *Bruhat decomposition.*

8.7 Real Inner Products

Recall that an *inner product* on the real vector space $\mathbb{R}^n$ is an $\mathbb{R}$-bilinear map $\beta\colon \mathbb{R}^n \times \mathbb{R}^n \longrightarrow \mathbb{R}$ which satisfies

$$\beta(\mathbf{x}, \mathbf{y}) = \beta(\mathbf{y}, \mathbf{x})$$

and so is *symmetric.* β is *non-degenerate* if for every non-zero $\mathbf{x} \in \mathbb{R}^n$ there is a $\mathbf{y} \in \mathbb{R}^n$ for which $\beta(\mathbf{x}, \mathbf{y}) \neq 0$.

Given the standard basis $\{\mathbf{e}_1, \dots, \mathbf{e}_n\}$ of $\mathbb{R}^n$, β is determined by the $n \times n$ matrix $B_\beta = [b_{ij}]$ for which

$$b_{ij} = \beta(\mathbf{e}_i, \mathbf{e}_j) = b_{ji}.$$

Hence B_β is a symmetric matrix and

$$\beta(\mathbf{x}, \mathbf{y}) = \mathbf{x}^T B_\beta \mathbf{y}.$$

If we express an arbitrary vector $\mathbf{x} \in \mathbb{R}^n$ as $\mathbf{x} = x_1\mathbf{e}_1 + \cdots + x_n\mathbf{e}_n$ for $x_i \in \mathbb{R}$, then

$$\begin{aligned}
\beta(\mathbf{x}, \mathbf{x}) &= \sum_{i,j=1}^{n} b_{ij}x_ix_j \\
&= \sum_{i=1}^{n} b_{ii}x_i^2 + 2 \sum_{1\leqslant i<j\leqslant n}^{n} b_{ij}x_ix_j \\
&= Q_\beta(\mathbf{x}),
\end{aligned}$$

which is a real symmetric *quadratic form.* The following is a well-known result on real inner products and quadratic forms; see [5, 16].

Theorem 8.18

Given an inner product β on $\mathbb{R}^n$, there is a matrix $P \in \mathrm{GL}_n(\mathbb{R})$ for which

$$(P^T)^{-1}B_\beta P^{-1} = \mathrm{diag}(\underbrace{1, \dots, 1}_{r_+}, \underbrace{0, \dots, 0}_{r_0}, \underbrace{-1, \dots, -1}_{r_-})$$

with r_+ 1's, r_0 0's and r_- -1's. Moreover, the numbers r_+, r_0 and r_- are independent of the matrix P.

We will say that two inner products β_1 and β_2 on $\mathbb{R}^n$ (or equivalently the symmetric matrices B_{β_1} and B_{β_2}) are *related* and write $\beta_1 \approx \beta_2$ (or $B_{\beta_1} \approx B_{\beta_2}$) if

$$B_{\beta_2} = (P^T)^{-1}B_{\beta_1}P^{-1}$$

for some $P \in \mathrm{GL}_n(\mathbb{R})$.

Corollary 8.19

β is non-degenerate if and only if $\det B \neq 0$, or equivalently if $r_0 \neq 0$.

Certain combinations of the numbers occurring here are given special names:

$$\begin{aligned} \text{the } \textit{rank} \text{ of } \beta &= \operatorname{rank}\beta = n = r_+ + r_-, \\ \text{the } \textit{index} \text{ or } \textit{signature} \text{ of } \beta &= \operatorname{sign}\beta = r_+ - r_-, \\ \text{the } \textit{nullity} \text{ of } \beta &= \operatorname{null}\beta = r_0. \end{aligned}$$

These three numbers determine the diagonal matrix $(P^T)^{-1}B_\beta P$. We can also sensibly refer to $\operatorname{rank} B$, $\operatorname{sign} B$ and $\operatorname{null} B$ for a real symmetric matrix.

The $n \times n$ real symmetric matrices form a subspace $\mathrm{Sym}_n(\mathbb{R}) \subseteq \mathrm{M}_n(\mathbb{R})$ of dimension $\binom{n}{2}$. We introduce a continuous action of $\mathrm{GL}_n(\mathbb{R})$ on $\mathrm{Sym}_n(\mathbb{R})$ by

$$\mathrm{GL}_n(\mathbb{R}) \times \mathrm{Sym}_n(\mathbb{R}) \longrightarrow \mathrm{Sym}_n(\mathbb{R}); \quad P \cdot B = (P^T)^{-1}BP^{-1}.$$

Notice that for each $P \in \mathrm{GL}_n(\mathbb{R})$, P acts as a linear transformation on $\mathrm{Sym}_n(\mathbb{R})$.

We can of course consider the stabiliser and orbit of an element $B \in \mathrm{Sym}_n(\mathbb{R})$. For example, if $B = I_n$, then

$$\begin{aligned} \mathrm{Stab}_{\mathrm{GL}_n(\mathbb{R})}(I_n) &= \{P \in \mathrm{GL}_n(\mathbb{R}) : (P^T)^{-1}I_nP^{-1} = I_n\} = \mathrm{O}(n), \\ \mathrm{Orb}_{\mathrm{GL}_n(\mathbb{R})}(I_n) &= \{B \in \mathrm{Sym}_n(\mathbb{R}) : \operatorname{sign} B = n\}. \end{aligned}$$

Hence we see that the homogenous space of $\mathrm{O}(n)$ is

$$\mathrm{GL}_n(\mathbb{R})/\,\mathrm{O}(n) = \{B \in \mathrm{Sym}_n(\mathbb{R}) : \operatorname{sign} B = n\},$$

which can also be interpreted as the set of all *positive definite* inner products.

More generally, following the ideas of Section 1.5, considering any $n \times n$ real symmetric matrix Q, we find that its stabiliser is

$$\mathrm{Stab}_{\mathrm{GL}_n(\mathbb{R})}(Q) = \mathrm{O}_Q = \{P \in \mathrm{GL}_n(\mathbb{R}) : P^TQP = Q\},$$

the generalised orthogonal group associated with Q. In particular, when

$$Q = \operatorname{diag}(\underbrace{1,\ldots,1}_{r_+},\underbrace{0,\ldots,0}_{r_0},\underbrace{-1,\ldots,-1}_{r_-}),$$

the homogeneous space $\mathrm{GL}_n(\mathbb{R})/\,\mathrm{O}_Q$ can be interpreted as the space of all inner products with rank n, signature $(r_+ - r_-)$ and nullity r_0.

8.8 Symplectic Forms

Definition 8.20

A *symplectic form* on $\mathbb{R}^n$ is a bilinear form

$$\omega: \mathbb{R}^n \times \mathbb{R}^n \longrightarrow \mathbb{R}$$

which is *skew symmetric*, *i.e.*, for all $\mathbf{x}, \mathbf{y} \in \mathbb{R}^n$,

$$\omega(\mathbf{y}, \mathbf{x}) = -\omega(\mathbf{x}, \mathbf{y}),$$

and non-degenerate, *i.e.*, for every non-zero vector $\mathbf{u} \in \mathbb{R}^n$, there is a vector $\mathbf{v} \in \mathbb{R}^n$ for which $\omega(\mathbf{u}, \mathbf{v}) \neq 0$.

Suppose that $\mathbf{v} = \{\mathbf{v}_1, \ldots, \mathbf{v}_n\}$ is a basis for $\mathbb{R}^n$. Then associated with a skew symmetric bilinear form ω on $\mathbb{R}^n$ is a matrix $S_{\omega,\mathbf{v}} = [s_{ij}]$ given by

$$s_{ij} = \omega(\mathbf{v}_i, \mathbf{v}_j).$$

It is easy to see that $S_{\omega,\mathbf{v}}$ is skew symmetric. Conversely, a skew symmetric matrix $S = [s_{ij}]$ gives rise to a skew symmetric bilinear form ω for which

$$s_{ij} = \omega(\mathbf{v}_i, \mathbf{v}_j).$$

An alternative way to see this connection is using the formula

$$\omega(\mathbf{x}, \mathbf{y}) = \mathbf{x}^T S_{\omega,\mathbf{v}} \mathbf{y} \quad (\mathbf{x}, \mathbf{y} \in \mathbb{R}^n). \tag{8.1}$$

Notice that for any vector $\mathbf{x} \in \mathbb{R}^n$,

$$\omega(\mathbf{x}, \mathbf{x}) = -\omega(\mathbf{x}, \mathbf{x}),$$

hence $\omega(\mathbf{x}, \mathbf{x}) = 0$.

Lemma 8.21

i) A skew symmetric bilinear form ω on $\mathbb{R}^n$ is a symplectic form if and only if the associated matrix $S_{\omega,\mathbf{v}}$ is non-singular, *i.e.*, $\det S_{\omega,\mathbf{v}} \neq 0$.
ii) If ω is a symplectic form on $\mathbb{R}^n$, then n must be even.

Proof

(i) Suppose that ω is a symplectic form. If $S_{\omega,\mathbf{v}}$ is singular, there is a non-zero vector $\mathbf{u}$ for which $S_{\omega,\mathbf{v}}\mathbf{u} = \mathbf{0}$. If $\mathbf{v} \in \mathbb{R}^n$ is any vector for which $\omega(\mathbf{u}, \mathbf{v}) \neq 0$, then

$$\omega(\mathbf{u}, \mathbf{v}) = -\omega(\mathbf{v}, \mathbf{u}) = \mathbf{v}^T S_{\omega,\mathbf{v}} \mathbf{u} = 0,$$

which contradicts the assumption on $\mathbf{v}$.

If $S_{\omega,\mathbf{v}}$ is non-singular, then for any non-zero vector $\mathbf{u} \in \mathbb{R}^n$, $S_{\omega,\mathbf{v}}\mathbf{u} \neq \mathbf{0}$, hence there must be some vector $\mathbf{v} \in \mathbb{R}^n$ for which

$$\mathbf{v}^T(S_{\omega,\mathbf{v}}\mathbf{u}) \neq 0.$$

But then we have $\omega(\mathbf{v}, \mathbf{u}) \neq 0$. So ω is non-degenerate.
(ii) If $\det S_{\omega,\mathbf{v}} \neq 0$ then n must be even by Equation (1.5). □

Theorem 8.22 (Darboux's Theorem)

Let ω be a symplectic form on $\mathbb{R}^{2m}$. Then there is a basis $\mathbf{w}_1, \ldots, \mathbf{w}_{2m}$ of $\mathbb{R}^{2m}$ for which

$$\begin{aligned} \omega(\mathbf{w}_i, \mathbf{w}_j) &= 0 \quad \text{if } (i-j) \neq \pm 1, \\ \omega(\mathbf{w}_{2k}, \mathbf{w}_{2k+1}) &= 0, \\ \omega(\mathbf{w}_{2k-1}, \mathbf{w}_{2k}) &= 1. \end{aligned}$$

Proof

Start with any basis $\mathbf{v} = \{\mathbf{v}_1, \ldots, \mathbf{v}_{2m}\}$ of $\mathbb{R}^{2m}$. Taking $\mathbf{w}_1 = \mathbf{v}_1$, notice that since ω is non-degenerate, at least one of the $\mathbf{v}_j$ with $j \neq 1$ must satisfy

$$\omega(\mathbf{w}_1, \mathbf{v}_j) \neq 0.$$

By reordering the vectors $\mathbf{v}_j$ if necessary, we can assume that $j = 2$. Setting

$$\mathbf{w}_2 = \frac{1}{\omega(\mathbf{w}_1, \mathbf{v}_2)}\mathbf{v}_2,$$

we have $\omega(\mathbf{w}_1, \mathbf{w}_2) = 1$. Notice that

$$\{\mathbf{w}_1, \mathbf{w}_2, \mathbf{v}_3, \ldots, \mathbf{v}_{2m}\}$$

is a basis. Replace each of the vectors $\mathbf{v}_3, \ldots, \mathbf{v}_{2m}$ by

$$\mathbf{v}'_j = \mathbf{v}_j - \omega(\mathbf{v}_j, \mathbf{w}_2)\mathbf{w}_1 + \omega(\mathbf{v}_j, \mathbf{w}_1)\mathbf{w}_2.$$

It is easy to see that

$$\{\mathbf{w}_1, \mathbf{w}_2, \mathbf{v}'_3, \ldots, \mathbf{v}'_{2m}\}$$

is a basis for which

$$\omega(\mathbf{w}_i, \mathbf{v}'_j) = 0.$$

We can keep repeating this process to replace this basis by another of the form

$$\{\mathbf{w}_1, \mathbf{w}_2, \mathbf{w}_3, \mathbf{w}_4, \ldots, \mathbf{w}_{2k-1}, \mathbf{w}_{2k}, \mathbf{v}''_{2k} \ldots, \mathbf{v}''_{2m}\}$$

for which

$$\begin{aligned}
\omega(\mathbf{w}_i, \mathbf{w}_j) &= 0 \quad \text{if } (i-j) \neq \pm 1,\\
\omega(\mathbf{w}_{2r}, \mathbf{w}_{2r+1}) &= 0,\\
\omega(\mathbf{w}_{2r-1}, \mathbf{w}_{2r}) &= 1,\\
\omega(\mathbf{w}_i, \mathbf{v}_j'') &= 0.
\end{aligned}$$

Continuing with this we obtain a basis

$$\{\mathbf{w}_1, \mathbf{w}_2, \mathbf{w}_3, \mathbf{w}_4, \ldots, \mathbf{w}_{2m-1}, \mathbf{w}_{2m}\}$$

with $\omega(\mathbf{w}_i, \mathbf{w}_j)$ with properties as above. □

Corollary 8.23

If ω is a symplectic form on $\mathbb{R}^{2m}$, there is a basis

$$\mathbf{w} = \{\mathbf{w}_1, \mathbf{w}_2, \mathbf{w}_3, \mathbf{w}_4, \ldots, \mathbf{w}_{2m-1}, \mathbf{w}_{2m}\}$$

for which $S_{\omega,\mathbf{w}} = J_{2m}$.

Corollary 8.24

Let S be a non-singular $2m \times 2m$ skew symmetric matrix. Then there is a matrix $P \in \mathrm{GL}_{2m}(\mathbb{R})$ for which $S = P^T J_{2m} P$.

It is sometimes useful to reorder such a basis $\mathbf{w}$ with $S_{\omega,\mathbf{w}} = J_{2m}$ to give another basis

$$\widetilde{\mathbf{w}} = \{\mathbf{w}_1, \mathbf{w}_3, \ldots, \mathbf{w}_{2m-1}, \mathbf{w}_2, \mathbf{w}_4, \ldots, \mathbf{w}_{2m}\}$$

for which

$$S_{\omega,\widetilde{\mathbf{w}}} = \begin{bmatrix} O_m & I_n \\ -I_n & O_m \end{bmatrix} = J_{2m}'.$$

This leads to another canonical form for a skew symmetric matrix.

Corollary 8.25

Let S be a non-singular $2m \times 2m$ skew symmetric matrix. Then there is a matrix $Q \in \mathrm{GL}_{2m}(\mathbb{R})$ for which $S = Q^T J_{2m}' Q$.

Let $\Sigma_{2m} \subseteq M_{2m}(\mathbb{R})$ be the subspace of all $2m \times 2m$ non-singular real skew symmetric matrices. There is a continuous action of $GL_{2m}(\mathbb{R})$ on Σ_{2m} given by

$$P \cdot S = (P^{-1})^T S P^{-1}.$$

Then using Corollary 8.24, the homogeneous space $GL_{2m}(\mathbb{R})/\operatorname{Symp}_{2m}(\mathbb{R})$ can be identified with Σ_{2m}. A similar result is true for $GL_{2m}(\mathbb{R})/\operatorname{Symp}'_{2m}(\mathbb{R})$ and Proposition 1.42 can be viewed as a special case of part (iii) of Theorem 1.60. The details are left as an exercise.

We end by giving another interpretation of the *Pfaffian function* introduced in the exercises of Chapter 5.

Proposition 8.26

There is a continuous function

$$\widetilde{\operatorname{pf}}\colon GL_{2m}(\mathbb{R})/\operatorname{Symp}_{2m}(\mathbb{R}) \longrightarrow \mathbb{R}^{\times}; \quad \widetilde{\operatorname{pf}}(P\operatorname{Symp}_{2m}(\mathbb{R})) = \det P.$$

Proof

The point is that the coset $P\operatorname{Symp}_{2m}(\mathbb{R}) \in GL_{2m}(\mathbb{R})/\operatorname{Symp}_{2m}(\mathbb{R})$ corresponds to a skew symmetric matrix $P^T J_{2m} P$ and then

$$\widetilde{\operatorname{pf}}(P\operatorname{Symp}_{2m}(\mathbb{R})) = \operatorname{pf}(P^T J_{2m} P).$$

We leave the reader to work out the details. □

EXERCISES

8.1. Let $\mathbf{k} = \mathbb{R}$ or $\mathbb{C}$ and $n \geqslant 1$. From the proof of Proposition 8.9, recall that the projective space $\mathbf{k}\mathrm{P}^n$ contains an open subset

$$\mathbf{k}\mathrm{P}^n_{n+1} = \{[\mathbf{x}] : \mathbf{x} \neq \mathbf{0},\ x_{n+1} \neq 0\}$$

with a diffeomorphism $\sigma = \sigma_{n+1} \colon \mathbf{k}\mathrm{P}^n_{n+1} \longrightarrow \mathbf{k}^n$, where as usual we write

$$\mathbf{x} = \begin{bmatrix} x_1 \\ \vdots \\ x_{n+1} \end{bmatrix}.$$

Also recall the affine group $\operatorname{Aff}_n(\mathbf{k}) \leqslant GL_{n+1}(\mathbf{k})$.

a) Show that the action of $\operatorname{Aff}_n(\mathbf{k})$ on $\mathbf{k}^{n+1}_0$ gives rise to a continuous action of $\mathrm{Aff}_n(\Bbbk)$ on $\Bbbk\mathrm{P}^n$.

b) Show that the action of (a) restricts to an action of $\mathrm{Aff}_n(\Bbbk)$ on $\Bbbk\mathrm{P}^n_{n+1}$.

c) Using σ, obtain an action of $\mathrm{Aff}_n(\Bbbk)$ on $\Bbbk^n$. Explain how this action is related to that of Section 1.5.

d) How much of this would make sense with $\Bbbk = \mathbb{H}$, the skew field of quaternions?

8.2. For $n \geqslant 1$, consider the matrix group $\mathrm{Sp}(n) \times \mathrm{Sp}(1)$ and its quotient, the *quasi-symplectic group*

$$\mathrm{Sp}(n)\tilde{\times}\,\mathrm{Sp}(1) = \mathrm{Sp}(n) \times \mathrm{Sp}(1)/\{(I_n, 1), (-I_n, -1)\}.$$

a) Show that the Lie algebra of $\mathrm{Sp}(n)\tilde{\times}\,\mathrm{Sp}(1)$ is the direct product of two Lie ideals of dimensions $(2n^2 + n)$ and 3. Is the Lie group $\mathrm{Sp}(n)\tilde{\times}\,\mathrm{Sp}(1)$ a semi-direct product of groups of those dimensions?

b) Show that $\mathbb{H}\mathrm{P}^{n-1}$ is a homogeneous space of $\mathrm{Sp}(n)\tilde{\times}\,\mathrm{Sp}(1)$.

8.3. In the exercises for Chapter 7 we saw that the set of complex structures on $\mathbb{R}^{2n}$ is a submanifold

$$\mathcal{C}_{2n} = \{J \in \mathrm{GL}_{2n}(\mathbb{R}) : J^2 = -I_{2n}\} \subseteq \mathrm{GL}_{2n}(\mathbb{R})$$

and that $\mathrm{GL}_{2n}(\mathbb{R})$ acts smoothly on $\mathcal{C}_{2n}$ with a single orbit. Use this result to identify the homogeneous spaces $\mathrm{GL}_{2n}(\mathbb{R})/\rho_n\,\mathrm{GL}_{2n}(\mathbb{C})$ and $\mathrm{GL}_{2n}(\mathbb{R})/\rho'_n\,\mathrm{GL}_{2n}(\mathbb{C})$ of Section 1.6.

9
Connectivity of Matrix Groups

9.1 Connectivity of Manifolds

Definition 9.1

Let X be a topological space.

- X is *connected* if whenever $X = U \cup V$ with $U, V \neq \varnothing$ both open subsets, then $U \cap V \neq \varnothing$.
- X is *path connected* if whenever $x, y \in X$, there is a continuous path $p\colon [0,1] \longrightarrow X$ with $p(0) = x$ and $p(1) = y$.
- X is *locally path connected* if every point is contained in a path connected open neighbourhood.

The following is a fundamental result of Real Analysis.

Proposition 9.2

Every interval $[a,b], [a,b), (a,b], (a,b) \subseteq \mathbb{R}$ is path connected and connected. In particular, $\mathbb{R}$ is path connected and connected.

Proposition 9.3

If X is a path connected topological space then X is connected.

Proof

Suppose X is not connected. Then $X = U \cup V$ where $U, V \subseteq X$ are non-empty open subsets and $U \cap V = \varnothing$. Let $x \in U$ and $y \in V$. By path connectedness of X, there is a continuous map $p\colon [0,1] \longrightarrow X$ with $p(0) = x$ and $p(1) = y$. Then $[0,1] = p^{-1}U \cup p^{-1}V$ expresses $[0,1]$ as a union of open subsets with no common elements. But this contradicts the connectivity of $[0,1]$. So X must be connected. □

Proposition 9.4

Let X be a connected topological space which is locally path connected. Then X is path connected.

Proof

Let $x \in X$, and set

$$X_x = \{y \in X : \exists p\colon [0,1] \longrightarrow X \text{ continuous with } p(0) = x \text{ and } p(1) = y\}.$$

Then for each $y \in X_x$, there is a path connected open neighbourhood U_y. But for each point $z \in U_y$ there is a continuous path from x to z via y, hence $U_y \subseteq X_x$. This shows that

$$X_x = \bigcup_{y \in X_x} U_y \subseteq X$$

is open in X. Similarly, if $w \in X - X_x$, then $X_w \subseteq X - X_x$ and this is also open. But then so is

$$X - X_x = \bigcup_{w \in X - X_x} X_w.$$

Hence $X = X_x \cup (X - X_x)$, and so by connectivity, $X_x = \varnothing$ or $X - X_x = \varnothing$. Hence X is path connected. □

Proposition 9.5

If the topological spaces X and Y are path connected then their product $X \times Y$ is path connected.

Corollary 9.6

For $n \geqslant 1$, $\mathbb{R}^n$ is path connected and connected.

We also record the following standard results.

Proposition 9.7

i) Let $n \geqslant 2$. The unit sphere $\mathbb{S}^{n-1} \subseteq \mathbb{R}^n$ is path connected. In $\mathbb{S}^0 = \{\pm 1\} \subseteq \mathbb{R}$, the subsets $\{1\}$ and $\{-1\}$ are path connected. The set of non-zero vectors $\mathbb{R}^n_0 \subseteq \mathbb{R}^n$ is path connected.
ii) For $n \geqslant 1$, the sets of non-zero complex and quaternionic vectors $\mathbb{C}^n_0 \subseteq \mathbb{C}^n$ and $\mathbb{H}^n_0 \subseteq \mathbb{H}^n$ are path connected.

Proposition 9.8

Every manifold is locally path connected. Hence every connected manifold is path connected.

Proof

Every point is contained in an open neighbourhood homeomorphic to some open subset of $\mathbb{R}^n$ which can be taken to be an open disc which is path connected. The second statement now follows from Proposition 9.4. □

Theorem 9.9

Let M be a connected manifold and $N \subseteq M$ be a non-empty submanifold which is also a closed subset. If $\dim N = \dim M$ then $N = M$.

Proof

Since $N \subseteq M$ is closed, $M - N \subseteq M$ is open. But $N \subseteq M$ is also open since every element is contained in an open subset of M contained in N; hence $M - N \subseteq M$ is closed. Since M is connected, $M - N = \varnothing$. □

Proposition 9.10

Let G be a Lie group and $H \leqslant G$ be a closed subgroup. If H and G/H are connected, then so is G.

Proof

First we remark on the following: for any $g \in G$, the left translation map $\mathrm{L}_g \colon H \longrightarrow gH$ provides a homeomorphism between these spaces, hence gH is connected since H is.

Suppose that G is not connected, and let $U, V \subseteq G$ be non-empty open subsets for which $U \cap V = \varnothing$ and $U \cup V = G$. By Lemma 8.2, the projection $\pi\colon G \longrightarrow G/H$ is a surjective open mapping, so $\pi U, \pi V \subseteq G/H$ are open subsets for which $\pi U \cup \pi V = G/H$. As G/H is connected, there is an element gH say in $\pi U \cap \pi V$. In G we have

$$gH = (gH \cap U) \cup (gH \cap V),$$

where $(gH \cap U), (gH \cap V) \subseteq gH$ are open subsets in the subspace topology on gH since U, V are open in G. By connectivity of gH, this can only happen if $gH \cap U = \varnothing$ or $gH \cap V = \varnothing$, since these are subsets of U, V which have no common elements. As

$$\pi^{-1} gH = \{gh : h \in H\},$$

this is false, so $(gH \cap U) \cap (gH \cap V) \neq \varnothing$ which implies that $U \cap V \neq \varnothing$. This contradicts the original assumption on U, V. □

Propositions 9.8 and 9.10 provide a useful criterion for path connectedness of a Lie group which may need to be applied repeatedly to show a particular example is path connected. Recall that a closed subgroup of a Lie group is a submanifold by Theorem 7.31.

Proposition 9.11

Let G be a Lie group and $H \leqslant G$ be a closed subgroup. If H and G/H are connected, then G is path connected.

9.2 Examples of Path Connected Matrix Groups

In this section we apply Proposition 9.11 to show that many familiar matrix groups are path connected.

Example 9.12

For $n \geqslant 1$, $\mathrm{SL}_n(\mathbb{R})$ is path connected.

Proof

For the real case, we proceed by induction on n. Notice that $\mathrm{SL}_1(\mathbb{R}) = \{1\}$, which is certainly connected. Now suppose that $\mathrm{SL}_{n-1}(\mathbb{R})$ is path connected for some $n \geqslant 2$.

Recall that $\mathrm{SL}_n(\mathbb{R})$ acts continuously on $\mathbb{R}^n$ by matrix multiplication. Consider the continuous function

$$f\colon \mathrm{SL}_n(\mathbb{R}) \longrightarrow \mathbb{R}^n; \quad f(A) = A\mathbf{e}_n.$$

The image of f is $\operatorname{im} f = \mathbb{R}^n_0 = \mathbb{R}^n - \{0\}$ since every vector $\mathbf{v} \in \mathbb{R}^n_0$ can be extended to a basis

$$\mathbf{v}_1, \ldots, \mathbf{v}_{n-1}, \mathbf{v}_n = \mathbf{v}$$

of $\mathbb{R}^n$, and we can multiply $\mathbf{v}_1$ by a suitable scalar to ensure that the matrix $A_\mathbf{v}$ with these vectors as its columns has determinant 1. Then $A_\mathbf{v}\mathbf{e}_n = \mathbf{v}$.

Notice that $P\mathbf{e}_n = \mathbf{e}_n$ if and only if

$$P = \begin{bmatrix} Q & 0 \\ \mathbf{w} & 1 \end{bmatrix},$$

where Q is $(n-1) \times (n-1)$ with $\det Q = 1$, is the $(n-1) \times 1$ zero vector and $\mathbf{w}$ is an arbitrary $1 \times (n-1)$ vector. The set of all such matrices is the stabiliser of $\mathbf{e}_n$, $\mathrm{Stab}_{\mathrm{SL}_n(\mathbb{R})}(\mathbf{e}_n)$, which is a closed subgroup of $\mathrm{SL}_n(\mathbb{R})$. More generally, $A\mathbf{e}_n = \mathbf{v}$ if and only if $A = A_\mathbf{v}P$ for some $P \in \mathrm{Stab}_{\mathrm{SL}_n(\mathbb{R})}(\mathbf{e}_n)$. So the homogeneous space $\mathrm{SL}_n(\mathbb{R})/\mathrm{Stab}_{\mathrm{SL}_n(\mathbb{R})}(\mathbf{e}_n)$ is homeomorphic to $\mathbb{R}^n_0$.

Since $n \geqslant 2$, it is well known that $\mathbb{R}^n_0$ is path connected, hence it is connected. This implies that $\mathrm{SL}_n(\mathbb{R})/\mathrm{Stab}_{\mathrm{SL}_n(\mathbb{R})}(\mathbf{e}_n)$ is connected.

The subgroup $\mathrm{SL}_{n-1}(\mathbb{R}) \leqslant \mathrm{Stab}_{\mathrm{SL}_n(\mathbb{R})}(\mathbf{e}_n)$ is closed and the well-defined map

$$\mathrm{Stab}_{\mathrm{SL}_n(\mathbb{R})}(\mathbf{e}_n)/\mathrm{SL}_{n-1}(\mathbb{R}) \longrightarrow \mathbb{R}^{n-1}; \quad \begin{bmatrix} Q & 0 \\ \mathbf{w} & 1 \end{bmatrix} \mathrm{SL}_{n-1}(\mathbb{R}) \longmapsto (\mathbf{w}Q^{-1})^T$$

is a homeomorphism so the homogeneous space $\mathrm{Stab}_{\mathrm{SL}_n(\mathbb{R})}(\mathbf{e}_n)/\mathrm{SL}_{n-1}(\mathbb{R})$ is homeomorphic to $\mathbb{R}^{n-1}$. Hence by Corollary 9.6 together with the inductive assumption, $\mathrm{Stab}_{\mathrm{SL}_n(\mathbb{R})}(\mathbf{e}_n)$ is path connected. We can combine this with the connectivity of $\mathbb{R}^n_0$ to deduce that $\mathrm{SL}_n(\mathbb{R})$ is path connected, thus demonstrating the inductive step. □

Example 9.13

For $n \geqslant 1$, $\mathrm{GL}^+_n(\mathbb{R})$ is path connected.

Proof

Since $\mathrm{SL}_n(\mathbb{R}) \leqslant \mathrm{GL}_n^+(\mathbb{R})$, it suffices to show that $\mathrm{GL}_n^+(\mathbb{R})/\mathrm{SL}_n(\mathbb{R})$ is path connected. But for this we can use the determinant to define a continuous map

$$\det\colon \mathrm{GL}_n^+(\mathbb{R}) \longrightarrow \mathbb{R}^+ = (0,\infty),$$

which is surjective onto a path connected space. Then the homogeneous space $\mathrm{GL}_n^+(\mathbb{R})/\mathrm{SL}_n(\mathbb{R})$ is diffeomorphic to $\mathbb{R}^+$ and hence is path connected. So $\mathrm{GL}_n^+(\mathbb{R})$ is path connected. □

This result implies that

$$\mathrm{GL}_n(\mathbb{R}) = \mathrm{GL}_n^+(\mathbb{R}) \cup \mathrm{GL}_n^-(\mathbb{R})$$

is the decomposition of $\mathrm{GL}_n(\mathbb{R})$ into two path connected components.

Example 9.14

For $n \geqslant 1$, $\mathrm{SO}(n)$ is path connected. Hence

$$\mathrm{O}(n) = \mathrm{SO}(n) \cup \mathrm{O}(n)^-$$

is the decomposition of $\mathrm{O}(n)$ into two path connected components.

Proof

For $n = 1$, $\mathrm{SO}(1) = \{1\}$. So we will assume that $n \geqslant 2$ and proceed by induction on n. So assume that $\mathrm{SO}(n-1)$ is path connected.

Consider the continuous action of $\mathrm{SO}(n)$ on $\mathbb{R}^n$ by left multiplication. The stabiliser of $\mathbf{e}_n$ is $\mathrm{SO}(n-1) \leqslant \mathrm{SO}(n)$ thought of as the closed subgroup of matrices of the form

$$\begin{bmatrix} P & \mathbf{0} \\ \mathbf{0}^T & 1 \end{bmatrix}$$

with $P \in \mathrm{SO}(n-1)$ and $\mathbf{0}$ the $(n-1)\times 1$ zero matrix. The orbit of $\mathbf{e}_n$ is the unit sphere $\mathbb{S}^{n-1}$ which is path connected. Since the orbit space is also diffeomorphic to $\mathrm{SO}(n)/\mathrm{SO}(n-1)$ we have the inductive step. □

Example 9.15

For $n \geqslant 1$, $\mathrm{U}(n)$ and $\mathrm{SU}(n)$ are path connected.

Proof

For $n = 1$, $\mathrm{U}(1)$ is the unit circle in $\mathbb{C}$ while $\mathrm{SU}(1) = \{1\}$, so both of these are path connected. Assume that $\mathrm{U}(n-1)$ and $\mathrm{SU}(n-1)$ are path connected for some $n \geqslant 2$.

Then $\mathrm{U}(n)$ and $\mathrm{SU}(n)$ act on $\mathbb{C}^n$ by matrix multiplication and by arguments of Chapter 8,

$$\mathrm{Stab}_{\mathrm{U}(n)}(\mathbf{e}_n) = \mathrm{U}(n-1), \quad \mathrm{Stab}_{\mathrm{SU}(n)}(\mathbf{e}_n) = \mathrm{SU}(n-1).$$

We also have

$$\mathrm{Orb}_{\mathrm{U}(n)}(\mathbf{e}_n) = \mathrm{Orb}_{\mathrm{SU}(n)}(\mathbf{e}_n) = \mathbb{S}^{2n-1},$$

where $\mathbb{S}^{2n-1} \subseteq \mathbb{C}^n \cong \mathbb{R}^{2n}$ denotes the unit sphere consisting of unit vectors. Since $\mathbb{S}^{2n-1}$ is path connected, we can deduce that $\mathrm{U}(n)$ and $\mathrm{SU}(n)$ are too, which gives the inductive step. □

9.3 The Path Components of a Lie Group

Let G be a Lie group. We say that two elements $x, y \in G$ are *connected by a path in G* if there is a continuous path $p\colon [0,1] \longrightarrow G$ with $p(0) = x$ and $p(1) = y$; we will then write $x \underset{G}{\sim} y$.

Lemma 9.16

$\underset{G}{\sim}$ is an equivalence relation on G.

For $g \in G$, we can consider the equivalence class of g, the *path component* of g in G,

$$G_g = \{x \in G : x \underset{G}{\sim} g\}.$$

Proposition 9.17

i) The path component of the identity is a clopen normal subgroup of G, $G_1 \triangleleft G$; hence it is a closed Lie subgroup of dimension $\dim G$.
ii) The path component G_g agrees with the coset of g with respect to G_1, $G_g = gG_1 = G_1 g$ and is a closed submanifold of G.

Proof

By Proposition 9.8, G_g contains an open neighbourhood of g in G. This shows that every component is actually a submanifold of G with dimension equal to $\dim G$. The argument used in the proof of Proposition 9.4 shows that each G_g is actually clopen in G.

Let $x, y \in G_1$. Then there are continuous paths $p, q\colon [0,1] \longrightarrow G$ with $p(0) = 1 = q(0)$, $p(1) = x$ and $q(1) = y$. The product path

$$r\colon [0,1] \longrightarrow G; \quad r(t) = p(t)q(t)$$

has $r(0) = 1$ and $r(1) = xy$. So $G_1 \leqslant G$. For $g \in G$, the path

$$s\colon [0,1] \longrightarrow G; \quad s(t) = gp(t)g^{-1}$$

has $s(0) = 1$ and $s(1) = gxg^{-1}$; hence $G_1 \lhd G$. If $z \in gG_1 = G_1 g$, then $g^{-1}z \in G_1$ and so there is a continuous path $h\colon [0,1] \longrightarrow G$ with $h(0) = 1$ and $h(1) = g^{-1}z$. Then the path

$$gh\colon [0,1] \longrightarrow G; \quad gh(t) = g(h(t))$$

has $gh(0) = g$ and $gh(1) = z$. So each coset gG_1 is path connected, and $gG_1 \subseteq G_g$. To show equality, suppose that g is connected by a path $k\colon [0,1] \longrightarrow G$ in G to $w \in G_g$. Then the path $g^{-1}k$ connects 1 to $g^{-1}w$, so $g^{-1}w \in G_1$, giving $w \in gG_1$. This shows that $G_g \subseteq gG_1$.

The remaining details are left to the reader. □

The quotient group G/G_1 is the *group of path components* of G, which is denoted by $\pi_0 G$.

Example 9.18

We have the following groups of path components:

$$\pi_0\,\mathrm{SO}(n) = \pi_0\,\mathrm{SL}_n(\mathbb{R}) = \pi_0\,\mathrm{SU}(n) = \pi_0\,\mathrm{U}(n) = \pi_0\,\mathrm{SL}_n(\mathbb{C}) = \pi_0\,\mathrm{GL}_n(\mathbb{C}) = \{1\},$$
$$\pi_0\,\mathrm{O}(n) \cong \pi_0\,\mathrm{GL}_n(\mathbb{R}) \cong \{\pm 1\}.$$

Example 9.19

Let

$$T = \left\{ \begin{bmatrix} \cos\theta & -\sin\theta & 0 \\ \sin\theta & \cos\theta & 0 \\ 0 & 0 & 1 \end{bmatrix} : \theta \in \mathbb{R} \right\} \leqslant \mathrm{SO}(3)$$

and let $G = \mathrm{N}_{\mathrm{SO}(3)}(T) \leqslant \mathrm{SO}(3)$ be the normaliser of T. Then T and G are Lie subgroups of $\mathrm{SO}(3)$ and $\pi_0 G \cong \{\pm 1\}$.

Proof

A straightforward computation shows that

$$N_{SO(3)}(T) = T \cup \left\{ \begin{bmatrix} -\cos\theta & \sin\theta & 0 \\ \sin\theta & \cos\theta & 0 \\ 0 & 0 & -1 \end{bmatrix} : \theta \in \mathbb{R} \right\} = T \cup ZT,$$

where

$$Z = \begin{bmatrix} -1 & 0 & 0 \\ 0 & 1 & 0 \\ 0 & 0 & -1 \end{bmatrix}.$$

Notice that T is isomorphic to the unit circle,

$$T \cong \mathbb{T}; \quad \begin{bmatrix} \cos\theta & -\sin\theta & 0 \\ \sin\theta & \cos\theta & 0 \\ 0 & 0 & 1 \end{bmatrix} \longleftrightarrow e^{\theta i}.$$

This implies that T is path connected and abelian since $\mathbb{T}$ is. The function

$$\varphi\colon G \longrightarrow \mathbb{R}^{\times}; \quad \varphi([a_{ij}]) = a_{33}$$

is continuous and

$$\varphi^{-1}\mathbb{R}^{+} = T, \quad \varphi^{-1}\mathbb{R}^{-} = ZT,$$

hence these are clopen subsets. This shows that the path components of G are $G_I = T$ and ZT, so $\pi_0 G \cong \{\pm 1\}$.

Notice that $N_{SO(3)}(T)$ acts by conjugation on T and every element of $T \lhd N_{SO(3)}(T)$ acts trivially since T is abelian. Then $\pi_0 G$ acts on T with the action of the non-trivial coset arising from conjugation by the matrix Z, *i.e.*,

$$Z \begin{bmatrix} \cos\theta & -\sin\theta & 0 \\ \sin\theta & \cos\theta & 0 \\ 0 & 0 & 1 \end{bmatrix} Z^{-1} = \begin{bmatrix} \cos\theta & \sin\theta & 0 \\ -\sin\theta & \cos\theta & 0 \\ 0 & 0 & 1 \end{bmatrix} = \begin{bmatrix} \cos\theta & -\sin\theta & 0 \\ \sin\theta & \cos\theta & 0 \\ 0 & 0 & 1 \end{bmatrix}^{-1}$$

which agrees with the inverse homomorphism on the unit circle $\mathbb{T} \cong T$. □

Example 9.20

Let

$$T = \{x1 + yi : x, y \in \mathbb{R},\ x^2 + y^2 = 1\} \leqslant \mathrm{Sp}(1),$$

the group of unit quaternions and let $G = N_{Sp(1)}(T) \leqslant \mathrm{Sp}(1)$ be its normaliser. Then T and G are Lie subgroups of $\mathrm{Sp}(1)$ and $\pi_0 G \cong \{\pm 1\}$.

Proof

By a straightforward calculation,

$$G = T \cup \{xj - yk : x, y \in \mathbb{R},\ x^2 + y^2 = 1\} = T \cup jT.$$

T is isomorphic to the unit circle so is path connected and abelian. The function

$$\theta\colon G \longrightarrow \mathbb{R}; \quad \theta(t1 + xi + yj + zk) = y^2 + z^2,$$

is continuous and

$$\theta^{-1}0 = T, \quad \theta^{-1}1 = jT.$$

Hence the path components of G are T, jT, while the group of path components is

$$\pi_0 G = G/T \cong \{\pm 1\}.$$

Under the conjugation action of G on T, every element of T acts trivially, so $\pi_0 G$ acts on T. The action of the non-trivial coset is given by conjugation with j,

$$j(x1 + yi)j^{-1} = x1 - yi,$$

corresponding to the inversion map on the unit circle $\mathbb{T} \cong T$. □

The significance of such examples will become clearer when we discuss maximal tori and their normalisers in Chapter 10.

9.4 Another Connectivity Result

The following result will be used in Chapter 10.

Proposition 9.21

Let G be a connected Lie group and $H \leqslant G$ be a subgroup which contains an open neighbourhood of 1 in G. Then $H = G$.

Proof

Let $U \subseteq H$ be an open neighbourhood of 1 in G. Since the inverse map $\mathrm{inv}\colon G \longrightarrow G$ is a homeomorphism and maps H into itself, by replacing U with $U \cap \mathrm{inv}\, U$ if necessary, we may assume that inv maps the open neighbourhood into itself, *i.e.*, $\mathrm{inv}\, U = U$.

For $k \geqslant 1$, consider

$$U^k = \{u_1 \cdots u_k \in G : u_j \in U\} \subseteq H.$$

Notice that $\operatorname{inv} U^k = U^k$. Also, $U^k \subseteq G$ is open since for $u_1, \ldots, u_k \in U$,

$$u_1 \cdots u_k \in \mathrm{L}_{u_1 \cdots u_{k-1}} U \subseteq U^k$$

where

$$\mathrm{L}_{u_1 \cdots u_{k-1}} U = \mathrm{L}^{-1}_{(u_1 \cdots u_{k-1})^{-1}} U$$

is an open subset of G. Then

$$V = \bigcup_{k \geqslant 1} U^k \subseteq H$$

satisfies $\operatorname{inv} V = V$.

V is closed in G since given $g \in G - V$, for the open set $gV \subseteq G$, if $x \in gV \cap V$ there are $u_1, \ldots, u_r, v_1 \ldots, v_s \in U$ such that

$$gu_1 \cdots u_r = v_1 \cdots v_s,$$

implying that $g = v_1 \cdots v_s u_1^{-1} \cdots u_r^{-1} \in V$, contradicting the assumption on g.

So V is a non-empty clopen subset of G, which is connected. Hence $G - V = \varnothing$, and therefore $V = G$, which also implies that $H = G$. □

EXERCISES

9.1. Let G be a Lie group.
a) Using the fact that a manifold is separable, show that $\pi_0 G$ is a countable group.
b) If G is compact, show that $\pi_0 G$ is a finite group.

9.2. Let $\mathrm{SL}_n(\mathbb{R})$ act smoothly on $\mathbb{R}^n$ by matrix multiplication.
a) Find $\operatorname{Stab}_{\mathrm{SL}_n(\mathbb{R})}(\mathbf{e}_n)$ and $\operatorname{Orb}_{\mathrm{SL}_n(\mathbb{R})}(\mathbf{e}_n)$.
b) Identify the homogeneous space $\mathrm{SL}_n(\mathbb{R})/\operatorname{Stab}_{\mathrm{SL}_n(\mathbb{R})}(\mathbf{e}_n)$ and show that it is path connected if $n \geqslant 2$. Use this to give another proof that $\mathrm{GL}_n(\mathbb{R})$ has two path components.

9.3. Let $\mathrm{SL}_n(\mathbb{C})$ act smoothly on $\mathbb{C}^n$ by matrix multiplication.
a) Find $\operatorname{Stab}_{\mathrm{SL}_n(\mathbb{C})}(\mathbf{e}_n)$ and $\operatorname{Orb}_{\mathrm{SL}_n(\mathbb{C})}(\mathbf{e}_n)$.
b) Identify the homogeneous space $\mathrm{SL}_n(\mathbb{C})/\operatorname{Stab}_{\mathrm{SL}_n(\mathbb{C})}(\mathbf{e}_n)$ and show that it is path connected. Use this to prove that $\mathrm{SL}_n(\mathbb{C})$ is path connected.
c) By making use of the determinant $\det\colon \mathrm{GL}_n(\mathbb{C}) \longrightarrow \mathbb{C}^\times$, deduce that $\mathrm{GL}_n(\mathbb{C})$ is path connected.

9.4. Let $A \in \mathrm{GL}_n(\mathbb{R})$.
a) Show that the symmetric matrix $S = AA^T$ is positive definite, *i.e.*, its eigenvalues are all positive real numbers. Deduce that S has a positive definite real symmetric square root, *i.e.*, there is a positive definite real symmetric matrix S_1 satisfying $S_1^2 = S$.
b) Show that $S_1^{-1}A$ is orthogonal.
c) If $PR = QS$ where P, Q are positive definite real symmetric and $R, S \in \mathrm{O}(n)$, show that $P^2 = Q^2$.
d) Let S_2 be a positive definite real symmetric matrix for which $S_2^2 = \mathrm{diag}(\lambda_1, \ldots, \lambda_n)$. Show that $S_2 = \mathrm{diag}(\sqrt{\lambda_1}, \ldots, \sqrt{\lambda_n})$.
e) Show that A can be uniquely expressed as $A = PR$ where P is positive definite real symmetric and $R \in \mathrm{O}(n)$. If $\det A > 0$, show that $R \in \mathrm{SO}(n)$.
f) Show that the homogeneous space $\mathrm{GL}_n^+(\mathbb{R})/\mathrm{SO}(n)$ is path connected. Using Example 9.14, deduce that $\mathrm{GL}_n^+(\mathbb{R})$ is path connected.
g) Let $B \in \mathrm{GL}_n(\mathbb{C})$. By suitably modifying the details of the real case, show that B can be uniquely expressed as $B = QT$ with Q positive definite hermitian and $T \in \mathrm{U}(n)$. Using Example 9.15, deduce that $\mathrm{GL}_n(\mathbb{C})$ is path connected.
[Such factorisations are called polar decompositions of A and B.]

9.5. For $\Bbbk = \mathbb{R}, \mathbb{C}$ and $n \geqslant 1$, the affine group $\mathrm{Aff}_n(\Bbbk)$ acts on $\Bbbk^n$ as explained in Chapter 1.
a) Find $\mathrm{Stab}_{\mathrm{Aff}_n(\Bbbk)}(\mathbf{0})$ and $\mathrm{Orb}_{\mathrm{Aff}_n(\Bbbk)}(\mathbf{0})$.
b) Show that the affine group $\mathrm{Aff}_n(\mathbb{R})$ has two path components, while $\mathrm{Aff}_n(\mathbb{C})$ is path connected.

9.6. a) If $n \geqslant 1$, use the Gram–Schmidt orthogonalisation process to show that every matrix $A \in \mathrm{GL}_n(\mathbb{R})$ can be expressed in the form $A = QR$, where $Q \in \mathrm{O}(n)$ and $R \in \mathrm{UT}_n(\mathbb{R}) \leqslant \mathrm{GL}_n(\mathbb{R})$ the subgroup of upper triangular matrices. Deduce that the homogeneous space $\mathrm{GL}_n(\mathbb{R})/\mathrm{UT}_n(\mathbb{R})$ is compact. Is it connected?
b) If $A \in \mathrm{SL}_n(\mathbb{R})$, show that we can take $Q \in \mathrm{SO}(n)$ and $\det R = 1$. Deduce that the homogeneous space $\mathrm{SL}_n(\mathbb{R})/(\mathrm{UT}_n(\mathbb{R}) \cap \mathrm{SL}_n(\mathbb{R}))$ is compact. Is it connected?
c) What is the relationship between these two homogeneous spaces?

9.7. Work through the analogue of the previous question for $\mathrm{GL}_n(\mathbb{C})$ after first showing that a matrix $A \in \mathrm{GL}_n(\mathbb{C})$ can be expressed in the form $A = QR$, where $Q \in \mathrm{U}(n)$ and $R \in \mathrm{UT}_n(\mathbb{C})$. Discuss the compactness and connectedness of the homogeneous spaces $\mathrm{GL}_n(\mathbb{C})/\mathrm{UT}_n(\mathbb{C})$ and $\mathrm{SL}_n(\mathbb{C})/(\mathrm{UT}_n(\mathbb{C}) \cap \mathrm{SL}_n(\mathbb{C}))$.

9.8. a) Let $m \geqslant 1$. Using Corollary 8.24, show how to identify the homogeneous spaces $\mathrm{GL}_{2m}(\mathbb{R})/\operatorname{Symp}_{2m}(\mathbb{R})$ and $\mathrm{GL}_{2m}(\mathbb{R})/\operatorname{Symp}'_{2m}(\mathbb{R})$ with the subspace $\Sigma_{2m} \subseteq \mathrm{M}_{2m}(\mathbb{R})$ of all $2m \times 2m$ non-singular real skew symmetric matrices. Use these identifications to prove Proposition 1.42.

b) Use the Pfaffian function $\mathrm{pf}\colon \Sigma_{2m} \longrightarrow \mathbb{R}^{\times}$ to prove that Σ_{2m} is not connected.

c) Fill in the details of Remark 1.43, in particular verify the inclusions $\operatorname{Symp}_{2m}(\mathbb{R}) \leqslant \mathrm{SL}_{2m}(\mathbb{R})$ and $\operatorname{Symp}'_{2m}(\mathbb{R}) \leqslant \mathrm{SL}_{2m}(\mathbb{R})$.

9.9. ⚠ Since $\operatorname{Symp}_2(\mathbb{R}) = \mathrm{SL}_2(\mathbb{R})$, this is a connected group. The aim of this exercise is to prove that $\operatorname{Symp}_{2n}(\mathbb{R})$ is connected for all $n \geqslant 1$. Let ω be the standard symplectic form on $\mathbb{R}^{2n}$, hence $\omega(\mathbf{x},\mathbf{y}) = \mathbf{x}^T J_{2n}\mathbf{y}$ where J_{2n} is defined in Section 1.5.

a) For $n \geqslant 1$, show that

$$\Psi_{2n} = \{(\mathbf{u},\mathbf{v}) \in \mathbb{R}^{2n} \times \mathbb{R}^{2n} : \omega(\mathbf{u},\mathbf{v}) = 1\} \subseteq \mathbb{R}^{2n} \times \mathbb{R}^{2n}$$

is a closed subset.

b) Show that the function

$$\varphi\colon \operatorname{Symp}_{2n}(\mathbb{R}) \longrightarrow \Psi_{2n}; \quad \varphi([\mathbf{a}_1,\ldots,\mathbf{a}_{2n}]) = (\mathbf{a}_{2n-1},\mathbf{a}_{2n})$$

is continuous and surjective, hence identify the homogeneous space $\operatorname{Symp}_{2n}(\mathbb{R})/\operatorname{Symp}_{2n-2}(\mathbb{R})$ with Ψ_{2n}.

c) Show that $\operatorname{Symp}_{2n}(\mathbb{R})$ acts continuously on Ψ_{2n} by

$$\begin{bmatrix} a & b \\ c & d \end{bmatrix} \cdot (\mathbf{u},\mathbf{v}) = (a\mathbf{u} + b\mathbf{v}, c\mathbf{u} + d\mathbf{v}).$$

Show that this action has only one orbit. Deduce that Ψ_{2n} is connected.

d) Prove by induction on n that $\operatorname{Symp}_{2n}(\mathbb{R})$ is connected.

9.10. Let

$$T = \{\operatorname{diag}(u,v) : |u| = |v| = 1\} \leqslant \mathrm{U}(2).$$

Find the normaliser $\mathrm{N}_{\mathrm{U}(2)}(T) \leqslant \mathrm{U}(2)$ and the group of path components $\pi_0 \mathrm{N}_{\mathrm{U}(2)}(T)$.

9.11. Let

$$G = \left\{ \begin{bmatrix} A & O_{2,2} \\ O_{2,2} & B \end{bmatrix} : A, B \in \mathrm{U}(2),\ \det A = \det B \right\} \leqslant \mathrm{SU}(4),$$

$$T' = \left\{ \begin{bmatrix} \operatorname{diag}(u,v) & O_{2,2} \\ O_{2,2} & \operatorname{diag}(w, uv\overline{w}) \end{bmatrix} : |u| = |v| = |w| = 1 \right\} \leqslant G.$$

Find the normaliser $\mathrm{N}_G(T') \leqslant G$ and the group of path components $\pi_0 \mathrm{N}_G(T')$.

Part III

Compact Connected Lie Groups and their Classification

10 Maximal Tori in Compact Connected Lie Groups

In this chapter we will describe some results on the structure of compact connected Lie groups, focusing on the important notion of a *maximal torus* which turns out to be central to the classification of simple compact connected Lie groups. From Chapter 9 we know that many familiar examples of compact matrix groups are path connected.

Although we state results for arbitrary Lie groups we will often give proofs for matrix groups. However, there is no loss of generality in assuming this because of the following important result whose proof requires *Haar measure* and integration on compact Lie groups; see [1, 27].

Theorem 10.1

Let G be a compact Lie group. Then for some $m, n \geqslant 1$, there are injective Lie homomorphisms $G \longrightarrow \mathrm{O}(m)$ and $G \longrightarrow \mathrm{U}(n)$, hence G is a matrix group.

10.1 Tori

The familiar *circle group*

$$\mathbb{T} = \{z \in \mathbb{C} : |z| = 1\} \leqslant \mathbb{C}^{\times}$$

is a matrix group since $\mathbb{C}^\times = \mathrm{GL}_1(\mathbb{C})$; its Lie algebra is the 1-dimensional real abelian subalgebra

$$\{ti \in \mathbb{C} : t \in \mathbb{R}\} \leqslant \mathbb{C}.$$

Notice that $\mathbb{T}$ is a closed and bounded subset of $\mathbb{C}$, hence it is compact; it is also path connected and abelian.

Definition 10.2

For each $r \geqslant 1$,

$$\mathbb{T}^r = \{\mathrm{diag}(z_1, \ldots, z_r) : |z_1| = \cdots = |z_r| = 1\} \leqslant \mathrm{GL}_r(\mathbb{C})$$

is the *standard torus of rank r*. More generally, a *torus of rank r* is any Lie group isomorphic to $\mathbb{T}^r$.

We will often view elements of $\mathbb{T}^r$ as sequences of complex numbers $(z_1, \ldots, z_r)$ with $|z_1| = \cdots = |z_r| = 1$, this corresponds to an identification

$$\mathbb{T}^r \cong \underbrace{\mathbb{T} \times \cdots \times \mathbb{T}}_{r \text{ factors}} \leqslant (\mathbb{C}^\times)^r.$$

$\mathbb{T}^r$ is an abelian Lie group of dimension r whose Lie algebra is the abelian Lie algebra

$$\{\mathrm{diag}(t_1 i, \ldots, t_r i) : t_1, \ldots, t_r \in \mathbb{R}\}$$

which can also be identified with

$$\{(t_1 i, \ldots, t_r i) : \forall k,\ t_k \in \mathbb{R}\}.$$

Proposition 10.3

Let T be a torus. Then T is a compact, path connected and abelian Lie group.

Proof

Since the circle $\mathbb{T}$ is compact and abelian the same is true for $\mathbb{T}^r$ and hence for any torus.

If $(z_1, \ldots, z_r) \in \mathbb{T}^r$, set $z_k = e^{\theta_k i}$. Then there is a continuous path

$$p \colon [0,1] \longrightarrow \mathbb{T}^r; \quad p(t) = (e^{t\theta_1 i}, \ldots, e^{t\theta_r i}),$$

with $p(0) = (1, \ldots, 1)$ and $p(1) = (z_1, \ldots, z_r)$. So $\mathbb{T}^r$ is path connected and therefore every torus is path connected. □

Our next result characterises which compact Lie groups are tori.

Theorem 10.4

Let H be a compact Lie group. Then H is a torus if and only if it is connected and abelian.

Proof

We know that H is a compact Lie group. Every torus is path connected and abelian by Proposition 10.3. So we need to show that when H is connected and abelian it is a torus since by Proposition 9.8 it would be path connected.

Suppose that $\dim H = r$ and let $\mathfrak{h}$ be the Lie algebra of H; then $\dim \mathfrak{h} = r$. Let $X, Y \in \mathfrak{h}$. From the definition of the Lie bracket in the proof of Theorem 3.18,

$$[X,Y] = \frac{\mathrm{d}}{\mathrm{d}\,s}\Big|_{s=0} \frac{\mathrm{d}}{\mathrm{d}\,t}\Big|_{t=0} \exp(sX)\exp(tY)\exp(-sX) = 0$$

since $\exp(sX), \exp(tY) \in H$. Because H is abelian, this gives

$$\exp(sX)\exp(tY)\exp(-sX) = \exp(tY),$$

showing that all Lie brackets in $\mathfrak{h}$ are zero.

Now consider the exponential map $\exp\colon \mathfrak{h} \longrightarrow H$. Propositions 2.2 and 2.1 imply that for all $X, Y \in \mathfrak{h}$,

$$\exp(X)\exp(Y) = \exp(X+Y), \quad \exp(-X) = \exp(X)^{-1}.$$

So $\exp \mathfrak{h} = \operatorname{im}\exp \subseteq H$ is a subgroup. By Proposition 2.4, $\exp \mathfrak{h}$ also contains a neighbourhood of the identity so by Proposition 9.21 we have $\exp \mathfrak{h} = H$.

As exp is a continuous homomorphism, its kernel $K = \ker\exp$ must be discrete since otherwise $\dim \exp(\mathfrak{h}) < r$. This means that $K \subseteq \mathfrak{h}$ is a free abelian subgroup with basis $\{v_1, \ldots, v_s\}$ for some $s \leqslant r$. Extending this to an $\mathbb{R}$-basis $\{v_1, \ldots, v_s, v_{s+1}, \ldots, v_r\}$ of $\mathfrak{h}$ we obtain isomorphisms of Lie groups

$$\exp(\mathfrak{h}) \cong \mathfrak{h}/K \cong \mathbb{R}^r/\mathbb{Z}^s \times \mathbb{R}^{r-s}.$$

The right-hand term is only compact if $s = r$, therefore K contains a basis of $\mathfrak{h}$ and

$$\mathbb{R}^r/\mathbb{Z}^r \cong \mathfrak{h}/K \cong H.$$

Since $\mathbb{T} \cong \mathbb{R}/\mathbb{Z}$, this shows that H is a torus. □

In this proof we made use of the essential idea contained in the next result.

Proposition 10.5

Let T be a torus of rank r. Then the exponential map $\exp\colon \mathfrak{t} \longrightarrow T$ is a surjective homomorphism of Lie groups whose kernel is a discrete subgroup isomorphic to $\mathbb{Z}^r$. Hence there is an isomorphism of Lie groups $\mathbb{R}^r/\mathbb{Z}^r \cong T$.

In the proof Theorem 7.36, we met the idea of a topological generator of the circle group. It turns out that all tori have such generators.

Definition 10.6

Let G be a Lie group. Then an element $g \in G$ is called a *topological generator* or just a *generator* of G if the cyclic subgroup $\langle g \rangle \leqslant G$ is dense in G, *i.e.*, $\overline{\langle g \rangle} = G$.

Proposition 10.7

Every torus T has a generator.

Proof

Without loss of generality we can assume that $T = \mathbb{R}^r/\mathbb{Z}^r$ and will write elements in the form $[x_1, \ldots, x_r] = (x_1, \ldots, x_r) + \mathbb{Z}^r$. The group operation is then addition. Let $U_1, U_2, U_3, \ldots$ be a countable base for the topology on T.

A *cube of side* $\varepsilon > 0$ in T is a subset of the form

$$C([u_1, \ldots, u_r], \varepsilon) = \{[x_1, \ldots, x_r] \in T : \forall k \ |x_k - u_k| < \varepsilon/2\},$$

for some $[u_1, \ldots, u_r] \in T$. Such a cube is the image of a cube in $\mathbb{R}^r$ under the quotient map $\mathbb{R}^r \longrightarrow T$.

Let $C_0 \subseteq T$ be a cube of side $\varepsilon > 0$. Suppose that we have a decreasing sequence of cubes C_k of side ε_k,

$$C_0 \supseteq C_1 \supseteq \cdots \supseteq C_m,$$

where for each $1 \leqslant k \leqslant m$, there is an integer N_k satisfying $N_k\varepsilon_{k-1} > 1$ and $N_k C_k \subseteq U_k$. Now choose an integer N_{m+1} large enough to guarantee that $N_{m+1}C_m = T$. Next choose a small cube $C_{m+1} \subseteq C_m$ of side ε_{m+1} so that $N_{m+1}C_{m+1} \subseteq U_{m+1}$. Then if $\mathbf{z} = [z_1, \ldots, z_r] \in \bigcap_{k \geqslant 1} C_k$, we have $N_k\mathbf{z} \in U_k$ for each k, hence the multiples of $\mathbf{z}$ are dense in T, so $\mathbf{z}$ is a generator of T. $\square$

10.2 Maximal Tori in Compact Lie Groups

Definition 10.8

Let G be a Lie group and $T \leqslant G$ be a closed subgroup which is a torus. Then T is *maximal* or a *maximal torus* in G if the only torus $T' \leqslant G$ for which $T \leqslant T'$ is T.

Remark 10.9

It is easy to see that every torus $T \leqslant G$ is contained in a maximal torus $\overline{T} \leqslant G$ where $\dim T \leqslant \dim \overline{T} \leqslant \dim G$. Later we will also see that if G is connected then every element $g \in G$ is contained in a maximal torus.

We will now describe some important examples of maximal tori in compact connected matrix groups. For $\theta \in [0, 2\pi)$, let

$$R(\theta) = \begin{bmatrix} \cos\theta & -\sin\theta \\ \sin\theta & \cos\theta \end{bmatrix} \in \mathrm{SO}(2).$$

More generally, for each $n \geqslant 1$, and $\theta_i \in [0, 2\pi)$ $(i = 1, \ldots, n)$, let

$$R_{2n}(\theta_1, \ldots, \theta_n) = \begin{bmatrix} R(\theta_1) & O & \cdots & \cdots & \cdots & O \\ O & R(\theta_2) & O & \ddots & \ddots & \vdots \\ \vdots & \ddots & \ddots & \ddots & \ddots & \vdots \\ O & \cdots & \cdots & \cdots & O & R(\theta_n) \end{bmatrix},$$

$$R_{2n+1}(\theta_1, \ldots, \theta_n) = \begin{bmatrix} R(\theta_1) & O & \cdots & \cdots & \cdots & \cdots & O \\ O & R(\theta_2) & O & \ddots & \ddots & \ddots & \ddots \\ \vdots & \ddots & \ddots & \ddots & \ddots & \ddots & \vdots \\ \vdots & \ddots & \ddots & \ddots & O & R(\theta_n) & O \\ O & \cdots & \cdots & \cdots & \cdots & O & 1 \end{bmatrix},$$

where each entry marked O is an appropriately sized block so that these are matrices of sizes $2n \times 2n$ and $(2n + 1) \times (2n + 1)$ respectively. Then $R_{2n}(\theta_1, \ldots, \theta_n) \in \mathrm{SO}(2n)$ and $R_{2n+1}(\theta_1, \ldots, \theta_n) \in \mathrm{SO}(2n + 1)$.

It is obvious that $\mathbb{T} \cong \mathrm{U}(1)$ as Lie groups. After first identifying $\mathbb{C}$ with $\mathbb{R}^2$ as real vector spaces using the bases $\{1, i\}$ and $\{e_1, e_2\}$, we also obtain an isomorphism of Lie groups

$$\mathbb{T} \longrightarrow \mathrm{SO}(2); \quad e^{\theta i} \longmapsto R_2(\theta).$$

The maximal tori listed in the next result are usually regarded as the *standard maximal tori* for the corresponding groups. For details see [1, 7, 13].

Proposition 10.10

Each of the following is a maximal torus in the stated group:

$$\begin{aligned}
&\{R_{2n}(\theta_1,\ldots,\theta_n) : \forall k\ \theta_k \in [0,2\pi)\} \leqslant \mathrm{SO}(2n),\\
&\{R_{2n+1}(\theta_1,\ldots,\theta_n) : \forall k\ \theta_k \in [0,2\pi)\} \leqslant \mathrm{SO}(2n+1),\\
&\{\mathrm{diag}(z_1,\ldots,z_n) : \forall k\ |z_k| = 1\} \leqslant \mathrm{U}(n),\\
&\{\mathrm{diag}(z_1,\ldots,z_n) : \forall k\ |z_k| = 1,\ z_1\cdots z_n = 1\} \leqslant \mathrm{SU}(n),\\
&\{\mathrm{diag}(z_1,\ldots,z_n) : \forall k\ z_k \in \mathbb{C},\ |z_k| = 1\} \leqslant \mathrm{Sp}(n),
\end{aligned}$$

We now begin to study the structure of compact Lie groups in terms of their maximal tori. Throughout the rest of the section, let G be a compact connected Lie group and $T \leqslant G$ be a maximal torus. The next result is fundamental to understanding the relationship between the different maximal tori in G.

Theorem 10.11

If $g \in G$, there is an $x \in G$ such that $g \in xTx^{-1}$, *i.e.*, g is conjugate to an element of T. Equivalently,

$$G = \bigcup_{x \in G} xTx^{-1}.$$

Proof

The proof this depends on the powerful and important *Lefschetz Fixed Point Theorem* from Algebraic Topology and we only give a sketch indicating it is used. For details on this result, the interested reader might usefully consult an introductory book on Algebraic Topology such as [9, 20].

The quotient space G/T is a compact space and each element $g \in G$ gives rise to a continuous map

$$\mu_g \colon G/T \longrightarrow G/T; \quad \mu_g(xT) = (gx)T = gxT.$$

Since G is path connected, there is a continuous map

$$p\colon [0,1] \times G/T \longrightarrow G/T;$$

for which $p(0, xT) = xT$ and $p(1, xT) = gxT$, *i.e.*, p is a *homotopy* $\mathrm{Id}_{G/T} \simeq \mu_g$.

The *Lefschetz Fixed Point Theorem* asserts that μ_g has a fixed point provided that the *Euler characteristic* $\chi(G/T)$ is non-zero. It can indeed be shown that $\chi(G/T) \neq 0$, so this tells us that there is an $x \in G$ such that $gxT = xT$, or equivalently $g \in xTx^{-1}$. □

Theorem 10.12

If $T, T' \leqslant G$ are maximal tori then they are conjugate in G, *i.e.*, there is a $y \in G$ such that $T' = yTy^{-1}$.

Proof

By Proposition 10.7, T' has a generator t say. By Theorem 10.11, there is a $y \in G$ such that $T' \leqslant yTy^{-1}$. As T' is a maximal torus and yTy^{-1} is a torus, we must have $T' = yTy^{-1}$. □

Our next result gives some important special cases related to the examples of Proposition 10.10; it can be proved using Theorem 10.12 and also using elementary linear algebra. Notice that if $A \in \mathrm{SO}(m)$, $A^{-1} = A^T$, while if $B \in \mathrm{U}(m)$, $B^{-1} = B^*$.

Theorem 10.13 (Principal Axis Theorem)

In each of the following matrix groups every element is conjugate to one of the stated form.

$$\begin{array}{rll} \mathrm{SO}(2n): & R_{2n}(\theta_1, \ldots, \theta_n), & \forall k\ \theta_k \in [0, 2\pi); \\ \mathrm{SO}(2n+1): & R_{2n+1}(\theta_1, \ldots, \theta_n), & \forall k\ \theta_k \in [0, 2\pi); \\ \mathrm{U}(n): & \mathrm{diag}(z_1, \ldots, z_n), & \forall k\ z_k \in \mathbb{C},\ |z_k| = 1; \\ \mathrm{SU}(n): & \mathrm{diag}(z_1, \ldots, z_n), & \forall k\ z_k \in \mathbb{C},\ |z_k| = 1,\ z_1 \cdots z_n = 1; \\ \mathrm{Sp}(n): & \mathrm{diag}(z_1, \ldots, z_n), & \forall k\ z_k \in \mathbb{C},\ |z_k| = 1. \end{array}$$

There are related results on the Lie algebra $\mathfrak{g}$ of such a compact, connected matrix group G. Recall that for each $g \in G$, there is a linear transformation

$$\mathrm{Ad}_g \colon \mathfrak{g} \longrightarrow \mathfrak{g}; \quad \mathrm{Ad}_g(t) = gtg^{-1}.$$

Proposition 10.14

Suppose that $g \in G$ and $H, K \leqslant G$ are Lie subgroups with Lie algebras $\mathfrak{h}, \mathfrak{k} \leqslant \mathfrak{g}$. If $gHg^{-1} = K$, then $\mathrm{Ad}_g\, \mathfrak{h} = \mathfrak{k}$.

Proof

By definition, for each $x \in \mathfrak{h}$ there is a curve $\gamma\colon (-\varepsilon, \varepsilon) \longrightarrow H$ with $\gamma(0) = 1$

and $\gamma'(0) = x$. Since $t \mapsto g\gamma(t)g^{-1}$ is a curve in K,

$$\mathrm{Ad}_g(x) = \frac{\mathrm{d}}{\mathrm{d}t} g\gamma(t)g^{-1}_{|t=0} \in \mathfrak{k},$$

so $\mathrm{Ad}_g \mathfrak{h} \leqslant \mathfrak{k}$. Since Ad_g is injective and $\dim \mathfrak{h} = \mathfrak{k}$, we must have $\mathrm{Ad}_g \mathfrak{h} = \mathfrak{k}$. □

If $x, y \in \mathfrak{g}$ and $y = \mathrm{Ad}_g(x)$ we will sometimes say that x is *conjugate* to y with respect to G. Being conjugate in this sense gives an equivalence relation on $\mathfrak{g}$.

For $t \in \mathbb{R}$, let

$$R'(t) = \begin{bmatrix} 0 & -t \\ t & 0 \end{bmatrix}$$

and

$$R'_{2n}(t_1, \ldots, t_n) = \begin{bmatrix} R'(t_1) & O & \cdots & \cdots & \cdots & O \\ O & R'(t_2) & O & \ddots & \ddots & \vdots \\ \vdots & \ddots & \ddots & \ddots & \ddots & \vdots \\ O & \cdots & \cdots & \cdots & O & R'(t_n) \end{bmatrix},$$

$$R'_{2n+1}(t_1, \ldots, t_n) = \begin{bmatrix} R'(t_1) & O & \cdots & \cdots & \cdots & \cdots & O \\ O & R'(t_2) & O & \ddots & \ddots & \ddots & \ddots \\ \vdots & \ddots & \ddots & \ddots & \ddots & \ddots & \vdots \\ \vdots & \ddots & \ddots & \ddots & O & R'(t_n) & O \\ O & \cdots & \cdots & \cdots & \cdots & O & 1 \end{bmatrix}.$$

Then $R'_{2n}(t_1, \ldots, t_n) \in \mathfrak{so}(2n)$ and $R'_{2n+1}(t_1, \ldots, t_n) \in \mathfrak{so}(2n+1)$.

Theorem 10.15 (Principal Axis Theorem for Lie algebras)

For each of the following Lie algebras $\mathfrak{g}$, every element $x \in \mathfrak{g}$ is conjugate in G to one of the stated form.

$$\begin{aligned} \mathfrak{so}(2n):&\quad R'_{2n}(t_1, \ldots, t_n), && \forall k\ \theta_k \in [0, 2\pi); \\ \mathfrak{so}(2n+1):&\quad R'_{2n+1}(t_1, \ldots, t_n), && \forall k\ \theta_k \in [0, 2\pi); \\ \mathfrak{u}(n):&\quad \mathrm{diag}(t_1 i, \ldots, t_n i), && \forall k\ t_k \in \mathbb{R}; \\ \mathfrak{su}(n):&\quad \mathrm{diag}(t_1 i, \ldots, t_n i), && \forall k\ t_k \in \mathbb{R},\ t_1 + \cdots + t_n = 1; \\ \mathfrak{sp}(n):&\quad \mathrm{diag}(t_1 i, \ldots, t_n i), && \forall k\ t_k \in \mathbb{R}. \end{aligned}$$

We can now give an important result which we have already seen is true for many familiar examples.

Theorem 10.16

For a compact connected Lie group G, the exponential map $\exp\colon \mathfrak{g} \longrightarrow G$ is surjective.

Proof

Let $T \leqslant G$ be a maximal torus. By Theorem 10.11, every element $g \in G$ is conjugate to an element $xgx^{-1} \in T$. By Proposition 10.5, $xgx^{-1} = \exp(t)$ for some $t \in \mathfrak{t}$, hence

$$g = x^{-1}\exp(t)x = \exp(\mathrm{Ad}_x(t)),$$

where $\mathrm{Ad}_x(t) \in \mathfrak{g}$. So $g \in \exp\mathfrak{g}$. Therefore $\exp\mathfrak{g} = G$. □

10.3 The Normaliser and Weyl Group of a Maximal Torus

Given Theorem 10.11, we can continue to develop the general theory for a compact connected Lie group G.

Proposition 10.17

Let $A \leqslant G$ be a compact abelian Lie group and suppose that $A_1 \leqslant A$ is the connected component of the identity element. If A/A_1 is cyclic then A has a generator and hence A is contained in a torus in G.

Proof

Let $d = |A/A_1|$. As A_1 is connected and abelian, it is a torus by Theorem 10.4, hence it has a generator a_0 by Proposition 10.7. Let $g \in A$ be an element of A for which the coset gA_1 generates A/A_1. Notice that $g^d \in A_1$ and therefore $a_0 g^{-d} \in A_1$. Now choose $b \in A_1$ so that $a_0 g^{-d} = b^d$. Then $a_0 = (gb)^d$, so the powers $(gb)^{kd}$ are dense in A_1. More generally, the powers of the form $(gb)^{kd+r}$ are dense in the coset $g^r A_1$. Hence the powers of gb are dense in A, which shows that this element is a generator of A.

Let $T \leqslant G$ be a maximal torus. By Theorem 10.11, any generator u of A is contained in a maximal torus xTx^{-1} conjugate to T. Hence $\langle u \rangle$ and its closure A are contained in xTx^{-1} which completes the proof. □

Proposition 10.18

Let $A \leqslant G$ be a connected abelian subgroup and let $g \in G$ commute with all the elements of A. Then there is a torus $T \leqslant G$ containing the subgroup $\langle A, g\rangle \leqslant G$ generated by A and g.

Proof

By replacing A by its closure which is also connected, we can assume that A is closed in G, hence compact and so a torus, by Theorem 10.4. Now consider the abelian subgroup $\langle A, g\rangle \leqslant G$ generated by A and g, whose closure $B \leqslant G$ is again compact and abelian. If the connected component of the identity is $B_1 \leqslant B$ then B_1 has finitely many cosets by compactness, and these are of the form $g^r B_1$ $(r = 0, 1, \dots, d-1)$ for some d. By Proposition 10.17, $\langle A, g\rangle$ is contained in a torus. □

Theorem 10.19

Let $T \leqslant G$ be a maximal torus and let $T \leqslant A \leqslant G$ where A is abelian. Then $A = T$. Equivalently, every maximal torus is a maximal abelian subgroup.

Proof

For each element $g \in A$, Proposition 10.18 implies that there is a torus containing $\langle T, g\rangle$, but by the maximality of T this must equal T. Hence $A = T$. □

We have now established that every maximal torus is also a maximal abelian subgroup, and that any two maximal tori are conjugate in G. Next we focus on the relationship of a maximal torus $T \leqslant G$ to the rest of G.

Recall that for a subgroup $H \leqslant G$, the *normaliser* of H in G is the largest subgroup of G in which H is normal,

$$\mathrm{N}_G(H) = \{g \in G : gHg^{-1} = H\}.$$

Then $\mathrm{N}_G(H) \leqslant G$ is a closed subgroup of G and so is also compact. It also contains H and its closure in G as normal subgroups. There is a continuous left action of $\mathrm{N}_G(H)$ on H by conjugation, *i.e.*, for $g \in \mathrm{N}_G(H)$ and $h \in H$, the action is given by

$$g \cdot h = ghg^{-1}.$$

If $H = T$ is a maximal torus in G, the quotient group $\mathrm{W}_G(T) = \mathrm{N}_G(T)/T$ is known as the *Weyl group* of T in G. Notice that the connected component of the identity in $\mathrm{N}_G(T)$ contains T; in fact it agrees with T as we will soon see.

Lemma 10.20

Let $T \leqslant G$ be a torus and let $Q \leqslant \mathrm{N}_G(T)$ be a connected subgroup acting on T by conjugation. Then Q acts trivially, *i.e.*, for $g \in Q$ and $x \in T$,

$$g \cdot x = gxg^{-1} = x.$$

Proof

Recall that $T \cong \mathbb{R}^r/\mathbb{Z}^r$ as Lie groups. By Proposition 10.5, the exponential map is a surjective group homomorphism $\exp\colon \mathfrak{t} \longrightarrow T$ whose kernel is a discrete subgroup. In fact, there is a commutative diagram

$$\begin{array}{ccccc} \ker\exp & \longrightarrow & \mathfrak{t} & \longrightarrow & T \\ \cong\downarrow & & \cong\downarrow & & \cong\downarrow \\ \mathbb{Z}^r & \longrightarrow & \mathbb{R}^r & \longrightarrow & \mathbb{R}^r/\mathbb{Z}^r \end{array}$$

in which all the maps are the evident ones.

Now a Lie group automorphism $\alpha\colon T \longrightarrow T$ lifts to homomorphism $\widetilde{\alpha}\colon \mathfrak{t} \longrightarrow \mathfrak{t}$ restricting to an isomorphism $\widetilde{\alpha}_0\colon \ker\exp \longrightarrow \ker\exp$. Indeed, since each element of $\ker\exp \cong \mathbb{Z}^r$ is uniquely divisible in $\mathfrak{t} \cong \mathbb{R}^r$, continuity implies that $\widetilde{\alpha}_0$ determines $\widetilde{\alpha}$ on $\mathfrak{t}$. But the automorphism group $\mathrm{Aut}(\ker\exp) \cong \mathrm{Aut}(\mathbb{Z}^r)$ of $\ker\exp \cong \mathbb{Z}^r$ is a discrete group.

From this we see that the action of Q on T by conjugation is determined by its restriction to the action on $\ker\exp$. As Q is connected, every element of Q gives rise to the identity automorphism of the discrete group $\mathrm{Aut}(\ker\exp)$. Hence the action of Q on T is trivial. □

This result shows that $\mathrm{N}_G(T)_1$, the connected component of the identity in $\mathrm{N}_G(T)$, acts trivially on the torus T. In fact, if $g \in \mathrm{N}_G(T)$ acts trivially on T then it commutes with all the elements of T, so by Theorem 10.19 g is in T. Thus T consists of all the elements of G with this property, *i.e.*,

$$T = \{g \in G : gxg^{-1} = x \ \forall x \in T\}. \tag{10.1}$$

In particular, we have $\mathrm{N}_G(T)_1 = T$. We can now state an important omnibus result on the Weyl group.

Theorem 10.21

Let $T \leqslant G$ be a maximal torus.
i) The Weyl group $\mathrm{W}_G(T)$ is isomorphic to the group of components of the normaliser,

$$\mathrm{W}_G(T) = \mathrm{N}_G(T)/T \cong \pi_0\,\mathrm{N}_G(T).$$

Hence $W_G(T)$ is finite.
ii) $W_G(T)$ acts on T by conjugation,

$$gT \cdot x = gxg^{-1}.$$

This action on T is faithful, *i.e.*, the coset $gT \in N_G(T)/T$ acts trivially on T if and only if $g \in T$.

Proof

i) $N_G(T)$ has finitely many cosets of T since it is closed, hence compact, so each coset is clopen.
ii) The faithfulness of the action follows from Equation (10.1). □

Proposition 10.22

Let $T \leqslant G$ be a maximal torus and $x, y \in T$. If x, y are conjugate in G then they are conjugate in $N_G(T)$, hence there is an element $w \in W_G(T)$ for which $y = w \cdot x$.

Proof

Suppose that $y = gxg^{-1}$. Then the centraliser $Z_G(y) \leqslant G$ of y is a closed subgroup containing T. It also contains the maximal torus gTg^{-1} since every element of this commutes with y. Let $H = Z_G(y)_1$, the connected component of the identity in $Z_G(y)$; this is a closed subgroup of G since it is closed in $Z_G(y)$. Then as T, gTg^{-1} are connected subgroups of $Z_G(y)$ they are both contained in H. So T, gTg^{-1} are tori in H and must be maximal since a torus in H containing one of these would be a torus in G where they are already maximal.

By Theorem 10.11 applied to the compact connected Lie group H, gTg^{-1} is conjugate to T in H, so for some $h \in H$ we have $gTg^{-1} = hTh^{-1}$ which gives

$$(h^{-1}g)T(h^{-1}g)^{-1} = T.$$

Thus $h^{-1}g \in N_G(T)$ and

$$(h^{-1}g)x(h^{-1}g)^{-1} = h^{-1}yh = y.$$

Now setting $w = h^{-1}gT \in W_G(T)$ we obtain the desired result. □

10.4 The Centre of a Compact Connected Lie Group

Recall that the centre of G is the closed normal subgroup

$$Z(G) = \{g \in G : \forall x \in G,\ gx = xg\} \triangleleft G.$$

Theorem 10.23

The centre of G is the intersection of all the maximal tori in G,

$$Z(G) = \bigcap_{\substack{T \leqslant G \\ \text{a maximal torus}}} T.$$

Proof

If $c \in Z(G)$ then $c \in T$ for some maximal torus $T \leqslant G$. So for $g \in G$,

$$c = gcg^{-1} \in gTg^{-1},$$

and by Theorem 10.12,

$$c \in \bigcap_{\substack{T \leqslant G \\ \text{a maximal torus}}} T.$$

Hence

$$Z(G) \leqslant \bigcap_{\substack{T \leqslant G \\ \text{a maximal torus}}} T.$$

Conversely, if

$$c \in \bigcap_{\substack{T \leqslant G \\ \text{a maximal torus}}} T,$$

then for any $x \in G$, $x \in T'$ for some maximal torus $T' \leqslant G$, hence c, x are both elements of the abelian group T'. So c and x commute. This shows that $c \in Z(G)$. □

Theorem 10.23 often proves useful when finding the centre of a compact connected Lie group. For example, the centres of $SO(2n)$ and $SO(2n+1)$ lie in the maximal tori of Proposition 10.10, and it is easy to find which elements of these are central in the whole groups. We illustrate this by determining the centre of $SO(4)$.

Consider the element

$$R_4(\theta_1,\theta_2) = \begin{bmatrix} \cos\theta_1 & -\sin\theta_1 & 0 & 0 \\ \sin\theta_1 & \cos\theta_1 & 0 & 0 \\ 0 & 0 & \cos\theta_2 & -\sin\theta_2 \\ 0 & 0 & \sin\theta_2 & \cos\theta_2 \end{bmatrix} \in \mathrm{Z}(\mathrm{SO}(4)).$$

For $R_4(\theta_1,\theta_2)$ to commute with $\begin{bmatrix} 0 & 1 & 0 & 0 \\ 1 & 0 & 0 & 0 \\ 0 & 0 & 0 & 1 \\ 0 & 0 & 1 & 0 \end{bmatrix}$ we must have $\sin\theta_1 = 0 = \sin\theta_2$, so $\cos\theta_1 = \pm 1$ and $\cos\theta_2 = \pm 1$. Hence

$$R_4(\theta_1,\theta_2) = \begin{bmatrix} \pm I_2 & O_2 \\ O_2 & \pm I_2 \end{bmatrix}.$$

Commutativity with $\begin{bmatrix} 0 & 0 & 0 & 1 \\ 1 & 0 & 0 & 0 \\ 0 & 1 & 0 & 0 \\ 0 & 0 & 1 & 0 \end{bmatrix}$ forces $R_4(\theta_1,\theta_2) = \pm I_4$. This also illustrates the case $n = 4$ of Proposition 5.31.

EXERCISES

10.1. Show that there are exactly two Lie isomorphisms $\mathbb{T} \longrightarrow \mathbb{T}$, but infinitely many Lie isomorphisms $\mathbb{T}^r \longrightarrow \mathbb{T}^r$ when $r \geqslant 2$.

10.2. Prove that in a Lie group G every torus $T \leqslant G$ is contained in a maximal torus in G, thus verifying the first part of Remark 10.9. Explain why the second part of this remark is false when G is not connected, *i.e.*, not every element of G is contained in a torus.

10.3. a) Show that the subgroup $T \leqslant \mathrm{U}(2)$ consisting of all the diagonal matrices is a maximal torus of $\mathrm{U}(2)$. Determine the normaliser $\mathrm{N}_{\mathrm{U}(2)}(T)$, its group of components $\pi_0\,\mathrm{N}_{\mathrm{U}(2)}(T)$ and its action by conjugation on T.
b) Show that the subgroup of diagonal matrices in $\mathrm{SU}(3)$ is a maximal torus and determine its normaliser and group of path components and describe the conjugation action of the latter on this torus.
c) Let

$$T_2 = \{\mathrm{diag}(u,v) \in \mathrm{Sp}(2) : u,v \in \mathbb{C}\} \leqslant \mathrm{Sp}(2).$$

Show that T_2 is a maximal torus of $\mathrm{Sp}(2)$. Making use of Example 9.20, determine $\mathrm{N}_{\mathrm{Sp}(2)}(T_2)$ and $\pi_0\,\mathrm{N}_{\mathrm{Sp}(2)}(T_2)$ and describe its conjugation action on T_2.

10.4. Show that the group

$$A = \{(\cos\theta_1 + \sin\theta_1 e_1 e_2)(\cos\theta_2 + \sin\theta_2 e_3 e_4) \cdots (\cos\theta_n + \sin\theta_n e_{2n-1} e_{2n}) : \theta_1, \ldots, \theta_n \in [0, 2\pi)\}$$

is a maximal torus in each of the spinor groups $\mathrm{Spin}(2n)$ and $\mathrm{Spin}(2n+1)$.

For small values of n, determine the normalisers and Weyl groups of A in $\mathrm{Spin}(2n)$ and $\mathrm{Spin}(2n+1)$. Find the conjugation action of each Weyl group on A.

How are these maximal tori related under the double covering maps $\rho\colon \mathrm{Spin}(n) \longrightarrow \mathrm{SO}(m)$ of Section 5.3 to the maximal tori of $\mathrm{SO}(2n)$ and $\mathrm{SO}(2n+1)$ given by Proposition 10.10?

10.5. a) For $n \geqslant 1$, show that

$$\{\mathrm{diag}(\varepsilon_1, \ldots, \varepsilon_n) : \varepsilon_1, \ldots, \varepsilon_n = \pm 1\} \leqslant \mathrm{O}(n)$$

is a maximal abelian subgroup of $\mathrm{O}(n)$.

b) Let $T_{2n} \leqslant \mathrm{SO}(2n)$ and $T_{2n+1} \leqslant \mathrm{SO}(2n+1)$ be the maximal tori of Proposition 10.10 and

$$T'_{2n} = T_{2n}, \quad T'_{2n+1} = T_{2n+1} \cup \mathrm{diag}(1, \ldots, 1, -1) T_{2n+1}.$$

Show that $T'_{2n} \leqslant \mathrm{O}(2n)$ and $T'_{2n+1} \leqslant \mathrm{O}(2n+1)$ are maximal abelian subgroups.

c) Explain why these results are compatible with those of Chapter 10.

11
Semi-simple Factorisation

In this chapter we begin our discussion of the classification theory of compact connected Lie groups. Our aim is to introduce the main ideas involved and eventually to describe the outcome of this classification in terms of root systems and Dynkin diagrams. The reader interested in seeing more details might usefully consult some of the books [1, 2, 4, 7, 13, 14, 26] which focus on various aspects of this theory.

Assumption 11.1

Throughout this chapter, unless otherwise specified, G will denote a compact connected Lie group with Lie algebra $\mathfrak{g}$. By Theorem 10.1 we may assume that $G \leqslant \mathrm{U}(n)$ for some n and $\mathfrak{g} \leqslant \mathfrak{u}(n)$ is a Lie subalgebra. This allows us to give proofs valid for matrix subgroups of unitary groups.

11.1 An Invariant Inner Product

In order to discuss the semi-simple decomposition of a compact connected Lie group and later of its adjoint representation, we first need to introduce an $\mathbb{R}$-bilinear inner product on its Lie algebra. We do this by first defining such an inner product on the real Lie algebra $\mathfrak{u}(n) = \text{Sk-Herm}_n$, then restricting it to a subalgebra.

We define $(\ |\ )$ by

$$(X \mid Y) = -\operatorname{tr}(XY) = \operatorname{tr}(X^*Y) \in \mathbb{C} \quad (X, Y \in \mathfrak{u}(n)). \tag{11.1}$$

Then $(\ |\ )$ is symmetric since $\operatorname{tr}(YX) = \operatorname{tr}(XY)$. Using standard properties of the trace together with the definition of the hermitian conjugate $(\ )^*$, we find that

$$\begin{aligned}\overline{(X \mid Y)} = \overline{\operatorname{tr}(XY)} &= \operatorname{tr}(\overline{X}\,\overline{Y}) \\ &= \operatorname{tr}(\overline{X}\,\overline{Y})^T \\ &= -\operatorname{tr}(Y^*X^*) \\ &= -\operatorname{tr}(YX) \\ &= (Y \mid X) \\ &= (X \mid Y),\end{aligned}$$

hence $(X \mid Y)$ is real. Finally,

$$(X \mid X) = -\operatorname{tr}(X^2) = \operatorname{tr}(X^*X),$$

and the latter is the sum of the eigenvalues of the positive definite hermitian matrix X^*X. Since these eigenvalues are real and non-negative, $(\ |\ )$ is real and positive definite. This is still true for the restriction of $(\ |\ )$ to an $\mathbb{R}$-Lie subalgebra of $\mathfrak{u}(n)$, in particular to the Lie subalgebra of a Lie subgroup of $\mathrm{U}(n)$. In the case $\mathrm{SU}(2) \leqslant \mathrm{U}(2)$, $(\ |\ )$ agrees with the inner product introduced in Section 3.5.

For $A \in \mathrm{U}(n)$,

$$\begin{aligned}(AXA^{-1} \mid AYA^{-1}) &= \operatorname{tr}(AXA^{-1}AYA^{-1}) \\ &= \operatorname{tr}(AXYA^{-1}) \\ &= \operatorname{tr}(XY),\end{aligned}$$

giving

$$(\mathrm{Ad}_A(X) \mid \mathrm{Ad}_A(Y)) = (X \mid Y). \tag{11.2}$$

This shows that $\mathrm{Ad}_A \colon \mathfrak{u}(n) \longrightarrow \mathfrak{u}(n)$ is a linear isometry with respect to the inner product $(\ |\ )$. If we choose an orthonormal basis for $\mathfrak{u}(n)$ then the adjoint representation becomes a homomorphism $\mathrm{Ad}\colon \mathrm{U}(n) \longrightarrow \mathrm{SO}(\dim \mathrm{U}(n))$ since $\mathrm{U}(n)$ is connected.

When $A = \exp(tH)$ with $t \in \mathbb{R}$ and $H \in \mathfrak{u}(n)$,

$$\begin{aligned}\frac{\mathrm{d}}{\mathrm{d}t}(AXA^{-1} \mid AYA^{-1})_{|t=0} &= \frac{\mathrm{d}}{\mathrm{d}t}\operatorname{tr}(AXYA^{-1})_{|t=0} \\ &= \operatorname{tr}(HXY - XYH) \\ &= \operatorname{tr}([H,X]Y + X[H,Y]) \\ &= ([H,X] \mid Y) + (X \mid [H,Y]),\end{aligned}$$

and so

$$(\mathrm{ad}_H(X) \mid Y) + (X \mid \mathrm{ad}_H(Y)) = 0. \tag{11.3}$$

Because of Equations (11.2) and (11.3), the inner product $(\ \mid\ )$ is said to be *invariant* with respect to Ad, ad or the Lie bracket $[\ ,\ ]$. On choosing an orthonormal basis for $\mathfrak{u}(n)$ with respect to the inner product $(\ \mid\ )$, each associated matrix $\mathrm{Ad}_H \in \mathrm{GL}_{\dim \mathrm{U}(n)}(\mathbb{R})$ is special orthogonal and $\mathrm{ad}_H \in \mathrm{M}_{\dim \mathrm{U}(n)}(\mathbb{R})$ is skew-symmetric, hence the adjoint representation becomes a homomorphism of Lie algebras

$$\mathrm{ad}\colon \mathfrak{u}(n) \longrightarrow \mathfrak{so}(\dim \mathrm{U}(n)) = \text{Sk-Sym}_{\dim \mathrm{U}(n)}(\mathbb{R}).$$

Here is the more general result obtained by restricting the inner product $(\ \mid\ )$ to the subspace $\mathfrak{g}$.

Theorem 11.2

Let $G \leqslant \mathrm{U}(n)$ be a compact connected Lie subgroup. On choosing an orthonormal $\mathbb{R}$-basis for $\mathfrak{g}$ with respect to the inner product $(\ \mid\ )$, the adjoint representations of G and $\mathfrak{g}$ take the form

$$\mathrm{Ad}\colon G \longrightarrow \mathrm{SO}(\dim G), \quad \mathrm{ad}\colon \mathfrak{g} \longrightarrow \mathfrak{so}(\dim G),$$

which are homomorphisms of groups and Lie algebras respectively.

Recall from Definition 3.31 the notion of complexification of a Lie algebra. We can extend the inner product $(\ \mid\ )$ to a $\mathbb{C}$-bilinear inner product on the complexification $\mathfrak{u}(n)_{\mathbb{C}} = \mathfrak{gl}_n(\mathbb{C})$. The adjoint action of $\mathrm{U}(n)$ on $\mathfrak{u}(n)$ extends to a $\mathbb{C}$-linear action on $\mathfrak{u}(n)_{\mathbb{C}}$ and this extension is Ad-invariant.

Proposition 11.3

Let $\mathfrak{g} \leqslant \mathfrak{u}(n)$. Then there is a unique extension of $(\ \mid\ )$ on $\mathfrak{g}$ to a $\mathbb{C}$-bilinear inner product $(\ \mid\ )_{\mathbb{C}}$ on the complexification $\mathfrak{g}_{\mathbb{C}}$. Furthermore, $(\ \mid\ )_{\mathbb{C}}$ is both Ad-invariant and ad-invariant.

Proof

Let $\{U_1, \ldots, U_d\}$ be an $\mathbb{R}$-basis of $\mathfrak{g}$, which is also a $\mathbb{C}$-basis of $\mathfrak{g}_{\mathbb{C}}$. Taking as the definition

$$(\sum_i z_i U_i \mid \sum_j w_j U_j)_{\mathbb{C}} = \sum_i \sum_j z_i w_j (U_i \mid U_j),$$

it is easily seen that $(\ \mid\ )_{\mathbb{C}}$ is indeed the unique $\mathbb{C}$-bilinear extension of $(\ \mid\ )$ to the complex vector space $\mathfrak{g}_{\mathbb{C}}$. The invariance is straightforward to check. □

Example 11.4

For $n \geqslant 1$, the $\mathbb{C}$-basis of $\mathfrak{u}(n)_{\mathbb{C}} = \mathfrak{gl}_n(\mathbb{C})$ of Equation (3.27) satisfies

$$(E^{k\ell} \mid E^{rs}) = -\delta_{kr}\delta_{\ell s},$$

while for $\mathfrak{su}(n)_{\mathbb{C}} = \mathfrak{sl}_n(\mathbb{C})$,

$$(E^{kk} - E^{(k-1)\,(k-1)} \mid E^{rs}) = 0 \qquad \text{if } r \neq s, \tag{11.4a}$$

$$(E^{kk} - E^{(k-1)\,(k-1)} \mid E^{rr} - E^{(r-1)\,(r-1)}) = \begin{cases} -2 & \text{if } r = k, \\ 1 & \text{if } |r-k| = 1. \end{cases} \tag{11.4b}$$

11.2 The Centre and its Lie Algebra

In Section 10.4 we described the centre $Z(G)$ of G in terms of the maximal tori of G. We can determine the dimension of $Z(G)$ by considering its Lie algebra.

Theorem 11.5

Let G be a compact connected Lie group and let $\mathfrak{z}$ be the Lie algebra of its centre $Z(G)$. Then $\mathfrak{z} = z(\mathfrak{g})$ and so $\dim Z(G) = \dim z(\mathfrak{g})$.

Proof

Let $\gamma\colon (-\varepsilon, \varepsilon) \longrightarrow Z(G)$ be a smooth curve with $\gamma(0) = I$. By Theorem 7.32, for all $x \in \mathfrak{g}$ and $s \in \mathbb{R}$, $\exp(sx) \in G$ and so for all $t \in \mathbb{R}$,

$$\gamma(t)\exp(sx)\gamma(t)^{-1} = \exp(sx).$$

Then we have

$$\begin{aligned}
[\gamma'(0), x] &= \frac{\mathrm{d}}{\mathrm{d}t}\gamma(t)x\gamma(t)^{-1}{}_{|t=0} \\
&= \frac{\mathrm{d}}{\mathrm{d}t}\gamma(t)\left(\frac{\mathrm{d}}{\mathrm{d}s}\exp(sx)_{|s=0}\right)\gamma(t)^{-1}{}_{|t=0} \\
&= \frac{\mathrm{d}}{\mathrm{d}t}\frac{\mathrm{d}}{\mathrm{d}s}\left(\gamma(t)\exp(sx)\gamma(t)^{-1}\right)_{|s=0|t=0} \\
&= \frac{\mathrm{d}}{\mathrm{d}t}\frac{\mathrm{d}}{\mathrm{d}s}\left(\exp(sx)\right)_{|s=0|t=0} \\
&= \frac{\mathrm{d}}{\mathrm{d}t}x_{|t=0} = 0,
\end{aligned}$$

so $\gamma'(0) \in \mathfrak{z}(\mathfrak{g})$. This shows that $\mathfrak{z} \subseteq \mathfrak{z}(\mathfrak{g})$.

Conversely, let $z \in \mathfrak{z}(\mathfrak{g})$ and $t \in \mathbb{R}$. Then by Theorem 10.16, every element $g \in G$ has the form $g = \exp(x)$ for some $x \in \mathfrak{g}$, therefore making use of Proposition 2.2 we have

$$\begin{aligned}\exp(tz)g\exp(tz)^{-1} &= \exp(tz)\exp(x)\exp(tz)^{-1}\\ &= \exp(tz+x)\exp(-tz)\\ &= \exp(tz+x-tz)\\ &= \exp(x),\end{aligned}$$

which shows that $\exp(tz) \in Z(G)$ and hence $z \in \mathfrak{z}$ by Equation (7.15). So $\mathfrak{z}(\mathfrak{g}) \subseteq \mathfrak{z}$.

Combining these two inclusions we obtain $\mathfrak{z}(\mathfrak{g}) = \mathfrak{z}$ and so

$$\dim Z(G) = \dim \mathfrak{z} = \dim \mathfrak{z}(\mathfrak{g}).$$ □

Corollary 11.6

$\dim Z(G) = 0$ if and only if $\mathfrak{g}$ is centreless.

A Lie group G for which Corollary 11.6 holds can still have non-trivial centre; this is illustrated in Propositions 5.31 and 5.32. Here is a more general observation in this vein.

Proposition 11.7

Let G be a compact Lie group and $H \leqslant G$ be a closed subgroup. Then $\dim H = 0$, if and only if H is a finite subgroup.

More generally, H has finitely many path components, each of which is compact.

Proof

When $\dim H = 0$, H is a discrete subgroup of the compact group G so it must be finite as shown in the exercises for Chapter 1.

If $H \leqslant G$ is a closed subset then it is compact by Proposition 1.26. The path components form an open cover of H by disjoint sets, and this must be finite by compactness. □

Now we can give another description of the centre of G.

Proposition 11.8

The adjoint action of G on $\mathfrak{g}$ gives a Lie homomorphism

$$\mathrm{Ad}\colon G \longrightarrow \mathrm{SO}(\dim G); \quad g \longmapsto \mathrm{Ad}_g,$$

with compact connected image and kernel $\ker \mathrm{Ad} = \mathrm{Z}(G)$.

Proof

Let $g \in \mathrm{Z}(G)$. Then if $x \in \mathfrak{g}$ and $t \in \mathbb{R}$, $g\exp(tx)g^{-1} = \exp(tx)$ and differentiating with respect to t at 0 gives $gxg^{-1} = x$. Hence $g \in \ker \mathrm{Ad}$.

Conversely, if $g \in \ker \mathrm{Ad}$ then for $x \in \mathfrak{g}$, $gxg^{-1} = x$ and by applying the exponential we obtain

$$g\exp(x)g^{-1} = \exp(gxg^{-1}) = \exp(x).$$

Since every element of G has the form $\exp(x)$, we have $g \in \mathrm{Z}(G)$. □

11.3 Lie Ideals and the Adjoint Action

We would like to generalise Theorem 11.5 to arbitrary closed normal subgroups. We will make use of ideas and notation introduced in Section 1.8. First we need to investigate the adjoint action further and in particular its behaviour with respect to Lie ideals in the Lie algebra $\mathfrak{g}$.

Proposition 11.9

Let G be a compact connected Lie group and $\mathfrak{n} \subseteq \mathfrak{g}$ be an $\mathbb{R}$-subspace. Then $\mathfrak{n} \lhd \mathfrak{g}$ if and only if the adjoint action of G on $\mathfrak{g}$ restricts to an action on $\mathfrak{n}$,

$$\mathrm{Ad}_g\colon \mathfrak{n} \longrightarrow \mathfrak{n}; \quad \mathrm{Ad}_g(x) = gxg^{-1}.$$

Proof

Let $\mathfrak{n} \lhd \mathfrak{g}$. Then for every $u \in \mathfrak{g}$,

$$\mathrm{ad}_u(x) = [u, x] \in \mathfrak{n} \quad (x \in \mathfrak{n}).$$

This shows that the adjoint action of $\mathfrak{g}$ on $\mathfrak{n}$ assigns to each $u \in \mathfrak{g}$ a linear transformation $\mathrm{ad}_u \in \mathrm{End}_{\mathbb{R}}(\mathfrak{n})$. Now consider the differential equation

$$\varphi'(t) = \varphi(t)\,\mathrm{ad}_u, \tag{11.5}$$

where $\varphi\colon \mathbb{R} \longrightarrow \mathrm{GL}_{\mathbb{R}}(\mathfrak{n})$ is supposed to be a smooth function and $\varphi(t)\,\mathrm{ad}_u$ means the composition of these elements $\mathrm{End}_{\mathbb{R}}(\mathfrak{n})$. By the general theory of such equations given in Theorem 2.12, this has a unique solution. To see this solution explicitly, for each $t \in \mathbb{R}$ consider the function

$$\mathrm{Ad}_{\exp(tu)}\colon \mathfrak{g} \longrightarrow \mathfrak{g}; \quad \mathrm{Ad}_{\exp(tu)}(x) = \exp(tu)x\exp(-tu).$$

On differentiating with respect to t we find

$$\begin{aligned}\frac{\mathrm{d}}{\mathrm{d}t}\exp(tu)x\exp(-tu) &= \exp(tu)[u,x]\exp(-tu)\\ &= \mathrm{Ad}_{\exp(tu)}(\mathrm{ad}_u(x)),\end{aligned}$$

hence this satisfies Equation (11.5) and so its unique solution is the function φ for which

$$\varphi(t) = \mathrm{Ad}_{\exp(tu)}\colon \mathfrak{n} \longrightarrow \mathfrak{n}.$$

Since G is compact and connected, by Theorem 10.16, every element $g \in G$ has the form $g = \exp(u)$ for some $u \in \mathfrak{g}$, hence Ad_g restricts to a map $\mathrm{Ad}_g\colon \mathfrak{n} \longrightarrow \mathfrak{n}$. This means that $\mathfrak{n}$ is indeed closed under the action of each Ad_g, *i.e.*, $\mathrm{Ad}_g \in \mathrm{GL}_{\mathbb{R}}(\mathfrak{n})$.

Conversely, suppose that $\mathfrak{n} \subseteq \mathfrak{g}$ is closed under the action of Ad_g for every $g \in G$, then for $u \in \mathfrak{g}$, $x \in \mathfrak{n}$ and $t \in \mathbb{R}$, $\mathrm{Ad}_{\exp(tu)}(x) \in \mathfrak{n}$ and differentiating at $t = 0$ gives $[u,x] \in \mathfrak{n}$. Hence $\mathfrak{n} \lhd \mathfrak{g}$. □

Now suppose that there is a connected normal Lie subgroup $N \lhd G$ with Lie algebra $\mathfrak{n} \leqslant \mathfrak{g}$. For each $x \in \mathfrak{n}$ and $u \in \mathfrak{g}$ we have $\mathrm{Ad}_{\exp(tu)}\exp(sx) \in G$ for all $s,t \in \mathbb{R}$. Differentiating with respect to s at 0, we obtain

$$\frac{\mathrm{d}}{\mathrm{d}s}\mathrm{Ad}_{\exp(tu)}\exp(sx)_{|s=0} = \mathrm{Ad}_{\exp(tu)}(x) \in \mathfrak{n}.$$

Differentiating this with respect to t at 0, now gives $[u,x] \in \mathfrak{n}$. Hence the Lie algebra $\mathfrak{n}$ of a normal subgroup of G is a Lie ideal, $\mathfrak{n} \lhd \mathfrak{g}$.

Proposition 11.10

Let G be a compact connected Lie group. Let $N \lhd G$ be a closed normal subgroup with Lie subalgebra $\mathfrak{n} \leqslant \mathfrak{g}$. Then $\mathfrak{n} \lhd \mathfrak{g}$ and the adjoint action of G on $\mathfrak{g}$ restricts to an action on $\mathfrak{n}$,

$$\mathrm{Ad}_g\colon \mathfrak{n} \longrightarrow \mathfrak{n}; \quad \mathrm{Ad}_g(x) = gxg^{-1}.$$

Corollary 11.11

Let G be a compact connected Lie group. If $\mathfrak{g}$ is simple then G has no closed normal subgroups of dimension greater than 0.

Definition 11.12

A compact connected Lie group with a simple Lie algebra is called *simple.*

Remark 11.13

This terminology conflicts with the usual usage in group theory; however, it does make sense in that much of the structure of a compact connected Lie group is captured by its Lie algebra. For example, as we will see, the classification of simple compact connected Lie groups is essentially accomplished by first classifying their Lie algebras up to isomorphism, then determining their associated groups which differ by surjective Lie homomorphisms with finite kernels, *i.e.*, *finite covering homomorphisms.*

The following converse result is true, but we will not give a proof.

Proposition 11.14

Let G be a compact connected Lie group. If G has no closed normal subgroups of dimension greater than 0, then the Lie algebra $\mathfrak{g}$ is simple.

Recall the *orthogonal complement* of $\mathfrak{n}$ in $\mathfrak{g}$,

$$\mathfrak{n}^{\perp} = \{x \in \mathfrak{g} : \forall y \in \mathfrak{n},\ (x \mid y) = 0\}.$$

Clearly $\mathfrak{n}^{\perp}$ is a vector subspace of $\mathfrak{g}$ and by standard arguments of linear algebra

$$\mathfrak{g} = \mathfrak{n} + \mathfrak{n}^{\perp}, \tag{11.6a}$$

$$\mathfrak{n} \cap \mathfrak{n}^{\perp} = \{0\}, \tag{11.6b}$$

$$(\mathfrak{n}^{\perp})^{\perp} = \mathfrak{n}. \tag{11.6c}$$

Proposition 11.15

Let $\mathfrak{g}$ be a finite dimensional $\mathbb{R}$-Lie algebra with a real positive definite *ad*-invariant inner product $(\ \mid\ )$. Let $\mathfrak{n} \triangleleft \mathfrak{g}$ be any Lie ideal.

i) The orthogonal complement $\mathfrak{n}^{\perp}$ of $\mathfrak{n}$ in $\mathfrak{g}$ is a Lie ideal, $\mathfrak{n}^{\perp} \triangleleft \mathfrak{g}$ which satisfies

the Equation (11.6).
ii) There is an isomorphism of Lie algebras $\mathfrak{n}^\perp \cong \mathfrak{g}/\mathfrak{n}$.

Proof

(i) If $x \in \mathfrak{n}^\perp$ and $y \in \mathfrak{n}$ and $z \in \mathfrak{g}$, then

$$([z, x] \mid y) = -(x \mid [z, y]) = 0,$$

by Equation (11.3) and the fact that $\mathfrak{n}$ is an ideal.
(ii) The quotient homomorphism $\mathfrak{g} \longrightarrow \mathfrak{g}/\mathfrak{n}$ restricts to a homomorphism of Lie algebras $\mathfrak{n}^\perp \longrightarrow \mathfrak{g}/\mathfrak{n}$ which is injective since $\mathfrak{n} \cap \mathfrak{n}^\perp = \{0\}$. Since

$$\dim \mathfrak{n}^\perp = \dim \mathfrak{g}/\mathfrak{n} = \dim \mathfrak{g} - \dim \mathfrak{n},$$

it is also surjective. □

Here are some further results of this type.

Proposition 11.16

Let $\mathfrak{g}$ be a finite dimensional $\mathbb{R}$-Lie algebra equipped with a real positive definite ad-invariant inner product $(\ \mid\ )$.
i) $\mathfrak{g}'$ and $\mathfrak{z}(\mathfrak{g})$ are mutually orthogonal, *i.e.*,

$$([x, y] \mid z) = 0 \quad (x, y \in \mathfrak{g}, z \in \mathfrak{z}(\mathfrak{g})).$$

Hence $\mathfrak{g}' \subseteq \mathfrak{z}(\mathfrak{g})^\perp$,
ii) There is an isomorphism of Lie algebras $\mathfrak{z}(\mathfrak{g})^\perp \cong \mathfrak{g}/\mathfrak{z}(\mathfrak{g})$.
iii) The centre of $\mathfrak{g}/\mathfrak{z}(\mathfrak{g})$ is trivial.

Proof

(i) If $x, y \in \mathfrak{g}$ and $z \in \mathfrak{z}(\mathfrak{g})$, then by Equation (11.3),

$$([x, y] \mid z) = -(y \mid [x, z]) = 0,$$

since $[x, z] = 0$. So $\mathfrak{g}'$ and $\mathfrak{z}(\mathfrak{g})$ are indeed mutually orthogonal.
(ii) This is a special case of Proposition 11.15(ii).
(iii) The isomorphism

$$\mathfrak{z}(\mathfrak{g}/\mathfrak{z}(\mathfrak{g})) \cong \mathfrak{z}(\mathfrak{g}^\perp)$$

and the identity

$$\mathfrak{z}(\mathfrak{g}^\perp) = \mathfrak{z}(\mathfrak{g}) \cap \mathfrak{z}(\mathfrak{g})^\perp = \{0\}$$

give this result. □

11.4 Semi-simple Decompositions

Let G be a compact connected Lie group. The centre $\mathrm{Z}(G)$ of G is a compact normal subgroup, which need not be connected. By Theorem 11.5, the centre of the Lie algebra $\mathfrak{g}$ is an ideal $\mathfrak{z}(\mathfrak{g}) \triangleleft \mathfrak{g}$. By Proposition 11.8, the quotient $G/\mathrm{Z}(G)$ is a compact connected Lie subgroup of $\mathrm{SO}(\dim G)$ and the quotient homomorphism $G \longrightarrow G/\mathrm{Z}(G)$ is a Lie homomorphism whose derivative coincides with the quotient homomorphism of Lie algebras $\mathfrak{g} \longrightarrow \mathfrak{g}/\mathfrak{z}(\mathfrak{g})$. This suggests that we should study G through $G/\mathrm{Z}(G)$ and its Lie algebra. In particular, by Proposition 11.16(iii), $\dim \mathrm{Z}(G/\mathrm{Z}(G)) = 0$ and $\mathrm{Z}(G/\mathrm{Z}(G))$ is a finite group. In fact the centre of $G/\mathrm{Z}(G)$ is trivial as the next result shows.

Proposition 11.17

Let G be a compact connected Lie group. Then the quotient group $G/\mathrm{Z}(G)$ has trivial centre, *i.e.*, $\mathrm{Z}(G/\mathrm{Z}(G)) = \{1\}$.

Proof

This follows from Proposition 11.8 together with Proposition 11.16(ii) and the factorisation of the adjoint map Ad as

$$\mathrm{Ad}\colon G \longrightarrow G/\mathrm{Z}(G) \xrightarrow{\overline{\mathrm{Ad}}} \mathrm{SO}(\dim G),$$

from which it follows that $\ker \mathrm{Ad}$ agrees with the kernel of the induced action of G on $\mathfrak{g}^{\perp} \cong \mathfrak{g}/\mathfrak{z}(\mathfrak{g})$. This in turn agrees with the inverse image of $\mathrm{Z}(G/\mathrm{Z}(G))$ under the quotient homomorphism $G \longrightarrow G/\mathrm{Z}(G)$. □

Assumption 11.18

Because of Proposition 11.17, we may as well replace G by $G/\mathrm{Z}(G)$, so from now on in this section, we will assume that $\mathrm{Z}(G) = \{1\}$. Such a Lie group is called *semi-simple.*

The adjoint action of G on $\mathfrak{g}$ gives rise to an orthogonal decomposition of $\mathfrak{g}$ into Lie ideals. The proof of this requires Haar measure on G and is closely related to Theorem 10.1. For readers familiar with the representation theory of finite groups, we remark that this kind of result is analogous to *Maschke's Theorem.*

Recall from Proposition 11.9 that a Lie ideal of $\mathfrak{g}$ is essentially the same thing as an Ad-closed $\mathbb{R}$-subspace of $\mathfrak{g}$.

Theorem 11.19

There are simple ideals $\mathfrak{g}_i \lhd \mathfrak{g}$ ($i = 1, \ldots, r$) which decompose $\mathfrak{g}$ as an orthogonal direct sum,

$$\mathfrak{g} = \mathfrak{g}_1 \oplus \cdots \oplus \mathfrak{g}_r, \quad \mathfrak{g}_i \subseteq \mathfrak{g}_j^{\perp} \quad \text{if } i \neq j.$$

For each $i = 1, \ldots, r$, the ideal

$$\mathfrak{g}(i) = \mathfrak{g}_1 \oplus \cdots \oplus \mathfrak{g}_{i-1} \oplus \mathfrak{g}_{i+1} \oplus \cdots \oplus \mathfrak{g}_r \lhd \mathfrak{g}$$

is the orthogonal complement of $\mathfrak{g}_i$, *i.e.*,

$$\mathfrak{g}(i) = \mathfrak{g}_i^{\perp}.$$

Notice that if $i \neq j$, $x \in \mathfrak{g}_i$ and $y \in \mathfrak{g}_j$, then since $[x, y] \in \mathfrak{g}_i \cap \mathfrak{g}_j = \{0\}$, we must have $[x, y] = 0$.

The adjoint action of G on $\mathfrak{g}$ restricts to an action on $\mathfrak{g}(i)$. Let

$$G_i = \{g \in G : \mathrm{Ad}_g = \mathrm{Id}_{\mathfrak{g}(i)} : \mathfrak{g}(i) \longrightarrow \mathfrak{g}(i)\}.$$

Then G_i is clearly a closed subgroup of G and so is a compact Lie subgroup. If $x \in \mathfrak{g}_i$ then for $y \in \mathfrak{g}(i)$, $[x, y] = 0$. Hence for all $t \in \mathbb{R}$,

$$\mathrm{Ad}_{\exp(tx)}(y) = y.$$

This shows that $\exp \mathfrak{g}_i \subseteq G_i$.

On the other hand, if $x \in \mathfrak{g}$ is in the Lie subalgebra of G_i, then for every $t \in \mathbb{R}$, $\exp(tx) \in G_i$ so then for every $y \in \mathfrak{g}(i)$,

$$\exp(tx) y \exp(-tx) = y.$$

On differentiating at $t = 0$ we see that $[x, y] = 0$. Writing $x = x_i + x_i'$ with $x_i \in \mathfrak{g}_i$ and $x_i' \in \mathfrak{g}(i)$ we find that $[x_i', y] = 0$ for every $y \in \mathfrak{g}(i)$, so $x_i' \in \mathfrak{z}(\mathfrak{g}) = \{0\}$. Therefore $x \in \mathfrak{g}_i$, which shows that the Lie algebra of G_i is $\mathfrak{g}_i$.

It easy to verify that $G_i \lhd G$ and in fact every pair of elements $g_i \in G_i$ and $g_j \in G_j$ commutes if $i \neq j$. This implies that $\mathrm{Z}(G_i) \leqslant \mathrm{Z}(G)$, giving $\mathrm{Z}(G_i) = \{1\}$.

This discussion leads to the following result.

Theorem 11.20

For each $i = 1, \ldots, r$ there is a closed normal subgroup $G_i \lhd G$ with Lie subalgebra $\mathfrak{g}_i \lhd \mathfrak{g}$. Moreover, there is a direct product decomposition

$$G = G_1 \times \cdots \times G_r.$$

So to understand the structure of an arbitrary compact connected Lie group G, we first factor out its centre $\mathrm{Z}(G)$ to form $G/\mathrm{Z}(G)$, which is semi-simple, so is a product of simple groups. This naturally leads us to consider the structure of simple compact connected Lie groups. In order to do this and to explain the classification of such simple groups we need to study the adjoint representation in greater detail since this provides the key ingredients.

11.5 Structure of the Adjoint Representation

Now let G be a compact connected Lie group of dimension $d = \dim G$ and $T \leqslant G$ be a maximal torus. We will use the notation $d = 2m$ if d is even and $d = 2m + 1$ if d is odd.

By Theorem 10.1, we may assume that G is a subgroup of some unitary group, say $G \leqslant \mathrm{U}(n)$. Also, by Proposition 10.7, T has a topological generator g_0 and on choosing an orthonormal basis for $\mathfrak{g}$, the adjoint action of g_0 gives a linear transformation $\mathrm{Ad}_{g_0}\colon \mathfrak{g} \longrightarrow \mathfrak{g}$ which is equivalent to a matrix $\mathrm{Ad}_{g_0} \in \mathrm{SO}(d)$; this matrix determines the adjoint action of T on $\mathfrak{g}$. The Principal Axis Theorem 10.13 implies that Ad_{g_0} is conjugate in $\mathrm{SO}(d)$ to a matrix of the form $R_d(\theta_1, \ldots, \theta_m)$. The eigenvalues of Ad_{g_0} (with repetitions) must then be of the form

$$\begin{cases} e^{\theta_1 i}, e^{-\theta_1 i}, \ldots, e^{\theta_m i}, e^{-\theta_m i} & \text{if } d \text{ is even,} \\ e^{\theta_1 i}, e^{-\theta_1 i}, \ldots, e^{\theta_m i}, e^{-\theta_m i}, 1 & \text{if } d \text{ is odd.} \end{cases}$$

If we view $\mathrm{SO}(d)$ as a subgroup of $\mathrm{U}(d)$, then Ad_{g_0} is conjugate in $\mathrm{U}(d)$ to the diagonal matrix

$$\begin{cases} \mathrm{diag}(e^{\theta_1 i}, e^{-\theta_1 i}, \ldots, e^{\theta_m i}, e^{-\theta_m i}) & \text{if } d \text{ is even,} \\ \mathrm{diag}(e^{\theta_1 i}, e^{-\theta_1 i}, \ldots, e^{\theta_m i}, e^{-\theta_m i}, 1) & \text{if } d \text{ is odd.} \end{cases}$$

These matrices give the action of Ad_{g_0} on the complexification $\mathfrak{g}_{\mathbb{C}}$ relative to the chosen basis of $\mathfrak{g}$ thought of as a $\mathbb{C}$-basis for $\mathfrak{g}_{\mathbb{C}}$.

For a power g_0^r ($r \in \mathbb{Z}$), $\mathrm{Ad}_{g_0^r} = (\mathrm{Ad}_{g_0})^r$ is conjugate to

$$\begin{cases} R_{\dim G}(r\theta_1, \ldots, r\theta_m) & \text{in } \mathrm{SO}(d), \\ \mathrm{diag}(e^{r\theta_1 i}, e^{-r\theta_1 i}, \ldots, e^{r\theta_m i}, e^{-r\theta_m i}) & \text{in } \mathrm{U}(d) \text{ if } d \text{ is even,} \\ \mathrm{diag}(e^{r\theta_1 i}, e^{-r\theta_1 i}, \ldots, e^{r\theta_m i}, e^{-r\theta_m i}, 1) & \text{in } \mathrm{U}(d) \text{ if } d \text{ is odd.} \end{cases} \tag{11.7}$$

This implies that there is an orthonormal $\mathbb{R}$-basis $\mathcal{B} = \{v_1, \ldots, v_d\}$ of $\mathfrak{g}$ for which

$$\begin{aligned} \mathrm{Ad}_{g_0^r}(v_{2k-1}) &= \cos(r\theta_k)v_{2k-1} + \sin(r\theta_k)v_{2k}, \\ \mathrm{Ad}_{g_0^r}(v_{2k}) &= -\sin(r\theta_k)v_{2k-1} + \cos(r\theta_k)v_{2k}, \end{aligned}$$

whenever $k \leqslant m$; if d is odd we also have

$$\mathrm{Ad}_{g_0^r}(v_d) = v_d.$$

So for $k = 1, \ldots, m$, there are continuous homomorphisms

$$\widetilde{\gamma}_k \colon T \longrightarrow \mathrm{SO}(2) \cong \mathbb{T}; \quad \widetilde{\gamma}_k(g_0^r) = R(r\theta_k).$$

For each $\widetilde{\gamma}_k$, the inverse map given by

$$\widetilde{\gamma}_k^{-1}(g_0^r) = R(r\theta_k)^{-1} = R(-r\theta_k)$$

corresponds to replacing the pair of basis vectors (v_{2k-1}, v_{2k}) by the pair $(-v_{2k-1}, v_{2k})$. The homomorphisms $\widetilde{\gamma}_k, \widetilde{\gamma}_k^{-1}$ $(k = 1, \ldots, m)$ are called the *roots* of the maximal torus T in G, the non-constant ones being called the *non-trivial roots.*

Notice that for $x \in \mathfrak{t} \subseteq \mathfrak{g}$, $\mathrm{Ad}_{g_0}(x) = x$, since this comes from the adjoint action of the abelian group T. Hence if we first choose an orthonormal basis for $\mathfrak{t}$, we can extend it to one for $\mathfrak{g}$ so that it has the properties above. From now on we will assume that $\mathcal{B}$ has been chosen in this way so that $v_1, \ldots, v_{\dim T}$ is an orthonormal basis for $\mathfrak{t}$.

For each root $\widetilde{\gamma} \colon T \longrightarrow \mathbb{T}$ we take its derivative and by identifying the Lie algebra of $\mathbb{T}$ with $\mathbb{R}$ this gives an $\mathbb{R}$-linear transformation $\gamma \colon \mathfrak{t} \longrightarrow \mathbb{R}$ which is also known as a *root.*

In general, if $\gamma \colon \mathfrak{t} \longrightarrow \mathbb{R}$ is a non-trivial root, we define $\mathfrak{g}(\pm\gamma) \subseteq \mathfrak{g}$ to be the maximal subspace for which there is a basis $\{u_1, \ldots, u_{2k}\}$ such that for $x \in \mathfrak{t}$ and $j = 1, \ldots, k$,

$$\mathrm{Ad}_{\exp(x)}(u_{2j-1}) = \cos\gamma(x)\, u_{2j-1} + \sin\gamma(x)\, u_{2j}, \tag{11.8a}$$

$$\mathrm{Ad}_{\exp(x)}(u_{2j}) = -\sin\gamma(x)\, u_{2j-1} + \cos\gamma(x)\, u_{2j}. \tag{11.8b}$$

This is called the *root space* associated to the *root pair* $\pm\gamma$ and by definition it is non-trivial. When γ is non-trivial we have $\dim \mathfrak{g}(\pm\gamma) = 2m_{\pm\gamma}$ for some natural number $m_{\pm\gamma} \geqslant 1$ called the *multiplicity* of the root pair $\pm\gamma$. We also set

$$\mathfrak{g}(0) = \{x \in \mathfrak{g} : \forall g \in T,\ \mathrm{Ad}_g(x) = x\},$$

which is the *null root space.* Then $\mathfrak{t} \subseteq \mathfrak{g}(0)$, so $\mathfrak{g}(0) \neq \{0\}$ if $\dim G > 0$; the number $m_0 = \dim \mathfrak{g}(0)$ is the multiplicity of the trivial root.

For practical purposes it is often more useful to work entirely with the Lie algebra and replace Equation (11.8) by the following equations which are obtained by differentiating the Ad-action at the identity,

$$\mathrm{ad}_x(u_{2j-1}) = \gamma(x)\, u_{2j}, \tag{11.9a}$$

$$\mathrm{ad}_x(u_{2j}) = -\gamma(x)\, u_{2j-1}. \tag{11.9b}$$

Theorem 11.21

There is an $\mathbb{R}$-linear direct sum decomposition which is orthogonal with respect to the invariant inner product $(\ |\ )$,

$$\mathfrak{g} = \bigoplus_{\pm\gamma \text{ a root pair}} \mathfrak{g}(\pm\gamma) = \mathfrak{g}(0) \oplus \bigoplus_{\substack{\pm\gamma \text{ a non-trivial} \\ \text{root pair}}} \mathfrak{g}(\pm\gamma).$$

Proof

This follows easily from the orthogonality of the eigenspaces associated with distinct conjugate pairs of eigenvalues of an orthogonal matrix. □

The decomposition in this result is called the *root space decomposition* of $\mathfrak{g}$ with respect to the maximal torus T. There is a complex version of this also called a *root space decomposition* or *Cartan decomposition* of the complexification $\mathfrak{g}_{\mathbb{C}}$. The adjoint action of T extends to a $\mathbb{C}$-linear action on $\mathfrak{g}_{\mathbb{C}}$ and each of the complexified root spaces $\mathfrak{g}(\pm\gamma)_{\mathbb{C}}$ for non-trivial γ splits into a direct sum of two $\mathbb{C}$-subspaces

$$\mathfrak{g}(\pm\gamma)_{\mathbb{C}} = \mathfrak{g}_{\gamma} \oplus \mathfrak{g}_{-\gamma},$$

where

$$\begin{aligned}\mathfrak{g}_{\gamma} &= \{v \in \mathfrak{g}_{\mathbb{C}} : \forall x \in \mathfrak{t},\ \mathrm{Ad}_{\exp(x)}(v) = e^{\gamma(x)i}v\} \\ &= \{v \in \mathfrak{g}_{\mathbb{C}} : \forall x \in \mathfrak{t},\ \mathrm{ad}_{x}(v) = \gamma(x)iv\}.\end{aligned}$$

We also set $\mathfrak{g}_0 = \mathfrak{g}(0)_{\mathbb{C}}$. When γ is non-trivial, $\dim_{\mathbb{C}} \mathfrak{g}_{\gamma} = m_{\pm\gamma}$, so we usually denote this by m_{γ}.

Theorem 11.22

There is a $\mathbb{C}$-linear direct sum decomposition,

$$\mathfrak{g}_{\mathbb{C}} = \mathfrak{g}_0 \oplus \bigoplus_{\substack{\gamma \text{ a non-trivial} \\ \text{root}}} \mathfrak{g}_{\gamma}.$$

Such root space decompositions contain all the essential information required to classify simple compact connected Lie groups. In order to explain the result of this classification we will need to investigate the geometry of root systems and their Weyl groups and we do this in Chapter 12. To finish this introduction we explain how the Weyl group acts on roots and then give some examples which are elucidated further in Section 12.3.

For a maximal torus $T \leqslant G$, we know that the Weyl group $\mathrm{W}_G(T) = \mathrm{N}_G(T)/T$ acts by conjugation on the Lie algebra $\mathfrak{t}$ of T. There is an associated *contragredient action* on the dual vector space $\mathfrak{t}^*$ consisting of all $\mathbb{R}$-linear transformations $\mathfrak{t} \longrightarrow \mathbb{R}$. This action is specified by taking ${}^{gT}\gamma = gT \cdot \gamma$ to be

$${}^{gT}\gamma \colon \mathfrak{t} \longrightarrow \mathbb{R}; \quad {}^{gT}\gamma(x) = \gamma(\mathrm{Ad}_{g^{-1}}(x)) = \gamma(\mathrm{Ad}_g^{-1}(x)).$$

It is easy to see that if γ is a root, then so is ${}^{gT}\gamma$, and this leads to the following result.

Proposition 11.23

Under the action of $\mathrm{W}_G(T)$ on $\mathfrak{t}^*$, the roots of G with respect to T are permuted.

Roots also have the properties given in the following result whose proof can be found in [13, 26].

Proposition 11.24

Let α, β be roots of G with respect to the maximal torus T.
i) For non-trivial α and $w \in \mathrm{W}_G(T)$, the multiplicities of ${}^{w}\alpha$ and α agree, *i.e.*, $m_{{}^w\alpha} = m_\alpha$; in fact these multiplicities are 1.
ii) For non-trivial α, the only roots which are non-zero multiples of α are $\pm\alpha$.
iii) $[\mathfrak{g}_\alpha, \mathfrak{g}_\beta] = \{0\}$ unless $\alpha + \beta$ is a root, and then

$$[\mathfrak{g}_\alpha, \mathfrak{g}_\beta] = \mathfrak{g}_{\alpha+\beta}$$

if $\alpha + \beta$ is non-trivial, while

$$\{0\} \neq [\mathfrak{g}_\alpha, \mathfrak{g}_{-\alpha}] \leqslant \mathfrak{g}_0.$$

By Theorems 10.23 and 11.5, we have $\mathfrak{z}(\mathfrak{g}) \leqslant \mathfrak{t}$. It is clear that for any root α we have $\mathfrak{z}(\mathfrak{g}) \subseteq \ker\alpha$; in fact this can be strengthened.

Proposition 11.25

Let T be a maximal torus of G. Then

$$\mathfrak{z}(\mathfrak{g}) = \bigcap_{\alpha \text{ a root}} \ker\alpha,$$

where the intersection is taken over roots of G with respect to T.

The roots of G with respect to T are best thought of as lying in the subspace

$$\mathfrak{z}(\mathfrak{g})^{\perp} = \{\gamma \in \mathfrak{t} : \mathfrak{z}(\mathfrak{g}) \subseteq \ker \gamma\} \subseteq \mathfrak{t}^*.$$

Recall that for $\theta \in \mathbb{R}$,

$$R(\theta) = \begin{bmatrix} \cos\theta & -\sin\theta \\ \sin\theta & \cos\theta \end{bmatrix}.$$

Example 11.26

$\mathrm{SO}(3)$ has the maximal torus

$$T = \left\{ \begin{bmatrix} R(\theta) & O_{2,1} \\ O_{1,2} & 1 \end{bmatrix} : \theta \in \mathbb{R} \right\} \leqslant \mathrm{SO}(3),$$

and the lie algebras of $\mathrm{SO}(3)$ and T are

$$\mathfrak{so}(3) = \left\{ \begin{bmatrix} 0 & -u & -v \\ u & 0 & -w \\ v & w & 0 \end{bmatrix} : u, v, w \in \mathbb{R} \right\},$$

$$\mathfrak{t} = \left\{ \begin{bmatrix} 0 & -t & 0 \\ t & 0 & 0 \\ 0 & 0 & 0 \end{bmatrix} : t \in \mathbb{R} \right\} \leqslant \mathfrak{so}(3).$$

The elements

$$U = \begin{bmatrix} 0 & -1 & 0 \\ 1 & 0 & 0 \\ 0 & 0 & 0 \end{bmatrix}, \quad V = \begin{bmatrix} 0 & 0 & -1 \\ 0 & 0 & 0 \\ 1 & 0 & 0 \end{bmatrix}, \quad W = \begin{bmatrix} 0 & 0 & 0 \\ 0 & 0 & -1 \\ 0 & 1 & 0 \end{bmatrix},$$

form a basis of $\mathfrak{so}(3)$ with $U \in \mathfrak{t}$. Then for $t \in \mathbb{R}$,

$$\exp(tU) = \begin{bmatrix} R(t) & O_{2,1} \\ O_{1,2} & 1 \end{bmatrix}$$

and

$$\begin{aligned} \exp(tU)U\exp(tU)^{-1} &= U, \\ \exp(tU)V\exp(tU)^{-1} &= \cos t\, V + \sin t\, W, \\ \exp(tU)W\exp(tU)^{-1} &= -\sin t\, V + \cos t\, W. \end{aligned}$$

In terms of the ad-action we have

$$[U, U] = 0, \quad [U, V] = W, \quad [U, W] = -V.$$

So there are two non-trivial roots $\pm\beta$, where

$$\beta(tP) = t \quad (t \in \mathbb{R}).$$

From Example 9.19, the normaliser of T in SO(3) is

$$\mathrm{N}_{\mathrm{SO}(3)}(T) = T \cup \left\{ \begin{bmatrix} -\cos\theta & \sin\theta & 0 \\ \sin\theta & \cos\theta & 0 \\ 0 & 0 & -1 \end{bmatrix} : \theta \in \mathbb{R} \right\} = T \cup ZT,$$

where $Z = \mathrm{diag}(-1, 1, -1)$. The Weyl group

$$\mathrm{W}_G(T) = \{IT, ZT\} \cong \{\pm 1\}$$

acts on the roots by

$$^{ZT}(\pm\beta) = \mp\beta.$$

Example 11.27

SU(2) has the maximal torus

$$T = \{\mathrm{diag}(z, \overline{z}) : |z| = 1\} \leqslant \mathrm{SU}(2),$$

which has Lie algebra

$$\mathfrak{t} = \{\mathrm{diag}(ti, -ti) : t \in \mathbb{R}\} \leqslant \mathfrak{su}(2).$$

Starting with the orthonormal basis $\{\hat{H}, \hat{E}, \hat{F}\}$ of $\mathfrak{su}(2)$ for which

$$\hat{H} = \frac{1}{\sqrt{2}} \begin{bmatrix} i & 0 \\ 0 & -i \end{bmatrix}, \quad \hat{E} = \frac{1}{\sqrt{2}} \begin{bmatrix} 0 & 1 \\ -1 & 0 \end{bmatrix}, \quad \hat{F} = \frac{1}{\sqrt{2}} \begin{bmatrix} 0 & i \\ i & 0 \end{bmatrix},$$

we find that for $t \in \mathbb{R}$,

$$\exp(t\hat{H}) = \mathrm{diag}(e^{ti}, e^{-ti})$$

and

$$\begin{aligned} \exp(t\hat{H})\hat{H}\exp(t\hat{H})^{-1} &= \hat{H}, \\ \exp(t\hat{H})\hat{E}\exp(t\hat{H})^{-1} &= \cos 2t\hat{E} + \sin 2t\hat{F}, \\ \exp(t\hat{H})\hat{F}\exp(t\hat{H})^{-1} &= -\sin 2t\hat{E} + \cos 2t\hat{F}. \end{aligned}$$

In terms of the ad-action we have

$$[\hat{H}, \hat{H}] = 0, \quad [\hat{H}, \hat{E}] = 2\hat{F}, \quad [\hat{H}, \hat{F}] = -2\hat{E}.$$

So here there are two non-trivial roots $\pm\alpha_1$ where

$$\alpha_1(\mathrm{diag}(ti, -ti)) = \alpha_1(t\hat{H}) = 2t \quad (t \in \mathbb{R}).$$

The normaliser of T in SU(2) is

$$N_{\mathrm{SU}(2)}(T) = T \cup \left\{ \begin{bmatrix} 0 & -\bar{z} \\ z & 0 \end{bmatrix} : |z| = 1 \right\} = T \cup JT,$$

where

$$J = \begin{bmatrix} 0 & -1 \\ 1 & 0 \end{bmatrix}.$$

The Weyl group

$$W_{\mathrm{SU}(2)}(T) = \{IT, JT\} \cong \{\pm 1\}$$

acts on the roots by

$$^{JT}(\pm\alpha_1) = \mp\alpha_1.$$

Example 11.28

U(2) has the maximal torus

$$T = \{\mathrm{diag}(z, w) : |z| = 1 = |w|\} \leqslant \mathrm{U}(2),$$

which has Lie algebra

$$\mathfrak{t} = \{\mathrm{diag}(si, ti) : s, t \in \mathbb{R}\} \leqslant \mathfrak{u}(2).$$

We have the orthonormal basis $\{\hat{H}_1, \hat{H}_2, \hat{E}, \hat{F}\}$ of $\mathfrak{u}(2)$ for which

$$\hat{H}_1 = \begin{bmatrix} i & 0 \\ 0 & 0 \end{bmatrix}, \quad \hat{H}_2 = \begin{bmatrix} 0 & 0 \\ 0 & i \end{bmatrix}, \quad \hat{E} = \frac{1}{\sqrt{2}} \begin{bmatrix} 0 & 1 \\ -1 & 0 \end{bmatrix}, \quad \hat{F} = \frac{1}{\sqrt{2}} \begin{bmatrix} 0 & i \\ i & 0 \end{bmatrix}.$$

Then for $s, t \in \mathbb{R}$,

$$\exp(s\hat{H}_1 + t\hat{H}_2) = \mathrm{diag}(e^{si}, e^{ti}).$$

Then

$$\begin{aligned} \exp(s\hat{H}_1 + t\hat{H}_2)(x\hat{H}_1 + y\hat{H}_2)\exp(s\hat{H}_1 + t\hat{H}_2)^{-1} &= (x\hat{H}_1 + y\hat{H}_2), \\ \exp(s\hat{H}_1 + t\hat{H}_2)\hat{E}\exp(s\hat{H}_1 + t\hat{H}_2)^{-1} &= \cos(s-t)\hat{E} + \sin(s-t)\hat{F}, \\ \exp(s\hat{H}_1 + t\hat{H}_2)\hat{F}\exp(s\hat{H}_1 + t\hat{H}_2)^{-1} &= -\sin(s-t)\hat{E} + \cos(s-t)\hat{F}. \end{aligned}$$

In terms of the ad-action we have

$$[s\hat{H}_1 + t\hat{H}_2, x\hat{H}_1 + y\hat{H}_2] = 0,$$
$$[s\hat{H}_1 + t\hat{H}_2, \hat{E}] = (s-t)\hat{F}, \quad [s\hat{H}_1 + t\hat{H}_2, \hat{F}] = -(s-t)\hat{E}.$$

So there are two non-trivial roots $\pm\alpha$ where

$$\alpha(s\hat{H}_1 + t\hat{H}_2) = s - t \quad (s, t \in \mathbb{R}).$$

Notice that

$$\ker \alpha = \{s(\hat{H}_1 + \hat{H}_2) : s \in \mathbb{R}\} = \mathfrak{z}(\mathfrak{u}(2)).$$

The normaliser of T in $\mathrm{U}(2)$ is

$$\mathrm{N}_{\mathrm{U}(2)}(T) = T \cup \left\{ \begin{bmatrix} 0 & -\bar{z} \\ z & 0 \end{bmatrix} : |z| = 1 \right\} = T \cup JT,$$

with J as in Example 11.27; the Weyl group

$$\mathrm{W}_{\mathrm{U}(2)}(T) = \{IT, JT\} \cong \{\pm 1\}$$

acts on the roots by

$$^{JT}(\pm\alpha) = \mp\alpha,$$

as indicated in Figure 11.1.

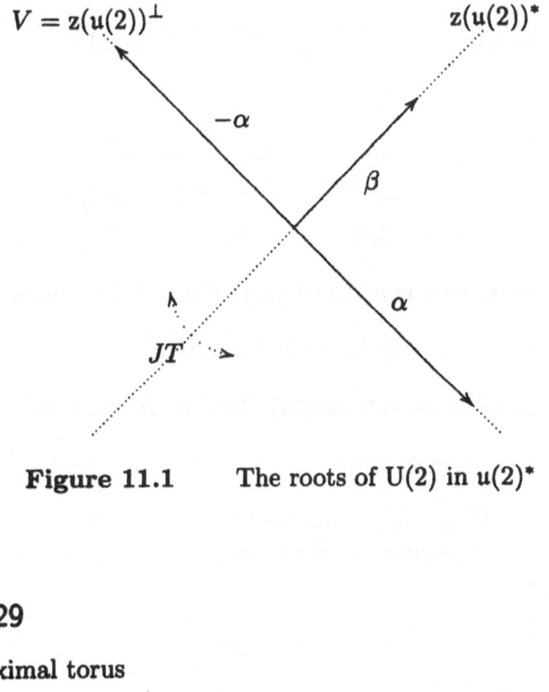

Figure 11.1 The roots of $\mathrm{U}(2)$ in $\mathfrak{u}(2)^*$

Example 11.29

$\mathrm{SU}(3)$ has maximal torus

$$T = \{\mathrm{diag}(z, w, \overline{zw}) : |z| = |w| = 1\} \leqslant \mathrm{SU}(3),$$

with Lie algebra

$$\mathfrak{t} = \{\mathrm{diag}(si, ti, -si - ti) : s, t \in \mathbb{R}\}.$$

In the Lie algebra $\mathfrak{su}(3)_{\mathbf{C}}$ we have

$$H_1 = \begin{bmatrix} i & 0 & 0 \\ 0 & -i & 0 \\ 0 & 0 & 0 \end{bmatrix}, \; H_2 = \begin{bmatrix} 0 & 0 & 0 \\ 0 & i & 0 \\ 0 & 0 & -i \end{bmatrix} \in \mathfrak{t},$$

and

$$E^{12} = \begin{bmatrix} 0 & 1 & 0 \\ 0 & 0 & 0 \\ 0 & 0 & 0 \end{bmatrix}, \quad E^{13} = \begin{bmatrix} 0 & 0 & 1 \\ 0 & 0 & 0 \\ 0 & 0 & 0 \end{bmatrix}, \quad E^{23} = \begin{bmatrix} 0 & 0 & 0 \\ 0 & 0 & 1 \\ 0 & 0 & 0 \end{bmatrix},$$
$$E^{21} = \begin{bmatrix} 0 & 0 & 0 \\ 1 & 0 & 0 \\ 0 & 0 & 0 \end{bmatrix}, \quad E^{31} = \begin{bmatrix} 0 & 0 & 0 \\ 0 & 0 & 0 \\ 1 & 0 & 0 \end{bmatrix}, \quad E^{32} = \begin{bmatrix} 0 & 0 & 0 \\ 0 & 0 & 0 \\ 0 & 1 & 0 \end{bmatrix}.$$

These six elements form a $\mathbb{C}$-basis of $\mathfrak{su}(3)_{\mathbf{C}}$ and with respect to the invariant inner product $(\;|\;)$ we have

$$(H_1 \mid H_1) = (H_2 \mid H_2) = 2, \quad (H_1 \mid H_2) = -1.$$

For the ad-action we have

$$\begin{aligned} [sH_1 + tH_2, E^{12}] &= (2s - t)iE^{12}, & [sH_1 + tH_2, E^{21}] &= -(2s - t)iE^{21}, \\ [sH_1 + tH_2, E^{13}] &= (s + t)iE^{13}, & [sH_1 + tH_2, E^{31}] &= -(s + t)iE^{13}, \\ [sH_1 + tH_2, E^{23}] &= (-s + 2t)iE^{23}, & [sH_1 + tH_2, E^{32}] &= -(-s + 2t)iE^{32}, \end{aligned}$$

showing that there are six non-trivial roots $\pm\alpha, \pm\beta, \pm\gamma$, where

$$\alpha(sH_1 + tH_2) = 2s - t, \quad \beta(sH_1 + tH_2) = s + t, \quad \gamma(sH_1 + tH_2) = -s + 2t.$$

Using the inner product we can identify these as elements of $\mathfrak{t} \cong \mathfrak{t}^*$,

$$H_1 \longleftrightarrow \alpha, \quad H_1 + H_2 \longleftrightarrow \beta, \quad H_2 \longleftrightarrow \gamma.$$

The Weyl group $W_{SU(3)}(T)$ consists of 6 elements which act on the roots as the symmetries of the regular hexagon whose edges are indicated with dotted lines in Figure 11.2.

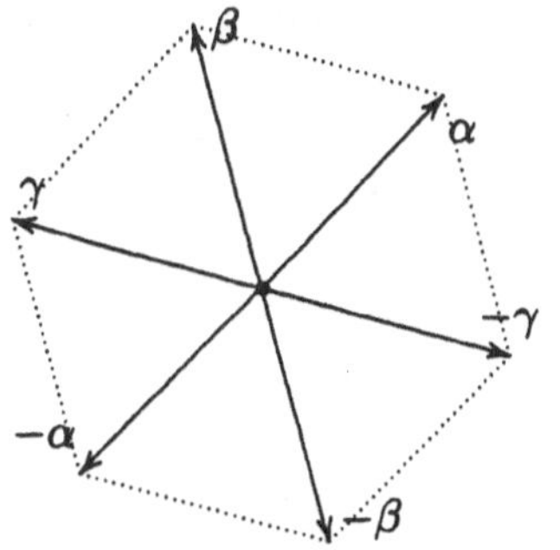

Figure 11.2 The roots of SU(3) in $\mathfrak{su}(3)^*$

EXERCISES

11.1. For $n \geqslant 2$, consider the unitary group $\mathrm{U}(n)$ and its Lie algebra $\mathfrak{u}(n)$ with its standard invariant inner product $(\ |\ )$.
a) Find the orthogonal complement of the centre of $\mathfrak{u}(n)$,

$$\mathfrak{z}(\mathfrak{u}(n))^{\perp} = \{X \in \mathfrak{u}(n) : \forall Z \in \mathfrak{z}(\mathfrak{u}(n))\ (X \mid Z) = 0\}$$

b) Show that there is a Lie homomorphism $\varphi\colon \mathrm{U}(n) \longrightarrow \mathrm{Z}(\mathrm{U}(n))/Z$ whose kernel $\ker\varphi \lhd \mathrm{U}(n)$ has Lie algebra $\mathfrak{z}(\mathfrak{u}(n))^{\perp} \leqslant \mathfrak{u}(n)$ and where $Z \lhd \mathrm{Z}(\mathrm{U}(n))$ is a finite normal subgroup.
c) Deduce that there is a surjective Lie homomorphism with finite kernel,

$$\mathrm{Z}(\mathrm{U}(n)) \times \ker\varphi \longrightarrow \mathrm{U}(n).$$

11.2. Derive Equations (11.9) from Equations (11.8).

11.3. Verify that the Weyl group of Example 11.29 acts on the roots as claimed.

11.4. Determine the roots of each of the groups Sp(2), and SO(4) for the standard maximal torus. In each case, find the action of the Weyl group on the roots.

11.5. a) Find a maximal torus T_1 for the group

$$G_1 = \left\{ \begin{bmatrix} A & O_{2,2} \\ O_{2,2} & B \end{bmatrix} : A, B \in \mathrm{U}(2),\ \det A \det B = 1 \right\} \leqslant \mathrm{SU}(4).$$

Determine the roots of G_1 with respect to T_1. Find the normaliser of T_1 in G_1 and the action of the Weyl group on the roots.

b) Do the same for the group

$$G_2 = \left\{ \begin{bmatrix} A & O_{2,2} \\ O_{2,2} & B \end{bmatrix} : A, B \in \mathrm{U}(2),\ \det A = \det B \right\} \leqslant \mathrm{U}(4).$$

12

Roots Systems, Weyl Groups and Dynkin Diagrams

In this chapter we discuss the theory of root systems and their Weyl groups and outline the classification of the irreducible ones. We indicate how these are associated with families of compact connected simple Lie groups. Accessible introductions to this material can be found in many books, *e.g.*, [4, 26, 10]; [14] contains a technical account of the study of Lie algebras from the perspective of root systems, including certain important infinite dimensional examples known as Kač-Moody algebras. We merely sketch the highlights without giving proofs.

12.1 Inner Products and Duality

In this section we recall some standard facts about inner products and duality. Throughout, let V be an ℓ-dimensional $\mathbb{R}$-vector space.

The $\mathbb{R}$-*linear dual* of V is the $\mathbb{R}$-vector space V^* consisting of all $\mathbb{R}$-linear transformations $V \longrightarrow \mathbb{R}$. Given a basis $\{v_1, \ldots, v_\ell\}$ for V, define elements $v_i^* \in V^*$ for $i = 1, \ldots, \ell$ by

$$v_i^*(v_j) = \delta_{ij}.$$

Proposition 12.1

If $\{v_1, \ldots, v_\ell\}$ is a basis of V, then $\{v_1^*, \ldots, v_\ell^*\}$ is a basis of V^*.

$\{v_1^*, \ldots, v_\ell^*\}$ is called the *dual basis* determined by $\{v_1, \ldots, v_\ell\}$.

Suppose that $\langle\ ,\ \rangle$ is a positive definite (but not necessarily symmetric) inner product on V. As usual, the length of a vector $\alpha \in V$ is defined to be $|\alpha| = \sqrt{\langle \alpha, \alpha \rangle}$. There is an $\mathbb{R}$-linear transformation

$$\nu\colon V \longrightarrow V^*; \quad \nu(u) = \langle u,\ \rangle,$$

where

$$\nu(u)(v) = \langle u, v \rangle \quad (v \in V).$$

The following result is standard.

Proposition 12.2

The $\mathbb{R}$-linear transformation $\nu\colon V \longrightarrow V^*$ is an isomorphism.

V^* inherits an inner product, which we will also write $\langle\ |\ \rangle$, defined by

$$\langle \alpha \mid \beta \rangle = \langle \nu^{-1}(\alpha) \mid \nu^{-1}(\beta) \rangle.$$

This is a positive definite inner product on V^*.

Proposition 12.3

Let $\{v_1, \ldots, v_\ell\}$ be a basis for V and

$$A = [\langle v_i \mid v_j \rangle] \in \mathrm{M}_\ell(\mathbb{R}).$$

Then the matrix $\breve{A} = [\langle v_i^* \mid v_j^* \rangle]$ associated with the inner product on $\langle\ |\ \rangle$ on V^* is given by

$$\breve{A} = (A^T)^{-1}.$$

Proof

This is a standard piece of linear algebra. Since

$$\nu(v_i) = \sum_{r=1}^{\ell} (v_i \mid v_r) v_r^*,$$

on setting $B = [b_{ij}] = A^{-1}$, we obtain

$$\nu^{-1}(v_i^*) = \sum_{r=1}^{\ell} b_{ir} v_r,$$

which yields

$$\begin{aligned}(v_i^* \mid v_j^*) = (\nu^{-1}(v_i^*) \mid \nu^{-1}(v_j^*)) &= \sum_{r=1}^{\ell}\sum_{s=1}^{\ell} b_{ir}b_{js}(v_r \mid v_s)\\ &= \sum_{s=1}^{\ell} \delta_{is}b_{js}\\ &= b_{ji}.\end{aligned}$$

This implies the matrix identity

$$[(v_i^* \mid v_j^*)] = (A^T)^{-1}$$

which is the desired result. □

12.2 Roots systems and their Weyl groups

Let V be an ℓ-dimensional real vector space with a positive definite (but not necessarily symmetric) inner product $\langle\ ,\ \rangle$; as usual, the length of a vector $\alpha \in V$ is defined to be $|\alpha| = \sqrt{\langle \alpha, \alpha \rangle}$.

If $u \in V - \{0\}$, there is an associated $\mathbb{R}$-linear transformation $s_u : V \longrightarrow V$ for which

$$s_u(x) = \begin{cases} -x & \text{if } x = tu, \\ x & \text{if } \langle u, x \rangle = 0. \end{cases}$$

It is easy to see that for $x \in V$,

$$s_u(x) = x - \frac{2\langle u, x \rangle}{\langle u, u \rangle} u. \tag{12.1}$$

Geometrically, s_u is a *hyperplane reflection* in the hyperplane

$$H_u^{\perp} = \{x \in V : \langle u, x \rangle = 0\} \subseteq V.$$

Proposition 12.4

If $u, v \in V - \{0\}$, the following identities are satisfied:

$$s_u^2 = s_u \circ s_u = \mathrm{Id}_V,$$

$$\begin{cases} s_v = s_u & \text{if } v = tu \text{ for some } t \in \mathbb{R}, \\ s_v \neq s_u & \text{otherwise.} \end{cases}$$

Definition 12.5

A subset $\Phi \subseteq V$ is a (*reduced*) *root system* if it has the following properties:

- Φ spans V and $0 \notin \Phi$;
- if $u \in \Phi$ then $-u \in \Phi$ and no other multiples of u are in Φ;
- if $u, v \in \Phi$ then $2\langle u, v\rangle / \langle u, u\rangle \in \mathbb{Z}$ and $s_u(v) \in \Phi$.

An element of Φ is called a *root.*

The final condition is equivalent to the requirement that for $u, v \in \Phi$ and some $k \in \mathbb{Z}$,

$$s_u(v) - v = ku.$$

Each s_u acts on Φ as a permutation and it can be viewed as an element of the group of all permutations of Φ, S_Φ.

Definition 12.6

The *Weyl group* of Φ is the subgroup $\mathrm{W}(\Phi) \leqslant S_\Phi$ generated by the s_u $(u \in \Phi)$.

Remark 12.7

By definition $\mathrm{W}(\Phi)$ is finite with order $|\,\mathrm{W}(\Phi)|$ dividing $|\Phi|!$.

Definition 12.8

A subset $\Pi \subseteq \Phi$ is a set of *fundamental* or *simple roots* if it satisfies

- Π is an $\mathbb{R}$-basis of V;
- every $u \in \Phi$ can be uniquely expressed in the form

$$u = \sum_{w \in \Pi} c_w w$$

where $c_w \in \mathbb{Z}$ either all satisfy $c_w \geqslant 0$ or all satisfy $c_w \leqslant 0$.

The choice of such a fundamental system is not unique, however given Π there is a partition $\Phi = \Phi^+ \cup \Phi^-$, where

$$\Phi^+ = \{\sum_{w \in \Pi} c_w w : c_w \geqslant 0\}, \quad \Phi^- = \{\sum_{w \in \Pi} c_w w : c_w \leqslant 0\}.$$

Notice that

$$-\Phi^+ = \{-u : u \in \Phi^+\} = \Phi^-, \quad -\Phi^- = \{-u : u \in \Phi^-\} = \Phi^+.$$

The elements of Π are called (*positive*) *simple roots*, while the elements of Φ^+ or Φ^- are called *positive* or *negative roots*.

Remark 12.9

It is easy to see that $W(\Phi)$ is generated by the reflections s_u ($u \in \Pi$).

12.3 Some Examples of Root Systems

We begin with an already familiar example from Section 11.5.

Example 12.10 (The root system of U(2))

This builds on Example 11.28. Consider $\mathfrak{t}^*$ with orthonormal basis $\{\varepsilon_1, \varepsilon_2\}$ dual to the orthonormal basis $\{\hat{H}_1, \hat{H}_2\}$ of $\mathfrak{t} \cong \mathfrak{t}^{**}$. Here we take $V = \mathfrak{z}(\mathfrak{u}(2))^\perp \subseteq \mathfrak{t}^*$. The root $\alpha = \varepsilon_1 - \varepsilon_2 \in V$ has length $|\alpha| = \sqrt{2}$. If $\beta = \varepsilon_1 + \varepsilon_2 \in \mathfrak{t}^*$, then $\alpha \cdot \beta = 0$, and $\mathbb{R}\{\beta\} = \mathfrak{z}(\mathfrak{u}(2))^*$. So for this root system we have

$$\Pi = \{\alpha\}, \quad \Phi = \{-\alpha, \alpha\},$$

with Weyl group $W(\Phi) = S_\Phi \cong S_2$. The generator JT of the Weyl group acts on $\mathfrak{t}^*$ by reflection in the line $\mathfrak{z}(\mathfrak{u}(2))^*$ as indicated in Figure 11.1.

In the remaining examples, $\mathbb{R}^\ell$ and its subspaces have the usual positive definite inner product $\langle \mathbf{x} \mid \mathbf{y} \rangle = \mathbf{x} \cdot \mathbf{y}$. In the diagrams, the solid arrows indicate a choice of positive simple roots, while dashed arrows indicate the remaining roots.

Example 12.11 (The root system A_1)

Let $V = \mathbb{R}$ and

$$\Pi = \{\alpha_1\}, \quad \Phi = \{-\alpha_1, \alpha_1\},$$

where $\alpha_1 \in \mathbb{R}$ is non-zero. This root system has Weyl group $W(\Phi) = S_\Phi \cong S_2$.

Figure 12.1 The roots of A_1

Example 12.12 (The root system A_2)

Let $V = \mathbb{R}^2$ and

$$\Pi = \{\alpha_1, \alpha_2\}, \quad \Phi = \{\alpha_1, -\alpha_1, \alpha_2, -\alpha_2, \alpha_1 + \alpha_2, -\alpha_1 - \alpha_2\},$$

where α_1, α_2 are two vectors for which

$$|\alpha_1| = |\alpha_2|, \quad \alpha_1 \cdot \alpha_2 = -\frac{|\alpha_1|\,|\alpha_2|}{2}.$$

This root system has the hyperplane reflections $s_{\alpha_1}, s_{\alpha_2}, s_{\alpha_1+\alpha_2}$ and Weyl group $W(\Phi) \cong S_3$.

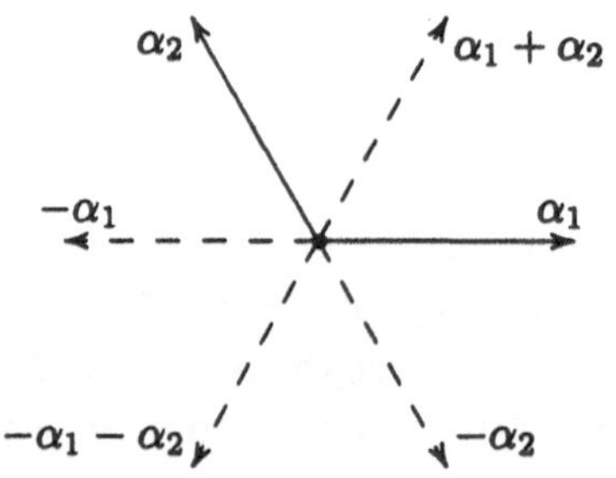

Figure 12.2 The roots of A_2

The preceding examples generalise to give the root systems A_n.

Example 12.13 (The root system A_n for $n \geqslant 1$)

Consider

$$V = \{x \in \mathbb{R}^{n+1} : x \cdot (e_1 + \cdots + e_{n+1}) = 0\} \subseteq \mathbb{R}^{n+1},$$

and $\alpha_k = e_k - e_{k+1}$ for $1 \leqslant k \leqslant n$. Then V is a vector space of dimension n with basis $\{\alpha_1, \dots, \alpha_n\}$. Notice that

$$\alpha_i \cdot \alpha_j = \begin{cases} 2 & \text{if } i = j, \\ -1 & \text{if } |i - j| = 1, \\ 0 & \text{otherwise.} \end{cases}$$

Taking

$$\begin{aligned} \Pi &= \{\alpha_1, \dots, \alpha_n\}, \\ \Phi &= \{\pm(\alpha_r + \alpha_{r+1} + \cdots + \alpha_s) : 1 \leqslant r \leqslant s \leqslant n\} \\ &= \{\pm(e_r - e_{s+1}) : 1 \leqslant r < s \leqslant n\}, \end{aligned}$$

we obtain a root system with Weyl group $W(\Phi) \cong S_{n+1}$.

Example 12.14 (The root system B_n for $n \geqslant 2$)

Take $V = \mathbb{R}^n$, $\alpha_i = e_i - e_{i+1}$ $(i = 1, \ldots, n-1)$ and $\alpha_n = e_n$. Then if

$$
\begin{aligned}
\Pi &= \{\alpha_1, \ldots, \alpha_n\}, \\
\Phi &= \{\pm(\alpha_r + \cdots + \alpha_s) : 1 \leqslant r \leqslant s \leqslant n\} \\
&\qquad \cup \{\pm(\alpha_r + \cdots + \alpha_{s-1} + 2\alpha_s + \cdots + 2\alpha_n) : 1 \leqslant r < s \leqslant n\} \\
&= \{\pm e_r : 1 \leqslant r \leqslant n\} \cup \{\pm(e_r + e_s), \pm(e_r - e_s) : 1 \leqslant r < s \leqslant n\},
\end{aligned}
$$

this is a root system whose Weyl group is the group of $n \times n$ generalised permutation matrices, which has order $|\,W(\Phi)| = n!\,2^n$.

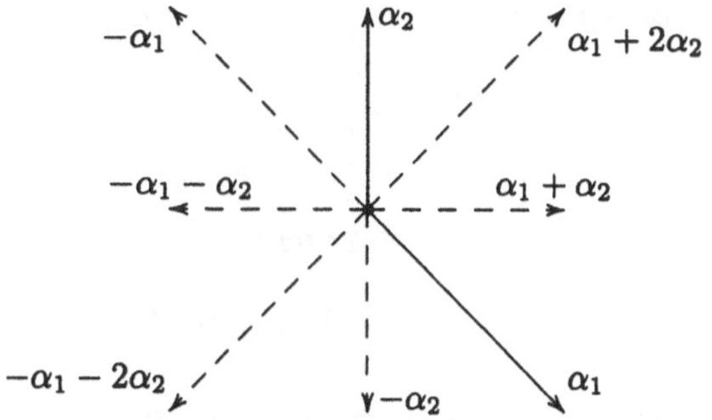

Figure 12.3 The roots of B_2

Example 12.15 (The root system C_n for $n \geqslant 2$)

Take $V = \mathbb{R}^n$, $\alpha_i = e_i - e_{i+1}$ $(i = 1, \ldots, n-1)$ and $\alpha_n = 2e_n$. If

$$
\begin{aligned}
\Pi &= \{\alpha_1, \ldots, \alpha_n\}, \\
\Phi &= \{\pm(\alpha_r + \cdots + \alpha_s) : 1 \leqslant r \leqslant s \leqslant n\} \\
&\qquad \cup \{\pm(\alpha_r + \cdots + \alpha_{s-1} + 2\alpha_s + \cdots + 2\alpha_{n-1} + \alpha_n) : 1 \leqslant r < s < n\} \\
&\qquad \cup \{\pm(2\alpha_r + \cdots + 2\alpha_{n-1} + \alpha_n) : 1 \leqslant r < n\} \\
&= \{\pm 2e_r : 1 \leqslant r \leqslant n\} \cup \{\pm(e_r + e_s), \pm(e_r - e_s) : 1 \leqslant r < s \leqslant n\},
\end{aligned}
$$

this is a root system whose Weyl group is again the group of $n \times n$ generalised permutation matrices, which has order $|\,W(\Phi)| = n!\,2^n$.

A careful examination of the roots for B_2 and C_2 shows that apart form the actual lengths of the root vectors and the way they are named, these are essentially the same root systems.

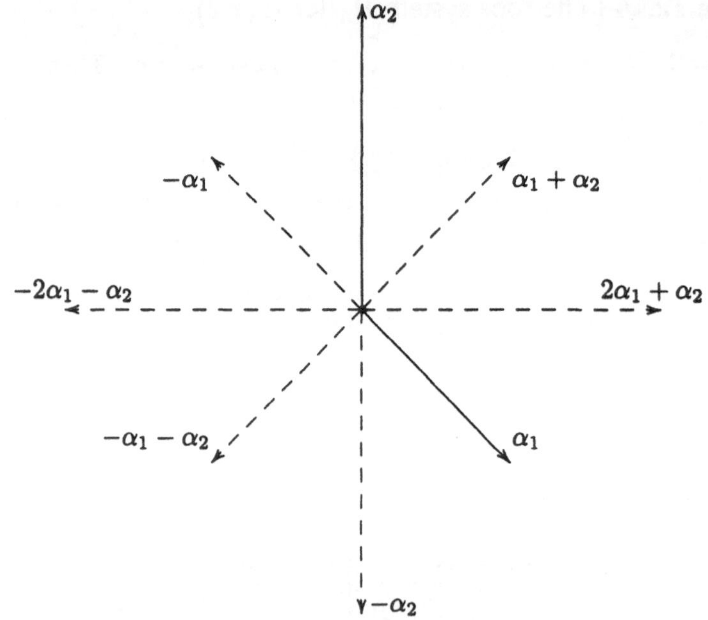

Figure 12.4 The roots of C_2

Example 12.16 (The root system D_n for $n \geqslant 4$)

Take $V = \mathbb{R}^n$, $\alpha_i = e_i - e_{i+1}$ $(i = 1, \dots, n-1)$ and $\alpha_n = e_{n-1} + e_n$. Then if

$$
\begin{aligned}
\Pi =& \{\alpha_1, \dots, \alpha_n\}, \\
\Phi =& \{\pm(\alpha_r + \cdots + \alpha_s) : 1 \leqslant r \leqslant s \leqslant n\} \\
& \cup \{\pm(\alpha_r + \cdots + \alpha_{n-2} + \alpha_n) : 1 \leqslant r \leqslant n-2\} \\
& \cup \{\pm(\alpha_r + \cdots + \alpha_{s-1} + 2\alpha_s + \cdots + 2\alpha_{n-2} + \alpha_{n-1} + \alpha_n : 1 \leqslant r < s \leqslant n-2\} \\
=& \{\pm(e_r + e_s), \pm(e_r - e_s) : 1 \leqslant r < s \leqslant n\},
\end{aligned}
$$

this is a root system whose Weyl group is the group of $n \times n$ generalised permutation matrices with an even number of sign changes, and this has order $|\mathrm{W}(\Phi)| = n!\, 2^{n-1}$.

12.4 The Dynkin Diagram of a Root System

Let Π be a fundamental system of the root system $\Phi \subseteq V$. Choosing an ordering $\alpha_1, \ldots, \alpha_\ell$ for the elements of Π, there are integers

$$a_{ij} = \frac{2\langle \alpha_i, \alpha_j \rangle}{\langle \alpha_i, \alpha_i \rangle} \quad (1 \leqslant i, j \leqslant \ell).$$

The $\ell \times \ell$ matrix $[a_{ij}]$ with integer entries is called the *Cartan matrix* of the root system Φ.

Defining the angle $\theta_{ij} \in [0, \pi]$ by

$$\theta_{ij} = \cos^{-1} \frac{\langle \alpha_i, \alpha_j \rangle}{|\alpha_i|\,|\alpha_j|},$$

we have

$$a_{ji} = \frac{2|\alpha_j|}{|\alpha_i|} \cos \theta_{ij}$$

and so

$$a_{ij} a_{ji} = 4 \cos^2 \theta_{ij}. \tag{12.2}$$

Because a_{ij}, a_{ji} are integers and $0 \leqslant \cos^2 \theta_{ij} \leqslant 1$, the expression $4\cos^2 \theta_{ij}$ has to take one of the values $0, 1, 2, 3, 4$, where the final case only occurs when $\cos \theta_{ij} = \pm 1$ and so $j = i$. In the situation where $j \neq i$, the following seven outcomes are the only ones possible.

$$\begin{array}{llll}
a_{ij} = 0, & a_{ji} = 0, & \theta_{ij} = \frac{\pi}{2}, & |\alpha_j| = |\alpha_i|. \\
a_{ij} = 1, & a_{ji} = 1, & \theta_{ij} = \frac{\pi}{3}, & |\alpha_j| = |\alpha_i|. \\
a_{ij} = -1, & a_{ji} = -1, & \theta_{ij} = \frac{2\pi}{3}, & |\alpha_j| = |\alpha_i|. \\
a_{ij} = 1, & a_{ji} = 2, & \theta_{ij} = \frac{\pi}{4}, & |\alpha_j| = \sqrt{2}|\alpha_i|. \\
a_{ij} = -1, & a_{ji} = -2, & \theta_{ij} = \frac{3\pi}{4}, & |\alpha_j| = \sqrt{2}|\alpha_i|. \\
a_{ij} = 1, & a_{ji} = 3, & \theta_{ij} = \frac{\pi}{6}, & |\alpha_j| = \sqrt{3}|\alpha_i|. \\
a_{ij} = -1, & a_{ji} = -3, & \theta_{ij} = \frac{5\pi}{6}, & |\alpha_j| = \sqrt{3}|\alpha_i|.
\end{array}$$

Notice that the value of θ_{ij} determines the (unordered) set of numbers

$$\left\{ \frac{|\alpha_i|}{|\alpha_j|}, \frac{|\alpha_j|}{|\alpha_i|} \right\}.$$

We will now show how to associate to a root system $\Pi \subseteq \Phi$ a *Dynkin diagram*. The diagram has one vertex for each simple root labelled by the root, with some bonds joining pairs of them. For distinct simple roots $\alpha_i, \alpha_j \in \Pi$, there are d_{ij} bonds between the corresponding vertices, where

$$d_{ij} = \begin{cases} 0 & \text{if } \varphi_{ij} = \dfrac{\pi}{2}, \\ 1 & \text{if } \varphi_{ij} = \dfrac{\pi}{3}, \dfrac{2\pi}{3}, \\ 2 & \text{if } \varphi_{ij} = \dfrac{\pi}{4}, \dfrac{3\pi}{4}, \\ 3 & \text{if } \varphi_{ij} = \dfrac{\pi}{6}, \dfrac{5\pi}{6}. \end{cases}$$

In the cases where $d_{ij} = 2, 3$, an inequality sign $>$ is added indicating which of the roots is larger. Here are the possible connections between nodes that can occur.

		Labels: α_i α_j
$d_{ij} = 0$		
$d_{ij} = 1$		
$d_{ij} = 2$	$\lvert\alpha_i\rvert < \lvert\alpha_j\rvert$	
$d_{ij} = 3$	$\lvert\alpha_i\rvert < \lvert\alpha_j\rvert$	

The whole Dynkin diagram is obtained by filling in the bonds between the vertices for all the simple roots. Such a diagram is *irreducible* or *connected* if it has only one path component.

12.5 Irreducible Dynkin Diagrams

The connected *irreducible Dynkin diagrams* are given in Tables 12.1 and 12.2. Each of the four classical series A_n, B_n, C_n and D_n consists of infinitely many distinct examples, while there are finitely many exceptional ones E_6, E_7, E_8, F_4, G_2 (the root system of G_2 is shown in Figure 12.5). The classical Dynkin diagrams are related to familiar matrix groups, where the entries at the bottom of Table 12.1 are consequences of certain special isomorphisms. Lie groups associated with the exceptional Dynkin diagrams are constructed in Adams [2]. The subscript n is the rank of the associated Lie group, and also the number of nodes in the Dynkin diagram.

Root system	Matrix group	Dynkin diagram
A_n $(n \geqslant 1)$	$SU(n)$	
B_n $(n \geqslant 2)$	$Spin(2n+1)$, $SO(2n+1)$	
C_n $(n \geqslant 3)$	$Sp(n)$	
D_n $(n \geqslant 4)$	$Spin(2n)$, $SO(2n)$	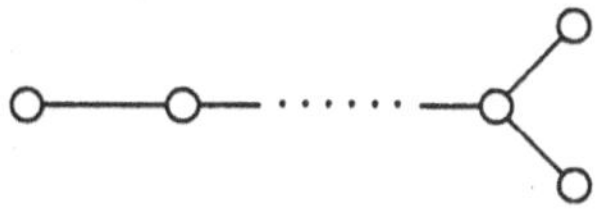
A_1	$SU(2) \cong Spin(3)$, $SO(3)$	
$B_2 = C_2$	$Sp(2) \cong Spin(5)$, $SO(5)$	

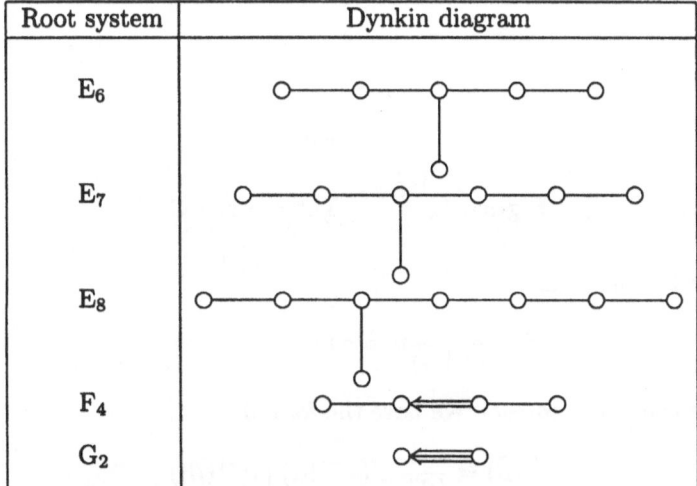

Table 12.1 Classical Dynkin diagrams and associated matrix groups

Root system	Dynkin diagram
E_6	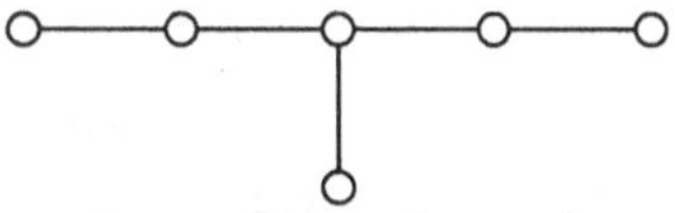
E_7	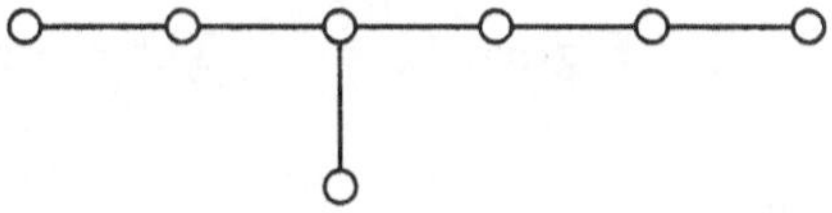
E_8	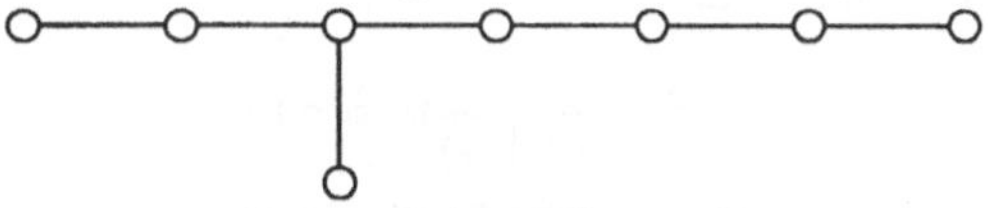
F_4	
G_2	

Table 12.2 Exceptional Dynkin diagrams

12.6 From Root Systems to Lie Algebras

Now assume that $\Pi = \{\alpha_1, \dots, \alpha_\ell\}$ is a set of simple roots for a root system $\Phi \subseteq \mathfrak{t}^*$, the linear dual of an $\mathbb{R}$-vector space $\mathfrak{t}$ of dimension ℓ, equipped with a positive definite inner product $(\ |\ )$. Using the duality isomorphism $\nu \colon \mathfrak{t} \longrightarrow \mathfrak{t}^*$,

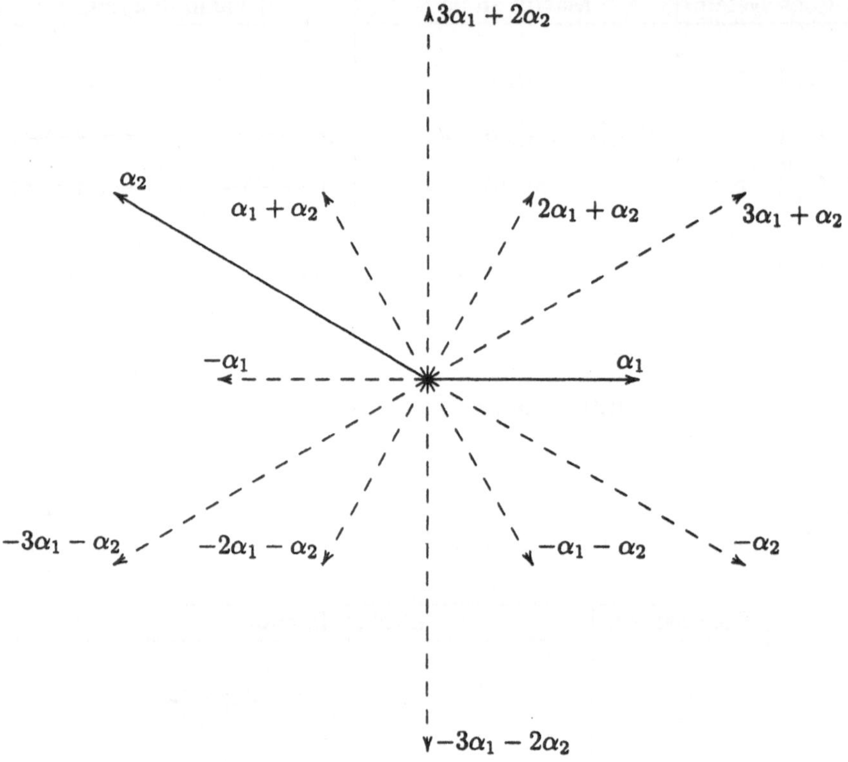

Figure 12.5 The roots of G_2

we can define elements

$$\check{\alpha} = \frac{2}{(\alpha \mid \alpha)} \nu^{-1}(\alpha) \in \mathfrak{t} \quad (\alpha \in \Phi).$$

These are called the *coroots*. We have the formula

$$\begin{aligned} \alpha(\check{\beta}) &= \frac{2}{(\beta \mid \beta)} (\nu^{-1}(\alpha) \mid \nu^{-1}(\beta)) \\ &= \frac{2}{(\beta \mid \beta)} (\alpha \mid \beta). \end{aligned} \tag{12.3}$$

We write

$$\check{\Phi} = \{\check{\alpha} : \alpha \in \Phi\}, \quad \check{\Pi} = \{\check{\alpha} : \alpha \in \Pi\},$$

for the sets of all coroots and *simple coroots* respectively. For each coroot $\check{\alpha} \in \check{\Phi}$, define $s_{\check{\alpha}}\mathfrak{t}$ by the formula analogous to Equation (12.1).

Proposition 12.17

$\check{\Phi} \subseteq \mathfrak{t}$ is a root system with $\check{\Pi}$ a set of simple roots.

The root system $\check{\Pi} \subseteq \check{\Phi}$ is called the *dual root system* to $\Pi \subseteq \Phi$.

Recall that for a finite dimensional vector space, $V \cong V^{**}$ and given a positive definite inner product such an isomorphism is obtained as the composite

$$V \xrightarrow{\nu} V^* \xrightarrow{\nu'} V^{**},$$

where $\nu' \colon V^* \longrightarrow V^{**}$ is obtained using the above inner product on V^*. It is easy to verify that given a root system in $\Phi' \subseteq V^{**}$, for each $\alpha \in \Phi'$,

$$\check{\check{\alpha}} = \nu^{-1}(\nu')^{-1}(\alpha). \tag{12.4}$$

This means that we can start with a root system in V or in V^* and pass back and forth by taking coroots.

The following important result is proved in [14]. We will illustrate it with the examples provided by the root systems A_n.

Theorem 12.18

Given a root system $\Phi \subseteq \mathfrak{t}^*$ with simple roots $\Pi = \{\alpha_1, \ldots, \alpha_\ell\}$, there is a $\mathbb{C}$-Lie algebra $\mathfrak{g}$ associated to Φ for which

- $\mathfrak{t}_\mathbb{C} \subseteq \mathfrak{g}$ is a maximal abelian $\mathbb{C}$-Lie subalgebra;
- for each root $\alpha \in \Phi$, there are elements $e_\alpha, f_\alpha \in \mathfrak{g}$ satisfying

  $$[e_\alpha, f_\beta] = \delta_{ij}\check{\alpha}, \quad [t, e_\alpha] = \alpha(t)e_\alpha, \quad [t, f_\alpha] = -\alpha(t)f_\alpha,$$

  for all $t \in \mathfrak{t}_\mathbb{C}$;
- The elements $\alpha_1, \ldots, \alpha_\ell, e_{\alpha_1}, f_{\beta_1}, \ldots, e_{\alpha_\ell}, f_{\beta_\ell}$ form a basis of $\mathfrak{g}$.

If $\mathfrak{g}'$ is a second such Lie algebra, there is a unique isomorphism of Lie algebras $\mathfrak{g} \longrightarrow \mathfrak{g}'$ which extends the identity function $\mathfrak{t}_\mathbb{C} \longrightarrow \mathfrak{t}_\mathbb{C} \subseteq \mathfrak{g}'$.

Because of the essential uniqueness provided by this result, we will write $\mathfrak{g}(\Phi)$ for the Lie algebra associated with a root system Φ.

Example 12.19

Consider the root system A_n of Example 12.13. Then the complexification of $\mathfrak{su}(n+1)$ is $\mathfrak{sl}_{n+1}(\mathbb{C})$. We take

$$\mathfrak{t} = \{\mathrm{diag}(t_1 i, \ldots, t_{n+1} i) : t_k \in \mathbb{R}\},$$

and

$$\breve{\alpha}_k = iE^{k\,k} - iE^{(k+1)\,(k+1)} \quad (k = 1, \ldots, n).$$

Then the coroots are the matrices of the form

$$\breve{\alpha} = \sum_{k=1}^{n} c_k \breve{\alpha}_k \quad (c_k = 0, \pm 1),$$

satisfying the inner product formulæ of Equation (11.4). The dual roots α satisfy

$$\alpha(\alpha_k) = (\alpha \mid \alpha_k) = (\breve{\alpha} \mid \breve{\alpha}_k).$$

For each root α, we take

$$e_\alpha = \sum_{k=1}^{n} c_k(E^{k\,(k+1)}), \quad f_\alpha = \sum_{k=1}^{n} c_k(E^{(k+1)\,k}).$$

Then there is an isomorphism of $\mathbb{C}$-Lie algebras $\mathfrak{g}(\mathrm{A}_n) \cong \mathfrak{sl}_{n+1}(\mathbb{C})$.

EXERCISES

12.1. Verify Remark 12.9.

12.2. Let Φ_{B_2} and Φ_{C_2} be the sets of roots for B_2 and C_2. Find a bijection $f\colon \Phi_{\mathrm{B}_2} \longrightarrow \Phi_{\mathrm{C}_2}$ which preserves angles but not lengths.

12.3. ⚠⚠ Using the ideas of Section 11.5, verify as many as you can of the identifications of root systems of Table 12.1.

12.4. ⚠⚠ For $n \geqslant 3$, determine the roots of $\mathrm{U}(n)$ with respect to the standard maximal torus

$$T_n = \{\mathrm{diag}(z_1, \ldots, z_n) : |z_1| = \cdots = |z_n| = 1\} \leqslant \mathrm{U}(n).$$

Find the normaliser of T_n in $\mathrm{U}(n)$ and the action of the Weyl group on the roots.

Hints and Solutions to Selected Exercises

Chapter 1

1.1. Using the formula of Remark 1.4, we obtain

$$\left\|\begin{bmatrix} u & 0 \\ 0 & v \end{bmatrix}\right\| = \max\{|u|, |v|\}, \qquad \left\|\begin{bmatrix} u & 1 \\ 0 & u \end{bmatrix}\right\| = \frac{\sqrt{2|u|^2 + 1 + \sqrt{4|u|^2 + 1}}}{\sqrt{2}},$$

$$\left\|\begin{bmatrix} \cos t & -\sin t \\ \sin t & \cos t \end{bmatrix}\right\| = 1, \qquad \left\|\begin{bmatrix} \cosh t & \sinh t \\ \sinh t & \cosh t \end{bmatrix}\right\| = \max\{e^t, e^{-t}\}.$$

1.2. (a) From the definition of the matrix norm,

$$\begin{aligned} \|BAB^{-1}\| &= \max\left\{\frac{|BAB^{-1}\mathbf{x}|}{|\mathbf{x}|} : \mathbf{x} \neq \mathbf{0}\right\} = \max\left\{\frac{|A\mathbf{x}|}{|B\mathbf{x}|} : \mathbf{x} \neq \mathbf{0}\right\} \\ &= \max\left\{\frac{|A\mathbf{x}|}{|\mathbf{x}|} : \mathbf{x} \neq \mathbf{0}\right\} = \|A\|. \end{aligned}$$

(b) The inequality $\|CAC^{-1}\| \leqslant \|C\|\,\|A\|\,\|C^{-1}\|$ always holds.

1.3. (a) We have $|\mathbf{x}|^2 = \mathbf{x}^*\mathbf{x}$ by definition, so

$$\|A\|^2 = \sup\{|A\mathbf{x}|^2 : \mathbf{x} \in \mathbb{C}^n,\ |\mathbf{x}| = 1\}.$$

Since the unit sphere $\{\mathbf{x} \in \mathbb{C}^n : |\mathbf{x}| = 1\}$ is closed and bounded, the continuous function $\mathbf{x} \longmapsto |A\mathbf{x}|^2$ defined on it attains its supremum, hence

$$\|A\|^2 = \max\{|A\mathbf{x}|^2 : \mathbf{x} \in \mathbb{C}^n,\ |\mathbf{x}| = 1\}.$$

(b) The matrix A^*A is hermitian and for any $\mathbf{x} \in \mathbb{C}^n$, $\mathbf{x}^*A^*A\mathbf{x} = |A\mathbf{x}|^2$, which is real and non-negative; it also vanishes precisely when $A\mathbf{x} = 0$. By the general theory of hermitian matrices, there is an orthonormal basis $\{\mathbf{u}_1, \ldots, \mathbf{u}_n\}$ consisting of eigenvalues $\lambda_1, \ldots, \lambda_n$ which are all real. For each j,

$$\mathbf{u}_j^*A^*A\mathbf{u}_j = \lambda_j\mathbf{u}_j^*\mathbf{u}_j = \lambda_j$$

since $\mathbf{u}_j^*\mathbf{u}_j = 1$; hence $\lambda_j \geqslant 0$. If $A \neq O$ there must be at least one non-zero eigenvalue of A^*A since otherwise $|A\mathbf{x}|^2$ would always vanish; so in this case we have a largest positive eigenvalue λ which we might as well assume is $\lambda = \lambda_1$. Now if $|\mathbf{x}| = 1$ we can write $\mathbf{x} = t_1\mathbf{u}_1 + \cdots + t_n\mathbf{u}_n$, where $t_j \in \mathbb{C}$ and $|t_1|^2 + \cdots + |t_n|^2 = 1$. Then

$$\begin{aligned}\mathbf{x}^*A^*A\mathbf{x} = \mathbf{x}^*(\lambda_1 t_1\mathbf{u}_1 + \cdots + \lambda_n t_n\mathbf{u}_n) &= \lambda_1|t_1|^2 + \cdots + \lambda_n|t_n|^2\\ &\leqslant \lambda_1|t_1|^2 + \cdots + \lambda_1|t_n|^2 = \lambda_1,\end{aligned}$$

hence $\|A\|^2 \leqslant \lambda_1$ and in fact we have equality since $\mathbf{u}_1^*A^*A\mathbf{u}_1 = \lambda_1$.

1.4. (a) Let

$$\ell = \frac{1}{2}\left(1 + \lim_{r\to\infty}\frac{\|A_{r+1}\|}{\|A_r\|}\right) < 1.$$

For large enough m we have

$$\begin{aligned}&\|A_m + \cdots + A_{m+k}\| \leqslant \|A_m\| + \cdots + \|A_{m+k}\|\\ &\quad = \|A_m\|\left(1 + \frac{\|A_{m+1}\|}{\|A_m\|} + \frac{\|A_{m+1}\|\|A_{m+2}\|}{\|A_m\|\|A_{m+1}\|} + \cdots + \frac{\|A_{m+1}\|\cdots\|A_{m+k}\|}{\|A_m\|\cdots\|A_{m+k-1}\|}\right)\\ &\quad < \|A_m\|\left(1 + \ell + \ell^2 + \cdots + \ell^{k-1}\right) = \|A_m\|\frac{(1-\ell^k)}{(1-\ell)} \to \frac{\|A_m\|}{1-\ell} \qquad \text{as } k \to \infty.\end{aligned}$$

Also, if m_0 is large enough, then for $m > m_0$, $\dfrac{\|A_m\|}{\|A_{m-1}\|} < \ell$, so

$$\|A_m\| = \frac{\|A_m\|\cdots\|A_{m_0+1}\|}{\|A_{m-1}\|\cdots\|A_{m_0}\|}\|A_{m_0}\| < \ell^{m-m_0}\|A_{m_0}\| \to 0 \quad \text{as } m \to \infty.$$

Combining these we see that the partial sums of the series form a Cauchy sequence and hence the series converges.
(b) Here $\|A_m\| \to \infty$ as $m \to \infty$ and the series diverges.
(c) Any of the usual tests based on absolute convergence works.

1.5. (a) For $m, n \geqslant 1$,

$$\begin{aligned}\left\|\sum_{r=m}^{m+n} A^r\right\| &= \|A^m + A^{m+1} + \cdots + A^{m+n}\| \leqslant \|A\|^m + \|A\|^{m+1} + \cdots + \|A\|^{m+n}\\ &= \|A\|^m\frac{(1-\|A\|^{n+1})}{(1-\|A\|)} \to 0 \quad \text{as } m, n \to \infty.\end{aligned}$$

So the sequence of partial sums is a Cauchy sequence. This can also be done using the ratio test of Exercise 1.4.
(b) For any $n \geqslant 1$,

$$(I - A)(I + A + A^2 + \cdots + A^n) = I - A^{n+1} = (I + A + A^2 + \cdots + A^n)(I - A),$$

so

$$\|(I - A)(I + A + A^2 + \cdots + A^n) - I\| = \|A^{n+1}\| \leqslant \|A\|^{n+1} \to 0 \quad \text{as } n \to \infty,$$

and similarly for $\|(I + A + A^2 + \cdots + A^n)(I - A) - I\|$. Hence $I - A$ has inverse

$$(I - A)^{-1} = \sum_{r=0}^{\infty} A^r.$$

(c) If $A^k = O$, then

$$(I - A)^{-1} = \sum_{r=0}^{k-1} A^r, \quad \exp(A) = \sum_{r=0}^{k-1} \frac{1}{r!} A^r.$$

1.6. (a) The subset

$$\mathrm{O}(n) = \{A \in \mathrm{M}_n(R) : A^T A - I = O\} \subseteq \mathrm{M}_n(R)$$

is bounded since for $A \in \mathrm{O}(n)$, $\|A\| = 1$. It is also closed since if a sequence $A_r \in \mathrm{O}(n)$ converges to $A \in \mathrm{M}_n(R)$ then

$$A^T A - I = \lim_{r \to \infty} (A_r^T A_r - I) = \lim_{r \to \infty} O = O.$$

Therefore $\mathrm{O}(n)$ is compact.
(b) The subset

$$\mathrm{U}(n) = \{A \in \mathrm{M}_n(\mathbb{C}) : A^* A - I = O\} \subseteq \mathrm{M}_n(\mathbb{C})$$

is compact by a similar argument to (a).
(c) Consider the $n \times n$ diagonal matrices

$$A_k = \mathrm{diag}(k, 1/k, 1, \ldots, 1) \quad (k \geqslant 1).$$

Then $\det A_k = 1$ so $A_k \in \mathrm{SL}_n(\mathbf{k}) \leqslant \mathrm{GL}_n(\mathbf{k})$. But $\|A_k\| = k \to \infty$ as $k \to \infty$, so this sequence is unbounded.

1.7. (a) If $A, B \in \overline{H}$ then there are sequences A_r, B_r in H with $A_r \to A$ and $B_r \to B$ as $r \to \infty$. Hence $A_r B_r \to AB$ and so $AB \in \overline{H}$ since each $A_r B_r \in H$. Similarly, $A_r^{-1} \to A^{-1}$, showing that $A^{-1} \in \overline{H}$.
(b) Here is an alternative proof in terms of open sets which applies to any topological group.
Let $u, v \in \overline{H}$ and consider uv. If $uv \notin \overline{H}$, the open set $G - \overline{H}$ contains uv; notice that $G - \overline{H}$ is the biggest open subset of G which does not intersect H. Since the product map mult is continuous, there is an open set of the form $U \times V \subseteq G \times G$ contained in the open set $\mathrm{mult}^{-1}(G - \overline{H})$ with $u \in U$ and $v \in V$. Notice that $U \cap H = \varnothing = V \cap H$ by the above remark. Now take $u' \in U \cap H$ and $v' \in V \cap H$ and note that $u'v' = \mathrm{mult}(u', v') \in H$ and also $\mathrm{mult}(u', v') \in G - \overline{H}$, giving a contradiction. So $uv \in \overline{H}$. Also, if $w \in \overline{H}$, suppose that $w^{-1} \in G - \overline{H}$. Then $\mathrm{inv}^{-1}(G - \overline{H}) \subseteq G$ is open, where inv is the inverse map. But as $w \in \mathrm{inv}^{-1}(G - \overline{H})$, there must be an element $h \in H \cap \mathrm{inv}^{-1}(G - \overline{H})$ and so $h^{-1} \in G - \overline{H}$ which is impossible. Hence $\overline{H} \leqslant G$.
(c) Consider any sequence of diagonal matrices

$$A_k = \mathrm{diag}(a_k, 1, \ldots, 1) \quad (k \geqslant 1)$$

where $a_k \leqslant 1$ and $a_k \to \sqrt{2}$ as $k \to \infty$ ($\sqrt{2}$ could be replaced by any other irrational number greater than 1). Then $A_k \to \mathrm{diag}(\sqrt{2}, 1, \ldots, 1) \notin \Gamma$. We must have $\overline{\Gamma} = \mathrm{GL}_n(\mathbb{R})$ since for every $B = [b_{ij}] \in \mathrm{GL}_n(\mathbb{R})$ we can find sequences $b_{ij}^{(k)} \in \mathbb{Q}$ satisfying $b_{ij}^{(k)} \to b_{ij}$ as $k \to \infty$.

1.8. By Proposition 1.26, $G \subseteq \mathrm{M}_n(\mathbf{k})$ is a closed subset of $\mathrm{GL}_n(\mathbb{R})$.

1.9. Suppose that $G \leqslant \mathrm{GL}_n(\mathbb{R}) \subseteq \mathrm{M}_n(\mathbb{R})$ for some n. Then $H \subseteq \mathrm{M}_n(\mathbb{R})$ is closed and bounded since G is compact. For each $h \in H$ the set $\{h\} \subseteq H$ is open and these form an open cover of H by disjoint open sets. By the Heine–Borel Theorem 1.20 there must be finitely many of these sets that cover H, so H is itself finite.

1.10. In each case, we have subgroups $G_{n+1} \leqslant \mathrm{GL}_{n+1}(\mathbf{k})$ (where $\mathbf{k} = \mathbb{R}$ or $\mathbb{C}$) and $G_n \leqslant \mathrm{GL}_n(\mathbf{k})$ where G_{n+1} is closed in $\mathrm{GL}_{n+1}(\mathbf{k})$ and $G_n = G_{n+1} \cap \mathrm{GL}_n(\mathbf{k})$. As $\mathrm{GL}_n(\mathbf{k})$ is closed in $\mathrm{GL}_{n+1}(\mathbf{k})$, G_n is closed in G_{n+1}.

1.11. (a) If $A \in \mathrm{Symp}_{2m}(\mathbb{R})$, then $A^T J_{2m} A = J_{2m}$ and as $\det J_{2m} = 1$, we have $\det A^2 = 1$, so $\det A = \pm 1$.

(b) Let $A = \begin{bmatrix} a & b \\ c & d \end{bmatrix} \in \mathrm{Symp}_2(\mathbb{R})$. By definition,

$$\begin{bmatrix} a & c \\ b & d \end{bmatrix} \begin{bmatrix} 0 & 1 \\ -1 & 0 \end{bmatrix} \begin{bmatrix} a & b \\ c & d \end{bmatrix} = \begin{bmatrix} 0 & ad-bc \\ -ad+bc & 0 \end{bmatrix} = \begin{bmatrix} 0 & 1 \\ -1 & 0 \end{bmatrix},$$

so $A \in \mathrm{Symp}_2(\mathbb{R})$ if and only if $\det A = 1$; hence $\mathrm{Symp}_2(\mathbb{R}) \leqslant \mathrm{SL}_2(\mathbb{R})$.

1.12. Clearly $\rho_n \mathrm{U}(n) \leqslant \rho_n \mathrm{GL}_n(\mathbb{C})$. If $Z = [z_{rs}] \in \mathrm{M}_n(\mathbb{C})$ where $z_{rs} = x_{rs} + y_{rs} i$ with $x_{rs}, y_{rs} \in \mathbb{R}$, then

$$\rho_n(Z^*) = \rho_n([x_{sr}] - [y_{sr} i]) = \rho_n(Z)^T.$$

Hence $Z \in \mathrm{U}(n)$ if and only if $\rho_n(Z) \in \mathrm{O}(2n)$. If $A \in \mathrm{O}(2n)$, then $A \in \mathrm{Symp}_{2n}(\mathbb{R})$ if and only if $AJ_{2n} = J_{2n}A$ and by Proposition 1.44 this is equivalent to $A \in \rho_n \mathrm{U}(n)$. This shows that $\rho_n \mathrm{U}(n) = \mathrm{O}(2n) \cap \mathrm{Symp}_{2n}(\mathbb{R})$. If $B \in \mathrm{Symp}_{2n}(\mathbb{R})$ then $BJ_{2n} = J_{2n}B$ if and only if $B^T = B$. This shows that $\rho_n \mathrm{GL}_n(\mathbb{C}) \cap \mathrm{Symp}_{2n}(\mathbb{R}) = \mathrm{O}(2n) \cap \mathrm{Symp}_{2n}(\mathbb{R})$. The second collection of equalities is proved in a similar way.

1.13. (a) The triangle inequality for ρ follows from the following sequence of inequalities valid for any $(x_1, x_2), (y_1, y_2), (z_1, z_2) \in X_1 \times X_2$,

$$\begin{aligned} \rho((x_1,x_2),(z_1,z_2)) &= \sqrt{\rho_1(x_1,z_1)^2 + \rho_2(x_2,z_2)^2} \\ &\leqslant \sqrt{(\rho_1(x_1,y_1) + \rho_1(y_1,z_1))^2 + (\rho_2(x_2,y_2) + \rho_2(y_2,z_2))^2} \\ &\qquad\qquad \text{[by the triangle inequalities for } \rho_1 \text{ and } \rho_2] \\ &\leqslant \sqrt{(\rho_1(x_1,y_1)^2 + \rho_2(x_2,y_2)^2} + \sqrt{\rho_1(y_1,z_1)^2 + \rho_2(y_2,z_2)^2} \\ &\qquad\qquad \text{[by the usual triangle inequality for } \mathbb{R}^2] \\ &= \rho((x_1,x_2),(y_1,y_2)) + \rho((y_1,y_2),(z_1,z_2)). \end{aligned}$$

The other properties required for ρ to be a metric are straightforward to check. If $U_1 \times U_2 \subseteq X_1 \times X_2$ with $U_1 \subseteq X_1$ and $U_2 \subseteq X_2$ open, then for each $x_1 \in U_1$ and $x_2 \in U_2$ there are open discs $\mathrm{N}_{X_1,\rho_1}(x_1; r_1) \subseteq U_1$ and $\mathrm{N}_{X_2,\rho_2}(x_2; r_2) \subseteq U_2$; for $r = \min\{r_1, r_2\}$, we have $\mathrm{N}_{X_1 \times X_2,\rho}((x_1,x_2); r) \subseteq U_1 \times U_2$, so $U_1 \times U_2$ is open with respect to ρ. Conversely, for an open disc $\mathrm{N}_{X_1 \times X_2,\rho}((x_1,x_2); r)$ we have

$$\mathrm{N}_{X_1,\rho_1}(x_1; r/\sqrt{2}) \times \mathrm{N}_{X_2,\rho_2}(x_2; r/\sqrt{2}) \subseteq \mathrm{N}_{X_1 \times X_2,\rho}((x_1,x_2); r).$$

Hence every subset of $X_1 \times X_2$ open with respect to ρ is a union of products of open discs in X_1 and X_2. So the topology associated with ρ has the same open sets as the product topology.

(b) If $(x_{1,r}, x_{2,r}) \to (x_1, x_2)$ say, then for $i = 1, 2$,

$$\rho_i(x_{i,r}, x_i) \leqslant \rho((x_{1,r}, x_{2,r}), (x_1, x_2)),$$

so $x_{i,r} \to x_i$ as $r \to \infty$. Conversely, if $x_{1,r} \to x_1$ and $x_{2,r} \to x_2$ then

$$\rho((x_{1,r}, x_{2,r}), (x_1, x_2)) = \sqrt{\rho_1(x_{1,r}, x_1)^2 + \rho_2(x_{2,r}, x_2)^2} \to 0,$$

so $(x_{1,r}, x_{2,r}) \to (x_1, x_2)$ as $r \to \infty$.

1.14. (a) If $g \notin \mathrm{Stab}_G(x)$ then $gx \in X - \{x\}$. Since $X - \{x\}$ is an open subset of X, $\mu^{-1}(X - \{x\}) \subseteq G \times X$ is open, so there are open sets $U \subseteq G$, $V \subseteq X$ with $g \in U$, $x \in V$ and $U \times V \subseteq \mu^{-1}(X - \{x\})$. But then for $h \in U$ we must have $hx \neq x$, hence $U \subseteq G - \mathrm{Stab}_G(x)$. This shows that $G - \mathrm{Stab}_G(x)$ is open in G.
(b) If $g \notin \mathrm{Stab}_G(W)$ then $gW \not\subseteq W$ or $g^{-1}W \not\subseteq W$. If $gW \not\subseteq W$ then for some $w \in W$, $gw \neq w$. Now a similar argument to (a) shows that there is an open set $U \subseteq G - \mathrm{Stab}_G(W)$ containing g. If $g^{-1}W \not\subseteq W$ then we find an open set $U' \subseteq G - \mathrm{Stab}_G(W)$ containing g^{-1}; applying the inverse map inv we obtain an open set $\mathrm{inv}\, U' \subseteq G - \mathrm{Stab}_G(W)$ containing g. Finally, since each stabiliser $\mathrm{Stab}_G(w)$ is closed the intersection of any collection of such stabilisers is also closed.

1.15. (a) Taking ρ_1 to be the norm metric on $\mathrm{M}_n(\mathbf{k})$ and ρ_2 to be the usual metric on $\mathbf{k}^n$, define ρ as in Exercise 1.13, *i.e.*, set $\rho(A, \mathbf{x}) = \sqrt{\|A\|^2 + |\mathbf{x}|^2}$. If $A, B \in \mathrm{M}_n(\mathbf{k})$ and $\mathbf{x}, \mathbf{y} \in \mathbf{k}^n$, then

$$|A\mathbf{x} - B\mathbf{y}| = |(A\mathbf{x} - B\mathbf{x}) + (B\mathbf{x} - B\mathbf{y})| \leqslant \|A - B\|\,|\mathbf{x}| + \|B\|\,|\mathbf{x} - \mathbf{y}|.$$

A routine ε-δ argument now establishes continuity of the product map at each $(A, \mathbf{x})$.
(b) These are consequences of results in the previous exercise which show that these stabiliser subgroups are closed.
(c) We have

$$\mathrm{Stab}_{\mathrm{GL}_n(\mathbb{R})}(\mathbf{e}_n) = \left\{ \begin{bmatrix} A & O_{n-1,1} \\ B & 1 \end{bmatrix} : A \in \mathrm{GL}_{n-1}(\mathbb{R}),\ B \in \mathrm{M}_{1,n-1}(\mathbb{R}) \right\},$$

$$\mathrm{Stab}_{\mathrm{GL}_n(\mathbb{R})}(X) = \left\{ \begin{bmatrix} A & O_{n-1,1} \\ B & t \end{bmatrix} : A \in \mathrm{GL}_{n-1}(\mathbb{R}),\ B \in \mathrm{M}_{1,n-1}(\mathbb{R}),\ t \in \mathbb{R} \right\},$$

$$\mathrm{Stab}_{\mathrm{SL}_n(\mathbb{R})}(\mathbf{e}_n) = \left\{ \begin{bmatrix} A & O_{n-1,1} \\ B & 1 \end{bmatrix} : A \in \mathrm{SL}_{n-1}(\mathbb{R}),\ B \in \mathrm{M}_{1,n-1}(\mathbb{R}) \right\},$$

$$\mathrm{Stab}_{\mathrm{SL}_n(\mathbb{R})}(X) = \left\{ \begin{bmatrix} A & O_{n-1,1} \\ B & \det A^{-1} \end{bmatrix} : A \in \mathrm{GL}_{n-1}(\mathbb{R}),\ B \in \mathrm{M}_{1,n-1}(\mathbb{R}) \right\},$$

$$\mathrm{Stab}_{\mathrm{O}(n)}(\mathbf{e}_n) = \left\{ \begin{bmatrix} A & O_{n-1,1} \\ O_{1,n-1} & 1 \end{bmatrix} : A \in \mathrm{O}(n-1), \right\},$$

$$\mathrm{Stab}_{\mathrm{O}(n)}(X) = \left\{ \begin{bmatrix} A & O_{n-1,1} \\ O_{1,n-1} & t \end{bmatrix} : A \in \mathrm{O}(n-1),\ t = \pm 1 \right\},$$

$$\mathrm{Stab}_{\mathrm{SO}(n)}(\mathbf{e}_n) = \left\{ \begin{bmatrix} A & O_{n-1,1} \\ O_{1,n-1} & 1 \end{bmatrix} : A \in \mathrm{SO}(n-1), \right\},$$

$$\mathrm{Stab}_{\mathrm{SO}(n)}(X) = \left\{ \begin{bmatrix} A & O_{n-1,1} \\ O_{1,n-1} & t \end{bmatrix} : A \in \mathrm{O}(n-1),\ t = \det A \right\}.$$

1.16. (b) We have

$$\begin{bmatrix} a & b \\ c & d \end{bmatrix} x^n = (ax + cy)^n,$$

so

$$\mathrm{Stab}_{\mathrm{SU}(2)}(x^n) = \left\{ \begin{bmatrix} u & 0 \\ 0 & \overline{u} \end{bmatrix} : u^n = 1 \right\}.$$

(c) We have

$$\begin{bmatrix} a & b \\ c & d \end{bmatrix} xy = (ax+cy)(bx+dy) = abx^2 + (ad+bc)xy + cdy^2,$$

so

$$\mathrm{Stab}_{\mathrm{SU}(2)}(xy) = \left\{ \begin{bmatrix} u & 0 \\ 0 & \overline{u} \end{bmatrix} : |u| = 1 \right\} \cup \left\{ \begin{bmatrix} 0 & v \\ -\overline{v} & 0 \end{bmatrix} : |v| = 1 \right\}.$$

(d) If $A \in \ker \varphi_n$, then $A \in \mathrm{Stab}_{\mathrm{SU}(2)}(x^n)$, so by (b),

$$A = \begin{bmatrix} u & 0 \\ 0 & \overline{u} \end{bmatrix} \quad \text{with } u^n = 1.$$

Also, for each $r = 1, \ldots, n$,

$$Ax^r y^{n-r} = (ux)^r (\overline{u}y)^{n-r} = u^{2r-n} x^r y^{n-r} = x^r y^{n-r},$$

so this can only happen if $u^{2r} = 1$ for all such r; hence $u^2 = 1$. If n is odd, this gives $u = 1$, while if n is even $u = \pm 1$.

1.17. (a) Arguing as in the discussion of the norm of a square matrix at the beginning of Section 1.2, there is a number $K \geqslant 0$ for which satisfies $\nu'(\varphi(v)) \leqslant K\nu(v)$ for every $v \in V$. Thus for $x, y \in V$ we have

$$\nu'(\varphi(y) - \varphi(x)) = \nu'(\varphi(y - x)) \leqslant K\nu(y - x) \to 0$$

as $y \to x$, hence φ is continuous.
(b) These projections are linear transformations, hence by (a) they are continuous. Now $\ker \varphi' = W$ and so $W = \varphi'^{-1}\{0\}$ is a closed subset of V.

Chapter 2

2.1. For $n \geqslant 0$ we have

$$\begin{bmatrix} 0 & 1 \\ -1 & 0 \end{bmatrix}^{2n} = (-1)^n \begin{bmatrix} 1 & 0 \\ 0 & 1 \end{bmatrix}, \quad \begin{bmatrix} 0 & 1 \\ -1 & 0 \end{bmatrix}^{2n+1} = (-1)^n \begin{bmatrix} 0 & 1 \\ -1 & 0 \end{bmatrix},$$

hence

$$\exp\left(\begin{bmatrix} 0 & t \\ -t & 0 \end{bmatrix}\right) = \begin{bmatrix} \sum_{n=0}^{\infty} \frac{(-1)^n t^{2n}}{(2n)!} & \sum_{n=0}^{\infty} \frac{(-1)^n t^{2n+1}}{(2n+1)!} \\ \sum_{n=0}^{\infty} \frac{(-1)^{n+1} t^{2n+1}}{(2n+1)!} & \sum_{n=0}^{\infty} \frac{(-1)^n t^{2n}}{(2n)!} \end{bmatrix} = \begin{bmatrix} \cos t & \sin t \\ -\sin t & \cos t \end{bmatrix}.$$

Similarly,

$$\begin{bmatrix} 0 & 1 \\ 1 & 0 \end{bmatrix}^{2n} = I_2, \quad \begin{bmatrix} 0 & 1 \\ 1 & 0 \end{bmatrix}^{2n+1} = \begin{bmatrix} 0 & 1 \\ 1 & 0 \end{bmatrix},$$

hence

$$\exp\left(\begin{bmatrix} 0 & t \\ t & 0 \end{bmatrix}\right) = \begin{bmatrix} \sum_{n=0}^{\infty} \frac{t^{2n}}{(2n)!} & \sum_{n=0}^{\infty} \frac{t^{2n+1}}{(2n+1)!} \\ \sum_{n=0}^{\infty} \frac{t^{2n+1}}{(2n+1)!} & \sum_{n=0}^{\infty} \frac{t^{2n}}{(2n)!} \end{bmatrix} = \begin{bmatrix} \cosh t & \sinh t \\ \sinh t & \cosh t \end{bmatrix}.$$

Finally, by an easy induction on n,

$$\begin{bmatrix} t & 0 \\ -2 & t \end{bmatrix}^n = \begin{bmatrix} t^n & 0 \\ -2nt^{n-1} & t^n \end{bmatrix},$$

hence

$$\exp\left(\begin{bmatrix} t & 0 \\ -2 & t \end{bmatrix}\right) = \begin{bmatrix} \sum_{n=0}^{\infty} \frac{t^n}{n!} & 0 \\ -2\sum_{n=0}^{\infty} \frac{nt^{n-1}}{n!} & \sum_{r=0}^{\infty} \frac{t^n}{n!} \end{bmatrix} = \begin{bmatrix} e^t & 0 \\ -2\sum_{n=0}^{\infty} \frac{t^n}{n!} & e^t \end{bmatrix} = \begin{bmatrix} e^t & 0 \\ -2e^t & e^t \end{bmatrix}.$$

2.2. (a) Each partial sum has the form

$$\sum_{r=0}^{n} \frac{1}{r!}(BAB^{-1})^r = B\left(\sum_{r=0}^{n} \frac{1}{r!}A^r\right)B^{-1}.$$

(b) Take $A = D$ and $B = C$ in (a) to obtain

$$\exp(D) = C\exp(\operatorname{diag}(\lambda_1, \dots, \lambda_n))C^{-1} = C\operatorname{diag}(e^{\lambda_1}, \dots, e^{\lambda_n})C^{-1}.$$

(c) The eigenvalues of $\begin{bmatrix} 0 & t \\ -t & 0 \end{bmatrix}$ are $\pm ti$ and the matrix $C = \begin{bmatrix} 1 & 1 \\ i & -i \end{bmatrix}$ satisfies

$$\begin{bmatrix} 0 & t \\ -t & 0 \end{bmatrix} = C\operatorname{diag}(ti, -ti)C^{-1}.$$

Now using the identities $\cos t = (e^{ti} + e^{-ti})/2$ and $\sin t = (e^{ti} - e^{-ti})/2i$ we obtain

$$\exp\left(\begin{bmatrix} 0 & t \\ -t & 0 \end{bmatrix}\right) = C\operatorname{diag}(e^{ti}, e^{-ti})C^{-1} = \begin{bmatrix} \cos t & \sin t \\ -\sin t & \cos t \end{bmatrix}.$$

Similarly,

$$\begin{bmatrix} 0 & t \\ t & 0 \end{bmatrix} = C\operatorname{diag}(t, -t)C^{-1},$$

where $C = \begin{bmatrix} 1 & -1 \\ 1 & 1 \end{bmatrix}$. Using the identities $\cosh t = (e^t + e^{-t})/2$ and $\sinh t = (e^t - e^{-t})/2$ we obtain

$$\exp\left(\begin{bmatrix} 0 & t \\ t & 0 \end{bmatrix}\right) = C\operatorname{diag}(e^t, e^{-t})C^{-1} = \begin{bmatrix} \cosh t & \sinh t \\ \sinh t & \cosh t \end{bmatrix}.$$

2.3. (a) Every positive power of N is strictly upper triangular and N is nilpotent, say $N^k = O$. So $\exp(N)$ is upper triangular with 1's down its main diagonal.
(b) Write $N = tI + U$, where $t \in \mathbf{k}$ and U is strictly upper triangular. Then U is nilpotent, say U^ℓ for some $\ell > 0$. So for $m \geqslant 1$ we have

$$N^m = (tI + U)^m = \sum_{r=0}^{m} \binom{m}{r} t^{m-r}U^r.$$

For $m \geqslant \ell$ this becomes

$$N^m = \sum_{r=0}^{\ell-1} \binom{m}{r} t^{m-r}U^r.$$

Thus

$$\exp(N) = \sum_{m=0}^{\infty} \frac{1}{m!} \sum_{r=0}^{\min\{\ell-1,m\}} \binom{m}{r} t^{m-r} U^r = \sum_{r=0}^{\ell-1} \left(\sum_{m=0}^{\infty} \frac{1}{m!} \binom{m}{r} t^{m-r} \right) U^r.$$

2.4. (a) The operation of transposition is a continuous function on $\mathrm{M}_n(\mathbb{R})$ so it commutes with taking limits of sequences. Alternatively, the partial sums of $\exp(S)$ satisfy

$$\left(\sum_{r=0}^{k} \frac{1}{k!} S^k \right)^T = \sum_{r=0}^{k} \frac{1}{k!} (S^T)^k = \sum_{r=0}^{k} \frac{1}{k!} (-S)^k = \exp(-S) = \exp(S)^{-1},$$

hence $\exp(S)^T = \exp(S)^{-1}$.
(b) Similarly, the hermitian conjugate of the exponential satisfies $\exp(S)^* = \exp(S)^{-1}$.

2.5. (a) Start by solving the differential equation

$$\alpha'(t) = \alpha(t) \begin{bmatrix} -1 & -2 \\ 0 & 1 \end{bmatrix}, \quad \alpha(0) = I.$$

The solution is

$$\alpha(t) = \begin{bmatrix} e^{-t} & 2(e^{-t} - e^t) \\ 0 & e^t \end{bmatrix}.$$

The desired solution is

$$\begin{bmatrix} x(t) \\ y(t) \end{bmatrix} = \begin{bmatrix} e^{-t} & 2(e^{-t} - e^t) \\ 0 & e^t \end{bmatrix} \begin{bmatrix} 1 \\ 2 \end{bmatrix} = \begin{bmatrix} 3e^{-t} - 2e^t \\ 2e^t \end{bmatrix}.$$

(b) The solutions are

$$\begin{bmatrix} x(t) \\ y(t) \end{bmatrix} = \begin{bmatrix} -2e^{-t} \sin t \\ e^{-t}(\cos t - \sin t) \end{bmatrix}, \quad \begin{bmatrix} x(t) \\ y(t) \end{bmatrix} = \begin{bmatrix} e^{-t} \\ e^{-t} \end{bmatrix}.$$

2.6. A solution has the form $\mathbf{x}(t) = \exp(tA)\mathbf{x}(0)$ for $t \in \mathbb{R}$. Since $A^* = -A$ we have $\exp(tA)^T = \exp(-tA) = \exp(tA)^{-1}$, hence $\exp(tA)$ is orthogonal. So

$$|\mathbf{x}(t)|^2 = \mathbf{x}(t) \cdot \mathbf{x}(t) = (\exp(tA)\mathbf{x}(0)) \cdot (\exp(tA)\mathbf{x}(0)) = \mathbf{x}(0) \cdot \mathbf{x}(0) = |\mathbf{x}(0)|^2.$$

We also have

$$\mathbf{x}(t) \cdot \mathbf{x}'(t) = \frac{1}{2} \frac{\mathrm{d}}{\mathrm{d}t} |\mathbf{x}(t)|^2 = 0,$$

showing that $\mathbf{x}(t)$ and $\mathbf{x}'(t)$ are orthogonal.

2.7. (a) Choose any sequence of matrices

$$A_n = \begin{bmatrix} \lambda_{1,n} & 1 & 0 & \cdots & 0 \\ 0 & \lambda_{2,n} & 1 & \cdots & \vdots \\ \ddots & \ddots & \ddots & \ddots & \ddots \\ \cdots & \cdots & 0 & \lambda_{r-1,n} & 1 \\ 0 & \cdots & \cdots & 0 & \lambda_{r,n} \end{bmatrix}$$

in which the r diagonal λ's in $J(\lambda, r)$ have been replaced by r sequences of non-zero terms $\lambda_{1,n}, \ldots, \lambda_{r,n}$ that satisfy

$$\lambda_{i,n} \neq \lambda_{j,n} \quad \text{if } i \neq j, \qquad \lambda_{j,n} \to \lambda \quad \text{as } n \to \infty.$$

Then each matrix A_n has r distinct eigenvalues so is diagonalisable and $A_n \to J(\lambda, r)$ as $n \to \infty$.
(b) This follows from (a) together with Theorem 2.9.

Chapter 3

3.1. (a) Any abelian Lie algebras of the same dimension are isomorphic since they are isomorphic as vector spaces. So let $\mathfrak{g}$ be any 2-dimensional $\mathbf{k}$-Lie algebra which is not abelian. Then there are elements $x, y \in \mathfrak{g}$ for which $[x, y] \neq 0$; these elements cannot be linearly dependent so they form a basis of the $\mathbf{k}$-vector space $\mathfrak{g}$. If $[x, y] = rx + sy$ with $r, s \in \mathbf{k}$, we can interchange x, y if necessary to ensure that $r \neq 0$, and then replace x by $(r^{-1})x$ to ensure that $[x, y] = x + sy$. Finally, since $[x + sy, y] = x + sy$, we can replace x by $x + sy$ to ensure that $[x, y] = x$. Notice that in $\mathfrak{b}$ we have the elements

$$U = \begin{bmatrix} 0 & 1 \\ 0 & 0 \end{bmatrix}, \quad V = \begin{bmatrix} -1 & 0 \\ 0 & 0 \end{bmatrix},$$

which have the bracket $[U, V] = U$. Then there is an obvious isomorphism of Lie algebras $\mathfrak{g} \longrightarrow \mathfrak{b}$ under which $x \longleftrightarrow U$ and $y \longleftrightarrow V$.
(b) Take

$$G = \left\{ \begin{bmatrix} z & w \\ 0 & 1 \end{bmatrix} \right\} \leqslant \mathrm{GL}_2(\mathbf{k}).$$

3.2. (a) If $\gamma\colon (a, b) \longrightarrow G$ is a differentiable curve then so are the curves defined by

$$(U\gamma)(t) = U\gamma(t),\ (\gamma U)(t) = \gamma(t)U,\ (U\gamma U^{-1})(t) = U\gamma(t)U^{-1},$$

and these have derivatives

$$(U\gamma)'(t) = U\gamma'(t),\ (\gamma U)'(t) = \gamma'(t)U,\ (U\gamma U^{-1})'(t) = U\gamma'(t)U^{-1}.$$

(b) Use the derivative maps at I, *i.e.*, $\lambda_U = \mathrm{d}\,\mathrm{L}_U$, $\rho_U = \mathrm{d}\,\mathrm{R}_U$ and $\chi_U = \mathrm{d}\,\mathrm{C}_U$. The required properties follow easily.

3.3. Follow the approach of Section 3.3.
The Lie algebra of G_1 is

$$\mathfrak{g}_1 = \left\{ \begin{bmatrix} u & 0 \\ w & v \end{bmatrix} : u, v, w \in \mathbb{R} \right\}.$$

The elements $U = \begin{bmatrix} 1 & 0 \\ 0 & 0 \end{bmatrix}$, $V = \begin{bmatrix} 0 & 0 \\ 0 & 1 \end{bmatrix}$ and $W = \begin{bmatrix} 0 & 0 \\ 1 & 0 \end{bmatrix}$ form a basis with the two non-trivial brackets $[U, W] = -W$ and $[V, W] = W$.
The Lie algebra of G_2 is

$$\mathfrak{g}_2 = \left\{ \begin{bmatrix} 0 & t \\ t & 0 \end{bmatrix} : t \in \mathbb{R} \right\},$$

and this is abelian.
The Lie algebra of G_3 is

$$\mathfrak{g}_3 = \left\{ \begin{bmatrix} r & 0 & u \\ v & s & w \\ -u & 0 & t \end{bmatrix} : r, s, t, u, v, w \in \mathbb{R} \right\}.$$

The Lie algebra of G_4 is

$$\mathfrak{g}_4 = \left\{ \begin{bmatrix} A & \mathbf{t} \\ \mathbf{0} & 0 \end{bmatrix} : A \in \mathrm{M}_n(\mathbf{k}),\ \mathbf{t} \in \mathbf{k}^n \right\},$$

with Lie bracket given by

$$\left[\begin{bmatrix} A_1 & \mathbf{t}_1 \\ 0 & 0 \end{bmatrix}, \begin{bmatrix} A_2 & \mathbf{t}_2 \\ 0 & 0 \end{bmatrix}\right] = \begin{bmatrix} [A_1, A_2] & A_1\mathbf{t}_2 - A_2\mathbf{t}_1 \\ 0 & 0 \end{bmatrix}.$$

Consider the $2m \times 2m$ real matrix

$$A = \begin{bmatrix} A_{11} & A_{12} & \cdots & A_{1m} \\ A_{21} & A_{22} & \cdots & \cdots \\ \vdots & \vdots & \ddots & \vdots \\ A_{m1} & \cdots & \cdots & A_{mm} \end{bmatrix}$$

where each $A_{rs} \in \mathrm{M}_2(\mathbb{R})$. Then $A^T J_{2m} + J_{2m} A = O$ if and only if the equations

$$A_{sr}^T J + J A_{rs} = O \quad (r, s = 1, \ldots, m)$$

are satisfied and these are equivalent to

$$A_{sr}^T = J A_{rs} J \quad (r, s = 1, \ldots, m).$$

If we write $A_{rs} = \begin{bmatrix} a_{rs} & b_{rs} \\ c_{rs} & d_{rs} \end{bmatrix}$ then

$$a_{sr} = -d_{rs}, \; b_{sr} = b_{rs}, \; c_{sr} = c_{rs}, \; d_{sr} = -a_{rs}.$$

If $s = r$, this gives $A_{rr} = \begin{bmatrix} a_{rr} & b_{rr} \\ c_{rr} & -a_{rr} \end{bmatrix}$.

3.4. We have $\mathfrak{b} \cong \mathfrak{a}/\ker\Phi$. Then $\mathfrak{b}$ is abelian if and only if for every commutator $[x, y]$ in $\mathfrak{a}$, $[x, y] + \ker\Phi = 0 + \ker\Phi$, *i.e.*, $[x, y] \in \Phi$. Hence $\mathfrak{b}$ is abelian if and only if $\mathfrak{a} \subseteq \ker\Phi$.

3.5. If $\exp(sX)\exp(tY) = \exp(tY)\exp(sX)$ for all $s, t \in \mathbb{R}$, then

$$[X, Y] = \frac{\mathrm{d}}{\mathrm{d}\,s}\bigg|_{s=0} \frac{\mathrm{d}}{\mathrm{d}\,t}\bigg|_{t=0} (\exp(sX)\exp(tY) - \exp(tY)\exp(sX)) = 0.$$

Conversely, if $[X, Y] = 0$ then for each $s \in \mathbb{R}$, consider the function

$$\gamma \colon \mathbb{R} \longrightarrow G; \quad \gamma(t) = \exp(sX)\exp(tY)\exp(-sX).$$

Then $\gamma(0) = I$ and

$$\gamma'(t) = \exp(sX)Y\exp(tY)\exp(-sX) = \exp(sX)\exp(tY)\exp(-sX)Y = \gamma(t)Y$$

since $\exp(sX)Y\exp(-sX) = Y$ by definition of the exponential function. But the differential equation

$$\gamma'(t) = \gamma(t)Y, \quad \gamma(0) = I$$

has the unique solution $\gamma(t) = \exp(tY)$, so for all s, t we have

$$\exp(sX)\exp(tY)\exp(-sX) = \exp(tY)$$

and therefore

$$\exp(sX)\exp(tY) = \exp(tY)\exp(sX).$$

3.6. (a) If $A \in U$ then

$$\Phi(A)^T = (I - A^T)(I + A^T)^{-1} = (I - A^{-1})(I + A^{-1})^{-1}$$
$$= (A - I)(A + I)^{-1} = -\Phi(A),$$

so $\Phi(A) \in \text{Sk-Sym}_n(\mathbb{R})$. Conversely, if $B \in \text{Sk-Sym}_n(\mathbb{R})$ then

$$\begin{aligned}\Phi((I-B)(I+B)^{-1}) &= \left(I - (I-B)(I+B)^{-1}\right)\left(I + (I-B)(I+B)^{-1}\right)^{-1} \\ &= \left(I - (I-B)(I+B)^{-1}\right)\left(I + (I-B)(I+B)^{-1}\right)^{-1} \\ &= ((I+B) - (I-B))\,((I+B) + (I-B))^{-1} \\ &= (2B)\,(2I)^{-1} = B,\end{aligned}$$

and

$$\begin{aligned}\left((I-B)(I+B)^{-1}\right)^T (I-B)(I+B)^{-1} &= (I - B^T)(I + B^T)^{-1}(I-B)(I+B)^{-1} \\ &= (I+B)(I-B)^{-1}(I-B)(I+B)^{-1} = I.\end{aligned}$$

Hence $(I - B)(I + B)^{-1} \in \mathrm{O}(n)$ and $\Phi((I - B)(I + B)^{-1}) = B$. In fact, there is a path $[0,1] \longrightarrow \mathrm{O}(n)$ given by

$$t \longmapsto (I - tB)(I + tB)^{-1},$$

hence $\det(I - tB)(I + tB)^{-1} = 1$ since $\det C = \pm 1$ if $C \in \mathrm{O}(n)$. So $(I - B)(I + B)^{-1} \in \mathrm{SO}(n)$ and therefore $\operatorname{im}\Phi = \text{Sk-Sym}_n(\mathbb{R})$.
(b) By the calculation in (a), $\Phi^{-1}(B) = (I - B)(I + B)^{-1}$.
(c) We have

$$\dim \mathrm{SO}(n) = \dim \text{Sk-Sym}_n(\mathbb{R}) = 1 + 2 + \cdots + (n-1) = \binom{n}{2}.$$

3.7. Parts (a)–(c) are very similar to the previous question and we obtain

$$\Theta^{-1}(B) = (I - B)(I + B)^{-1} \quad (B \in \text{Sk-Herm}_2(\mathbb{C})),$$

and

$$\dim \mathrm{U}(n) = \text{Sk-Herm}_n(\mathbb{C}) = n + 2\binom{n}{2} = n^2.$$

(d) In the case $n = 2$, if $A \in V \cap \mathrm{SU}(2)$, then the eigenvalues of A must have the form $\lambda, \overline{\lambda} \in \mathbb{C}$ where $|\lambda| = 1$. Then the eigenvalues of $\Theta(A)$ are $(1 - \lambda)/(1 + \lambda)$ and $(1 - \overline{\lambda})/(1 + \overline{\lambda})$, hence

$$\operatorname{tr}\Theta(A) = \frac{1-\lambda}{1+\lambda} + \frac{1-\overline{\lambda}}{1+\overline{\lambda}} = \frac{1-\lambda}{1+\lambda} + \frac{\lambda-1}{\lambda+1} = 0.$$

Conversely, if $B \in \text{Sk-Herm}_2^0(\mathbb{C})$ then B has imaginary eigenvalues $\pm ti$ for some $t \in \mathbb{R}$, so $\Theta^{-1}(B)$ has eigenvalues $(1 - ti)/(1 + ti)$ and $(1 + ti)/(1 - ti)$. Thus we have

$$\det \Theta^{-1}(B) = \frac{(1-ti)}{(1+ti)}\frac{(1+ti)}{(1-ti)} = 1.$$

Examples show that this can be false when $n > 2$.

3.8. (a) The surjection $\det\colon \mathrm{O}(n) \longrightarrow \{1,-1\}$ has kernel $\mathrm{SO}(n)$. The diagonal matrix $\mathrm{diag}(1,\ldots,1,-1) \in \mathrm{O}(n)$ generates a subgroup $C = \{I, \mathrm{diag}(1,\ldots,1,-1)\}$ of order 2 for which

$$\mathrm{O}(n) = C\,\mathrm{SO}(n), \quad C \cap \mathrm{SO}(n) = \{I\}.$$

(b) Let

$$\mathbb{T}' = \{\mathrm{diag}(1,\ldots,1,z) \in \mathrm{U}(n) : |z| = 1\} \leqslant \mathrm{U}(n).$$

Then

$$\mathrm{U}(n) = \mathbb{T}'\,\mathrm{SU}(n), \quad \mathbb{T}' \cap \mathrm{SU}(n) = \{I\}.$$

(c) Let

$$D = \{\mathrm{diag}(1,\ldots,1,t) \in \mathrm{GL}_n(\mathbb{R}) : t \in \mathbb{R}\} \leqslant \mathrm{GL}_n(\mathbb{R}).$$

Then

$$\mathrm{GL}_n(\mathbb{R}) = D\,\mathrm{SL}_n(\mathbb{R}), \quad D \cap \mathrm{SL}_n(\mathbb{R}) = \{I\}.$$

(d) Let

$$D' = \{\mathrm{diag}(1,\ldots,1,z) \in \mathrm{GL}_n(\mathbb{C}) : t \in \mathbb{C}\} \leqslant \mathrm{GL}_n(\mathbb{C}).$$

Then

$$\mathrm{GL}_n(\mathbb{C}) = D'\,\mathrm{SL}_n(\mathbb{C}), \quad D' \cap \mathrm{SL}_n(\mathbb{C}) = \{I\}.$$

Chapter 4

4.1. Some elements of order 2 in A_3' are -1 and

$$e_3(1\,2) = \frac{1}{6}[\,2(1\,2) - (1\,3) - (2\,3)\,],$$
$$e_3(1\,3) = \frac{1}{6}[\,2(1\,3) - (2\,3) - (1\,2)\,],$$
$$e_3(2\,3) = \frac{1}{6}[\,2(2\,3) - (1\,2) - (1\,3)\,].$$

As $\mathbb{H}$ has only the element -1 of order 2 these algebras cannot be isomorphic.

4.2. (a) The subset $A_1^\times \subseteq A$ is closed and bounded, so compact.
(b) Let $u, v \in A_1^\times$. Then we have the inequalities

$$\nu(uv) \leqslant \nu(u)\nu(v) = 1,$$
$$\nu((uv)^{-1}) = \nu(v^{-1}u^{-1}) \leqslant \nu(v^{-1})\nu(u^{-1}) \leqslant 1,$$
$$1 = \nu(1) = \nu(uv(uv)^{-1}) \leqslant \nu(uv)\nu((uv)^{-1}) \leqslant 1,$$

hence $\nu(uv) = \nu((uv)^{-1}) = 1$. So $A_1^\times \leqslant A^\times$.
(c) $\mathbb{C}_1^\times$ is the unit circle with abelian Lie algebra $\{ti : t \in \mathbb{R}\} \leqslant \mathbb{C}$.

$$\mathbb{H}_1^\times = \{u + vj : u, v \in \mathbb{C},\ |u|^2 + |v|^2 = 1\},$$

with Lie algebra

$$\{ri + sj + tk : u, v \in \mathbb{C},\ r, s, t \in \mathbb{R}\} \cong \mathfrak{su}(2).$$

For $\mathbf{k} = \mathbb{R}$ or $\mathbb{C}$, let $A = \mathrm{M}_n(\mathbf{k})_1^\times$. Suppose that $|A\mathbf{u}| < |\mathbf{u}|$ for some $\mathbf{u} \in \mathbf{k}^n$; then

$$1 = \frac{|\mathbf{u}|}{|\mathbf{u}|} = \frac{|A^{-1}(A\mathbf{u})|}{|\mathbf{u}|} < \frac{|A^{-1}(A\mathbf{u})|}{|A\mathbf{u}|} \leqslant \|A^{-1}\|,$$

contradicting the fact that $\|A^{-1}\| = 1$. This shows that for every $\mathbf{x} \in \mathbf{k}^n$, $|A\mathbf{x}| = |\mathbf{x}|$. By Proposition 1.38 and the analogous result for unitary groups, this shows that $A \in \mathrm{O}(n)$ if $\mathbf{k} = \mathbb{R}$ and $A \in \mathrm{U}(n)$ if $\mathbf{k} = \mathbb{C}$, while the corresponding Lie algebras are $\mathfrak{o}(n) \leqslant \mathrm{M}_n(\mathbb{R})$ and $\mathfrak{u}(n) \leqslant \mathrm{M}_n(\mathbb{C})$.

4.3. For the case $r > 0$, let $q \in \mathbb{H}$ satisfy $q^2 = r$. Using notation from the proof of Proposition 4.40, we find that $\mathbb{R}(q)$ is a field, so the number of roots of $x^2 - r$ is at most 2 and these are the real numbers $\pm\sqrt{r}$. The case where $r < 0$ can be dealt with using Proposition 4.42.

4.4. Here is what happens when $n = 1$. For $\mathbb{H}_{\mathbb{R}}$, $\Lambda\colon \mathbb{H} \longrightarrow \mathrm{M}_4(\mathbb{R})$ takes the values $\Lambda(1) = I_4$ and

$$\Lambda(i) = \begin{bmatrix} 0 & -1 & 0 & 0 \\ 1 & 0 & 0 & 0 \\ 0 & 0 & 0 & -1 \\ 0 & 0 & 1 & 0 \end{bmatrix}, \quad \Lambda(j) = \begin{bmatrix} 0 & 0 & -1 & 0 \\ 0 & 0 & 0 & 1 \\ 1 & 0 & 0 & 0 \\ 0 & -1 & 0 & 0 \end{bmatrix}, \quad \Lambda(k) = \begin{bmatrix} 0 & 0 & 0 & -1 \\ 0 & 0 & -1 & 0 \\ 0 & 1 & 0 & 0 \\ 1 & 0 & 0 & 0 \end{bmatrix},$$

and so

$$\Lambda(a1 + bi + cj + dk) = \begin{bmatrix} a & -b & -c & -d \\ b & a & -d & c \\ c & d & a & -b \\ d & -c & b & a \end{bmatrix}.$$

The reduced determinant $\mathrm{Rdet}_{\mathbb{R}}\colon \mathbb{H}^{\times} \longrightarrow \mathbb{R}^{\times}$ takes the value

$$\begin{aligned} \mathrm{Rdet}_{\mathbb{R}}(a1 + bi + cj + dk) &= \det \Lambda(a1 + bi + cj + dk) \\ &= a^4 + 2a^2b^2 + 2d^2a^2 + 2c^2a^2 + b^4 + 2d^2b^2 + 2c^2b^2 + c^4 + 2c^2d^2 + d^4. \end{aligned}$$

For $\mathbb{H}_{\mathbb{C}}$, $\Lambda\colon \mathbb{H} \longrightarrow \mathrm{M}_2(\mathbb{C})$ has the effect

$$\Lambda(1) = I_2, \quad \Lambda(j) = \begin{bmatrix} 0 & -1 \\ 1 & 0 \end{bmatrix},$$

so

$$\Lambda(a + bj) = \begin{bmatrix} a & -b \\ \bar{b} & \bar{a} \end{bmatrix}.$$

When $n = 1$, the reduced determinant $\mathrm{Rdet}_{\mathbb{C}}\colon \mathbb{H}^{\times} \longrightarrow \mathbb{C}^{\times}$ takes the value

$$\mathrm{Rdet}_{\mathbb{C}}(a1 + bj) = \det \Lambda(a1 + bj) = |a|^2 + |b|^2.$$

4.5. (b) We have

$$\exp(r + su) = \exp(r)\exp(su) = e^r(\cos s\, 1 + \sin s\, u).$$

4.6. (a) $\mathfrak{g} = \mathrm{M}_n(\mathbb{H})$. (b) $\mathfrak{g} = \{A \in \mathrm{M}_n(\mathbb{H}) : A^* = -A\}$.
(c) When $n = 1$, taking $\mathbf{k} = \mathbb{R}$ and using Exercise 4.4, we have $a1 + bi + cj + dk \in G$ with $a, b, c, d \in \mathbb{R}$ if and only if

$$a^4 + 2a^2b^2 + 2d^2a^2 + 2c^2a^2 + b^4 + 2d^2b^2 + 2c^2b^2 + c^4 + 2c^2d^2 + d^4 = 1,$$

so

$$\mathfrak{g} = \{ri + sj + tk : r, s, t \in \mathbb{R}\}.$$

For $\mathbf{k} = \mathbb{C}$, $u1 + vj \in G$ with $u, v \in \mathbb{C}$ if and only if $|u|^2 + |v|^2 = 1$, giving

$$\mathfrak{g} = \{ri + zj : r \in \mathbb{R},\ z \in \mathbb{C}\}.$$

(d) When $n = 1$,

$$\mathrm{Sp}(1) = \{a1 + bi + cj + dk : a, b, c, d \in \mathbb{R},\ a^2 + b^2 + c^2 + d^2 = 1\},$$

so again using Exercise 4.4, for $\mathbf{k} = \mathbb{R}$ or $\mathbb{C}$ we obtain

$$\mathfrak{g} = \{ri + zj : r \in \mathbb{R},\ z \in \mathbb{C}\}.$$

Chapter 5

5.1. (a) First write $u = r_1e_1 + \cdots + r_ne_n$ with $r_k \in \mathbb{R}$ and $r_1^2 + \cdots + r_n^2 = 1$. Then

$$u^2 = r_1^2e_1^2 + \cdots + r_n^2e_n^2 = -1.$$

Also, if $w \in \mathbb{R}^n$ with $w \cdot u = 0$, then writing $w = s_1e_1 + \cdots + s_ne_n$ with $s_k \in \mathbb{R}$, we have

$$\begin{aligned} wu &= (r_1s_1e_1^2 + \cdots + r_ns_n^2e_n^2) + \sum_{1 \leqslant i < j \leqslant n} (r_is_j - r_js_i)e_ie_j \\ &= -(u \cdot w) + \sum_{1 \leqslant i < j \leqslant n} (r_is_j - r_js_i)e_ie_j = \sum_{1 \leqslant i < j \leqslant n} (r_is_j - r_js_i)e_ie_j. \end{aligned}$$

A similar calculation shows that

$$uw = \sum_{1 \leqslant i < j \leqslant n} (r_js_i - r_is_j)e_ie_j = -wu.$$

So we have $wu = uw$ whenever $w \cdot u = 0$. Using this we obtain

$$uvu = uv_1u + uv_2u = v_1 - uv_2u.$$

(c) The elements $Ae_1, \ldots, Ae_n$ form an orthonormal basis of $\mathbb{R}^n \subseteq \mathrm{Cl}_n$, so using the universal property of Theorem 5.4, the function $\mathbb{R}^n \longrightarrow \mathrm{Cl}_n$ sending $\sum_{i=1}^n x_ie_i$ to $\sum_{i=1}^n x_iAe_i$ induces a homomorphism $A_*\colon \mathrm{Cl}_n \longrightarrow \mathrm{Cl}_n$. The inverse $A^{-1} = A^T$ induces $(A^{-1})_*\colon \mathrm{Cl}_n \longrightarrow \mathrm{Cl}_n$ which is actually the inverse of A_*. A straightforward calculation shows that $A_*(e_1 \cdots e_n) = (\det A)e_1 \cdots e_n$.

5.2. (a) The image of $i_n'\colon \mathrm{Cl}_n \longrightarrow \mathrm{Cl}_{n+1}$ is spanned by the monomials $e_{i_1} \cdots e_{i_r}$ with $i_k = 1, \ldots, n$.
(b) Since $x \cdot e_{n+1} = 0$,

$$(xe_{n+1})^2 = xe_{n+1}xe_{n+1} = -x^2e_{n+1}^2 = -(-|x|^2)(-1) = -|x|^2,$$

so the universal property gives the desired homomorphism. Also notice that $xe_{n+1} \in \mathrm{Cl}_n^+$ and it is easy to see that every monomial $e_{i_1} \cdots e_{i_{2s}}$ of even length with $i_k = 1, \ldots, n+1$ is in the image.

5.3. (a) For any $k = 1, \ldots, n$,

$$e_k\omega_n = e_ke_1 \cdots e_n = (-1)^{k-1}e_1 \cdots e_{k-1}e_k^2e_{k+1} \cdots e_n(-1)^ke_1 \cdots e_{k-1}e_{k+1} \cdots e_n,$$

while

$$\omega_ne_k = e_1 \cdots e_ne_k = (-1)^{n-k}e_1 \cdots e_{k-1}e_k^2e_{k+1} \cdots e_n(-1)^{n-k+1}e_1 \cdots e_{k-1}e_{k+1} \cdots e_n.$$

Since $n-k+1 \equiv k \bmod 2$, $e_k\omega_n = \omega_n e_k$. From this it follows that for every monomial in the e_k,

$$e_{i_1} \cdots e_{i_r}\omega_n = \omega_n e_{i_1} \cdots e_{i_r}.$$

(b) We have

$$\omega_n^2 = e_1 \cdots e_n e_1 \cdots e_n = (-1)^{(n-1)+\cdots+1} e_1^2 \cdots e_n^2 = (-1)^{\binom{n}{2}+n} = (-1)^{\binom{n+1}{2}}.$$

But $(n+1) \equiv 0 \bmod 4$, so $\binom{n+1}{2} \equiv 0 \bmod 2$ and therefore $\omega_n^2 = 1$. Now

$$\Omega_\pm^2 = \frac{1}{4}(1 \pm 2\omega_n + \omega_n^2) = \frac{1}{4}(2 \pm 2\omega_n) = \Omega_\pm,$$

so these are idempotents which are central and satisfy $\Omega_+ + \Omega_- = 1$.
(c) Set $\Omega_+ A_+ = \mathrm{Cl}_n$ and $\Omega_- A_- = \mathrm{Cl}_n$.
(d) By Exercise 5.2, $\mathrm{Cl}_3^+ = \mathrm{Cl}_2 = \mathbb{H}$ and $\mathrm{Cl}_7^+ = \mathrm{Cl}_6 = \mathrm{M}_8(\mathbb{R})$. In each case multiplication by $\Omega_\pm$ maps the simple algebra Cl_n^+ onto $A_\pm$ and therefore gives an isomorphism $\mathrm{Cl}_n^+ \cong A_\pm$.

5.4. (c) Send γ to -1 and ε_j to e_j.
(d) The elements $(1/2)(1 \pm \gamma) \in \mathbb{R}[\mathrm{ClGp}_n]$ are central idempotents. Put

$$A = (1/2)(1+\gamma)\mathbb{R}[\mathrm{ClGp}_n], \quad B = (1/2)(1-\gamma)\mathbb{R}[\mathrm{ClGp}_n].$$

5.6. (b) $\mathrm{Z}(\mathrm{O}(n)) = \mathrm{Z}(\mathrm{SO}(n))$.

Chapter 6

6.1. (b) For some $Q \in \mathrm{Lor}(3,1)$,

$$A_t = Q \begin{bmatrix} \cos t & -\sin t & 0 & 0 \\ \sin t & \cos t & 0 & 0 \\ 0 & 0 & \cosh(-t) & \sinh(-t) \\ 0 & 0 & \sinh(-t) & \cosh(-t) \end{bmatrix} Q^{-1}.$$

(c) With the same Q as in (b), take

$$U = Q \begin{bmatrix} 0 & -1 & 0 & 0 \\ 1 & 0 & 0 & 0 \\ 0 & 0 & 0 & -1 \\ 0 & 0 & -1 & 0 \end{bmatrix} Q^{-1}.$$

(d) A suitable equation is

$$\alpha'(t) = \alpha(t)U, \quad \alpha(0) = 1.$$

6.2. (a) We have

$$P_{24} + P_{34} = \begin{bmatrix} 0 & 0 & 0 & 0 \\ 0 & 0 & 0 & 1 \\ 0 & 0 & 0 & 0 \\ 0 & 1 & 0 & 0 \end{bmatrix} + \begin{bmatrix} 0 & 0 & 0 & 0 \\ 0 & 0 & 0 & 0 \\ 0 & 0 & 0 & 1 \\ 0 & 0 & 1 & 0 \end{bmatrix} = \begin{bmatrix} 0 & 0 & 0 & 0 \\ 0 & 0 & 0 & 1 \\ 0 & 0 & 0 & 1 \\ 0 & 1 & 1 & 0 \end{bmatrix}$$

and

$$\alpha(t) = \exp(t(P_{24} + P_{34})) = \begin{bmatrix} 1 & 0 & 0 & 0 \\ 0 & \dfrac{\cosh\sqrt{2}t + 2}{2} & \dfrac{\cosh\sqrt{2}t - 2}{2} & \dfrac{\sinh\sqrt{2}t}{\sqrt{2}} \\ 0 & \dfrac{\cosh\sqrt{2}t - 2}{2} & \dfrac{\cosh\sqrt{2}t + 2}{2} & \dfrac{\sinh\sqrt{2}t}{\sqrt{2}} \\ 0 & \dfrac{\sinh\sqrt{2}t}{\sqrt{2}} & \dfrac{\sinh\sqrt{2}t}{\sqrt{2}} & \dfrac{\cosh\sqrt{2}t}{2} \end{bmatrix}.$$

(b) We have

$$-i(E+F) = \begin{bmatrix} 0 & \frac{1}{2}(1-i) \\ \frac{1}{2}(1+i) & 0 \end{bmatrix}$$

and

$$\widetilde{\alpha}(t) = \exp(-ti(E+F)) = \begin{bmatrix} \cosh(t/\sqrt{2}) & \dfrac{\sinh(t/\sqrt{2})}{\sqrt{2}}(1-i) \\ \dfrac{\sinh(t/\sqrt{2})}{\sqrt{2}}(1+i) & \cosh(t/\sqrt{2}) \end{bmatrix}.$$

(c) The homomorphism $\widetilde{\mathrm{Ad}}\colon \mathrm{SL}_2(\mathbb{C}) \longrightarrow \mathrm{Lor}(3,1)$ relates these two curves through the equation $\alpha = \widetilde{\mathrm{Ad}} \circ \widetilde{\alpha}$.

Chapter 7

7.1. (a) Define the function

$$F\colon \mathrm{M}_n(\mathbb{R}) \times \mathbb{R} \longrightarrow \mathbb{R}; \quad F(A,b) = b\det A - 1.$$

This is smooth and $M = F^{-1}0 \subseteq \mathrm{M}_n(\mathbb{R}) \times \mathbb{R}$ is a closed subset. The derivative mapping $\mathrm{d} = \mathrm{d}_{(A,b)}$ at a point $(A,b) \in F^{-1}0$ has the form

$$\mathrm{d}(X,v) = (\det A)v + \mathrm{d}\det_A(X),$$

where we make the natural identifications $\mathrm{T}_{(A,b)}\,\mathrm{M}_n(\mathbb{R}) \times \mathbb{R} = \mathrm{M}_n(\mathbb{R}) \times \mathbb{R}$ and $\mathrm{T}_t\,\mathbb{R} = \mathbb{R}$ for any $t \in \mathbb{R}$. Since $\det A \neq 0$, d is surjective at such a point. Hence $M \subseteq \mathrm{M}_n(\mathbb{R}) \times \mathbb{R}$ is a smooth submanifold of dimension equal to $\dim \mathrm{M}_n(\mathbb{R}) = n^2$. We also have

$$\mathrm{T}_{(A,b)}\,M = \{(X, (\det A)^{-1}\,\mathrm{d}\det_A(X)) : X \in \mathrm{M}_n(\mathbb{R})\}.$$

(b) We have $M \cong \mathrm{GL}_n(\mathbb{R})$.

7.4. In each case we have to show that the subgroup is closed in G. In (a–c) we also use the continuous action (as introduced in Section 1.9, see the Exercises of Chapter 1 for details),

$$\mu\colon G \times G \longrightarrow G; \quad \mu(x,y) = xyx^{-1}.$$

(a) $\mathrm{Z}_G(g)$ agrees with the stabiliser of g, $\mathrm{Stab}_G(g)$ which is a closed subgroup of G.
(b) $\mathrm{Z}(G)$ is the intersection of all the closed sets $\mathrm{Z}_G(g)$ for $g \in G$ and so is closed.
(c) $\mathrm{N}_G(g)$ is identical with the generalised stabiliser $\mathrm{Stab}_G(H) \leqslant G$, which is a closed subgroup.
(d) Let $x \in G - \ker\varphi$. Then $\varphi(x) \neq 1$, so there is an open subset $U \subseteq H$ containing

$\varphi(x)$ but not 1. Then $\varphi^{-1}U \subseteq G$ is open since φ is continuous. Clearly $\ker\varphi \cap \varphi^{-1}U = \emptyset$ and $x \in \varphi^{-1}U$. This shows that $G - \ker\varphi$ is open in G, hence $\ker\varphi$ is closed.

7.5. This is similar to the previous exercise.

7.6. Use the smooth maps L_g, R_g and χ_g of Proposition 7.16. The tangent spaces are the images of the derivatives at g applied to the tangent space of H.

7.7. By Exercise 7.4(d), $\ker\varphi \leqslant G$ is closed. For $g \in \ker\varphi$ we have $\mathrm{T}_g \ker\varphi = \ker \mathrm{d}\,\varphi_g$.

7.8. (a) Use the continuous function

$$F\colon \mathrm{GL}_{2n}(\mathbb{R}) \longrightarrow \mathrm{M}_n(\mathbb{R}); \quad F(X) = X^2 + I_{2n}$$

for which $\mathcal{C}_{2n} = F^{-1}O$.

(b) The eigenvalues of an element $J \in \mathcal{C}_{2n}$ are $\pm i$. Since the minimal polynomial of J over $\mathbb{C}$ is clearly $X^2 + 1$, which has no repeated roots, J is diagonalisable over $\mathbb{C}$, and over $\mathbb{R}$ has the form

$$J = B \begin{bmatrix} J_2 & O & \cdots & \cdots & O \\ O & J_2 & O & \cdots & \vdots \\ \ddots & \ddots & \ddots & \ddots & \vdots \\ \cdots & \cdots & \cdots & J_2 & O \\ O & \cdots & \cdots & O & J_2 \end{bmatrix} B^{-1}, \quad \text{where } B \in \mathrm{GL}_{2n}(\mathbb{R}) \text{ and } J_2 = \begin{bmatrix} 0 & -1 \\ 1 & 0 \end{bmatrix}.$$

(c) These stabilisers are the subgroups $\rho_n \mathrm{GL}_n(\mathbb{C})$ and $\rho'_n \mathrm{GL}_n(\mathbb{C})$.

(d) For techniques useful for showing that $\mathcal{C}_{2n}$ is a submanifold, see Chapter 8. The tangent space to $\mathcal{C}_{2n}$ at J can be identified with

$$\ker \mathrm{d}\, F_J = \{X \in \mathrm{M}_{2n}(\mathbb{R}) : XJ + JX = O\}.$$

Chapter 8

8.2. (a) The obvious quotient homomorphism $\mathrm{Sp}(n) \times \mathrm{Sp}(1) \longrightarrow \mathrm{Sp}(n) \tilde{\times} \mathrm{Sp}(1)$ has the discrete subgroup $\{(I_n, 1), (-I_n, -1)\}$ as its kernel, so the derivative homomorphism is an isomorphism of Lie algebras; the Lie algebra of the domain is the direct product $\mathfrak{sp}(n) \times \mathfrak{sp}(1)$.

(b) We can identify $\mathbb{H}\mathrm{P}^{n-1}$ with $(\mathrm{Sp}(n) \tilde{\times} \mathrm{Sp}(1))/(\mathrm{Sp}(n-1) \tilde{\times} \mathrm{Sp}(1))$.

Chapter 9

9.1. (a) For each connected component U of G, choose an element $u \in U$. G has a countable open covering $\{U_1, U_2, \ldots\}$, hence for each u as above there is an open set $U_{j_u} \subseteq U$ containing u. This means that the components form a countable set.

(b) The components form an open covering of G. By the Heine–Borel Theorem 1.20, a finite subcollection also covers G. As these sets are disjoint there must be only finitely many of them.

9.2. (a) We have

$$\mathrm{Stab}_{\mathrm{SL}_n(\mathbb{R})}(\mathbf{e}_n) = \left\{ \begin{bmatrix} A & O_{n-1,1} \\ B & 1 \end{bmatrix} : A \in \mathrm{SL}_{n-1}(\mathbb{R}),\ B \in \mathrm{M}_{1,n-1}(\mathbb{R}) \right\},$$

$$\mathrm{Orb}_{\mathrm{SL}_n(\mathbb{R})}(\mathbf{e}_n) = \mathbb{R}^n - \{\mathbf{0}\}.$$

(b) This homogeneous space can be identified with $\mathrm{Orb}_{\mathrm{SL}_n(\mathbb{R})}(\mathbf{e}_n) = \mathbb{R}^n - \{\mathbf{0}\}$ and this is path connected since every pair of vectors $\mathbf{u}, \mathbf{v} \in \mathbb{R}^n - \{\mathbf{0}\}$ which are not parallel can be joined by a path of form

$$t \longmapsto t\mathbf{u} + (1-t)\mathbf{v} \quad (t \in [0,1]).$$

Now use induction on n as in Example 9.14 to show that $\mathrm{SL}_n(\mathbb{R})$ is path connected.

9.3. (a) We have

$$\mathrm{Stab}_{\mathrm{SL}_n(\mathbb{C})}(\mathbf{e}_n) = \left\{ \begin{bmatrix} A & O_{n-1,1} \\ B & 1 \end{bmatrix} : A \in \mathrm{SL}_{n-1}(\mathbb{C}),\ B \in \mathrm{M}_{1,n-1}(\mathbb{C}) \right\},$$
$$\mathrm{Orb}_{\mathrm{SL}_n(\mathbb{C})}(\mathbf{e}_n) = \mathbb{C}^n - \{\mathbf{0}\}.$$

(b) This is similar to part (b) of the previous exercise.
(c) Since $\mathrm{GL}_n(\mathbb{C})/\mathrm{SL}_n(\mathbb{C})$ is identified with $\mathbb{C}^\times$ using the determinant, Proposition 9.11 shows that $\mathrm{GL}_n(\mathbb{C})$ is path connected.

9.4. (a) If $\mathbf{x} \in \mathbb{R}^n$, then $\mathbf{x}^T AA^T \mathbf{x} = |A^T\mathbf{x}|^2 \geqslant 0$, and this vanishes only when $A^T\mathbf{x} = \mathbf{0}$, which happens only when $\mathbf{x} = \mathbf{0}$ since A^T is invertible. The eigenvalues of S are then positive real numbers and S is diagonalisable, say $S = P\,\mathrm{diag}(\lambda_1, \ldots, \lambda_n)P^T$ for some $P \in \mathrm{O}(n)$ and $\lambda_i > 0$. Then we take $S_1 = P\,\mathrm{diag}(\sqrt{\lambda_1}, \ldots, \sqrt{\lambda_n})P^T$.
(b) We have

$$\begin{aligned}(S_1^{-1}A)^T(S_1^{-1}A) &= A^T(S_1^T)^{-1}S_1^{-1}A = A^TS^{-1}A \\ &= A^T(AA^T)^{-1}A = A^T(A^T)^{-1}A^{-1}A = I.\end{aligned}$$

(c) Since $RR^T = I = SS^T$, we have

$$Q^2 = QSS^TQ^T = QS(QS)^T = PR(PR)^T = PRR^TP^T = P^2.$$

(d) S_2 must have the same eigenvalues as $\mathrm{diag}(\sqrt{\lambda_1}, \ldots, \sqrt{\lambda_n})$ and indeed

$$S_2 = U\,\mathrm{diag}(\sqrt{\lambda_1}, \ldots, \sqrt{\lambda_n})U^T$$

for some $U \in \mathrm{O}(n)$. So

$$U\,\mathrm{diag}(\lambda_1, \ldots, \lambda_n)U^T = \mathrm{diag}(\lambda_1, \ldots, \lambda_n)$$

and hence

$$U\,\mathrm{diag}(\lambda_1, \ldots, \lambda_n) = \mathrm{diag}(\lambda_1, \ldots, \lambda_n)U.$$

By comparing entries we find that this is only possible if either $\lambda_i = \lambda_j$ for all i, j or U is diagonal, in which case $S_2 = \mathrm{diag}(\sqrt{\lambda_1}, \ldots, \sqrt{\lambda_n})$.
(e) Use (c) and (d).
(f) First show that the set of positive definite symmetric matrices is a path connected subset of $\mathrm{GL}_n(\mathbb{R})$ by finding a path from every such matrix of the form $V\,\mathrm{diag}(\lambda_1, \ldots, \lambda_n)V^T$ with $V \in \mathrm{O}(n)$ to I.

9.5. (a) We have

$$\mathrm{Stab}_{\mathrm{Aff}_n(\mathbb{k})}(\mathbf{0}) = \left\{ \begin{bmatrix} A & 0 \\ 0 & 1 \end{bmatrix} : A \in \mathrm{GL}_n(\mathbb{k}) \right\}, \quad \mathrm{Orb}_{\mathrm{Aff}_n(\mathbb{k})}(\mathbf{0}) = \mathbb{k}^n.$$

Notice that $\mathrm{Stab}_{\mathrm{Aff}_n(\mathbf{k})}(\mathbf{0})$ can be identified with $\mathrm{GL}_n(\mathbf{k})$.
(b) This follows from the corresponding facts for $\mathrm{GL}_n(\mathbf{k})$.

9.6. See Section 8.5 for background on this. The spaces in (a) and (b) are connected and actually diffeomorphic.

9.7. This is similar to the previous exercise; the homogeneous spaces are diffeomorphic and connected.

9.8. (b) The Pfaffian function $\mathrm{pf}\colon \Sigma_{2m} \longrightarrow \mathbb{R}^{\times}$ is surjective, so Σ_{2m} has at least two path components. In fact it has exactly two components since the diffeomorphic homogeneous space $\mathrm{GL}_{2m}(\mathbb{R})/\mathrm{Symp}_{2m}(\mathbb{R})$ has at most two components; this implies that $\mathrm{Symp}_{2m}(\mathbb{R}) \leqslant \mathrm{SL}_{2m}(\mathbb{R})$.

9.10. By straightforward calculation we find that

$$\mathrm{N}_{\mathrm{U}(2)}(T) = T \cup \left\{ \begin{bmatrix} 0 & u \\ v & 0 \end{bmatrix} : |u| = |v| = 1 \right\} \leqslant \mathrm{U}(2).$$

This has the two path components T, $\left\{ \begin{bmatrix} 0 & u \\ v & 0 \end{bmatrix} : |u| = |v| = 1 \right\}$, so $\pi_0 \mathrm{N}_{\mathrm{U}(2)}(T) \cong \{1, -1\}$.

9.11. Using the previous exercise, we obtain

$$\mathrm{N}_G(T') = \left\{ \begin{bmatrix} P & O_{2,2} \\ O_{2,2} & Q \end{bmatrix} : P, Q \in \mathrm{N}_{\mathrm{U}(2)}(T),\ \det P = \det Q \right\},$$

which has 4 path components each of which contains one of the matrices

$$I_4, \quad \begin{bmatrix} J_2 & O_{2,2} \\ O_{2,2} & I_2 \end{bmatrix}, \quad \begin{bmatrix} I_2 & O_{2,2} \\ O_{2,2} & J_2 \end{bmatrix}, \quad \begin{bmatrix} J_2 & O_{2,2} \\ O_{2,2} & J_2 \end{bmatrix},$$

where $J_2 = \begin{bmatrix} 0 & -1 \\ 1 & 0 \end{bmatrix}$. Then $\pi_0 \mathrm{N}_G(T') \cong \{1, -1\} \times \{1, -1\}$.

Chapter 10

10.1. Given a Lie homomorphism $\varphi\colon \mathbb{T} \longrightarrow \mathbb{T}$, the curve $\gamma\colon \mathbb{R} \longrightarrow \mathbb{T}$ given by $\gamma(t) = \varphi(e^{2\pi i t})$ is a one-parameter subgroup satisfying $\gamma'(s+t) = \gamma(s)\gamma'(t)$ for all $s, t \in \mathbb{R}$. By our general results on such curves we have $\gamma(t) = e^{2\pi c i t}$ for some $c \in \mathbb{R}$. But whenever $n \in \mathbb{Z}$, $\gamma(n) = \varphi(1) = 1$, so $c \in \mathbb{Z}$. For φ to be an isomorphism we must have $c = \pm 1$.
For $r \geqslant 1$, for each integer matrix $A = [a_{ij}] \in \mathrm{M}_r(\mathbb{Z})$, there is a Lie homomorphism

$$\varphi_A\colon \mathbb{T}^r \longrightarrow \mathbb{T}^r; \quad \varphi_A(z_1, \ldots, z_r) = (z_1^{a_{11}} \cdots z_1^{a_{1r}}, \ldots, z_1^{a_{r1}} \cdots z_1^{a_{rr}})$$

with derivative $\mathrm{d} = A$ acting in the obvious way on $\mathfrak{t} = \mathbb{R}i \times \cdots \times \mathbb{R}i$. Isomorphisms come from those A with integer inverse $A^{-1} \in \mathrm{M}_r(\mathbb{Z})$.

10.2. If $T_0 = T$ is not maximal, then it is contained in another T_1 of greater dimension. Repeating this, for dimensional reasons eventually we obtain a sequence of tori $T_0 \leqslant T_1 \leqslant \cdots \leqslant T_k \leqslant G$ where there is no torus in G which properly contains T_k.

10.3. (a) We have

$$\mathrm{N}_{\mathrm{U}(2)}(T) = T \cup \left\{ \begin{bmatrix} 0 & u \\ v & 0 \end{bmatrix} : |u| = |v| = 1 \right\} = T \cup \begin{bmatrix} 0 & 1 \\ 1 & 0 \end{bmatrix} T, \quad \pi_0 \mathrm{N}_{\mathrm{U}(2)}(T) = \{1, -1\}.$$

The conjugation action is given by

$$\begin{bmatrix} 0 & 1 \\ 1 & 0 \end{bmatrix} T \cdot \operatorname{diag}(z, w) = \operatorname{diag}(w, z).$$

(b) and (c) are discussed in Chapters 11 and 12.

10.4. See Chapters 11 and 12.

10.5. (c) $\mathrm{O}(n)$ is not connected.

Chapter 11

11.1. (a) By Theorem 11.5, $\mathfrak{z}(\mathfrak{u}(n))$ consists of all scalar hermitian matrices tiI_n with $t \in \mathbb{R}$. Using the defining Equation (11.1), we have $X \in \mathfrak{z}(\mathfrak{u}(n))^{\perp}$ if and only if $\operatorname{tr}(tiX) = 0$ for all $t \in \mathbb{R}$. Since $\operatorname{tr}(tiX) = ti \operatorname{tr}(X)$, this gives

$$\mathfrak{z}(\mathfrak{u}(n))^{\perp} = \{X \in \mathfrak{u}(n) : \operatorname{tr} X = 0\} = \mathfrak{su}(n).$$

(b) Compose the determinant $\det\colon \mathrm{U}(n) \longrightarrow \mathbb{T} \cong Z(\mathrm{U}(n))$ with the projection $Z(\mathrm{U}(n)) \longrightarrow Z(\mathrm{U}(n))/Z$ for $Z = \{zI_N : z^n = 1\}$.
(c) Define this homomorphism by $(A, B) \longmapsto AB$.

11.4. See Chapter 12 for details.

11.5. (a) Take $T_1 = \{\operatorname{diag}(u, v, w, \overline{uvw}) : |u| = |v| = |w| = 1\}$. Then

$$\mathrm{N}_{G_1}(T_1) = \left\{ \begin{bmatrix} A & O_{2,2} \\ O_{2,2} & B \end{bmatrix} : A, B \in \mathrm{N}_{\mathrm{U}(2)}(T),\ \det A \det B = 1 \right\},$$

where $\mathrm{N}_{\mathrm{U}(2)}(T) \leqslant \mathrm{U}(2)$ is the normaliser of the maximal torus used in Exercise 10.3. $\mathrm{N}_{G_1}(T_1)$ has 4 components and the Weyl group $\mathrm{W}_{G_1}(T_1) = \mathrm{N}_{G_1}(T_1)/T_1$ consists of the cosets of the matrices

$$I_4, \quad W_1 = \begin{bmatrix} I_2 & O_{2,2} \\ O_{2,2} & J_2 \end{bmatrix}, \quad W_2 = \begin{bmatrix} J_2 & O_{2,2} \\ O_{2,2} & I_2 \end{bmatrix}, \quad W_3 = \begin{bmatrix} J_2 & O_{2,2} \\ O_{2,2} & J_2 \end{bmatrix},$$

where $J_2 = \begin{bmatrix} 0 & -1 \\ 1 & 0 \end{bmatrix}$; then $\mathrm{W}_{G_1}(T_1) \cong \{1, -1\} \times \{1, -1\}$. $\mathrm{W}_{G_1}(T_1)$ acts on T_1 by

$$\begin{aligned} W_1T_1 \cdot \operatorname{diag}(u, v, w, \overline{uvw}) &= \operatorname{diag}(u, v, \overline{uvw}, w), \\ W_2T_1 \cdot \operatorname{diag}(u, v, w, \overline{uvw}) &= \operatorname{diag}(v, u, w, \overline{uvw}), \\ W_3T_1 \cdot \operatorname{diag}(u, v, w, \overline{uvw}) &= \operatorname{diag}(v, u, \overline{uvw}, w). \end{aligned}$$

The Lie algebra of T_1 is $\mathfrak{t}_1 = \{\operatorname{diag}(ri, si, ti, -(r+s+t)i) : r, s, t \in \mathbb{R}\}$, and there are 4 non-trivial roots $\pm\alpha_1, \pm\alpha_2$, where

$$\alpha_1(ri, si, ti, -(r+s+t)i) = r - s, \quad \alpha_2(ri, si, ti, -(r+s+t)i) = r + s + 2t,$$

for $r, s, t \in \mathbb{R}$. $\mathrm{W}_{G_1}(T_1)$ acts on the roots by

$$\begin{aligned} W_1T_1 \cdot \alpha_1 &= \alpha_1, & W_2T_1 \cdot \alpha_1 &= -\alpha_1, & W_3T_1 \cdot \alpha_1 &= -\alpha_1, \\ W_1T_1 \cdot \alpha_2 &= -\alpha_2, & W_2T_1 \cdot \alpha_2 &= \alpha_2, & W_3T_1 \cdot \alpha_2 &= -\alpha_2. \end{aligned}$$

(b) This is similar to (a), see also Exercise 9.11.

Chapter 12

12.2. Rotate and rescale Figure 12.3 into Figure 12.4.

Bibliography

[1] J.F. Adams, Lectures on Lie Groups, University of Chicago Press (1969).

[2] J.F. Adams, Lectures on Exceptional Lie Groups, University of Chicago Press (1996).

[3] M.F. Atiyah, R. Bott & A. Shapiro, Clifford modules, Topology **3** (1964) suppl. 1, 3–38.

[4] R. Carter, G. Segal & I.G. Macdonald, Lectures on Lie groups and Lie algebras, Cambridge University Press (1995).

[5] P.M. Cohn, Classic Algebra, Wiley (2000).

[6] L. Conlon, Differentiable Manifolds: A First Course, Birkhäuser (1993).

[7] M.L. Curtis, Matrix Groups, Springer-Verlag (1984).

[8] R.W.R. Darling, Differential Forms and Connections, Cambridge University Press (1994).

[9] A. Dold, Lectures on Algebraic Topology, Springer-Verlag (1995).

[10] W. Ebeling, Lattices and Codes, Vieweg (1994).

[11] S. Helgason, Differential Geometry, Lie Groups, and Symmetric Spaces, American Mathematical Society (2001).

[12] R. Howe, Very basic Lie theory, American Mathematical Monthly **90** (1983), 600–623; Correction, Amer. Math. Monthly **91** (1984), 247.

[13] W.Y. Hsiang, Lectures on Lie Groups, World Scientific (2000).

[14] V.G. Kač, Infinite-dimensional Lie Algebras, Cambridge University Press (1990).

[15] S. Lang, Algebra, second edition, Addison-Wesley (1984).

[16] S. Lang, Linear Algebra, reprint of the third edition, Springer-Verlag, (1989).

[17] S. Lang, Differential and Riemannian Manifolds, third edition, Springer-Verlag (1995).

[18] S. Lang, Fundamentals of differential geometry, Springer-Verlag (1999).

[19] H.B. Lawson & M.-L. Michelsohn, Spin Geometry, Princeton University Press, (1989).

[20] C.R.F. Maunder, Algebraic Topology, Dover Publications (1996).

[21] M.A. Naĭmark, Normed Rings, translated by L.F. Boron, Wolters-Noordhoff Publishing (1970).

[22] G.K. Pedersen, Analysis Now, Springer-Verlag (1989).

[23] I.R. Porteous, Topological Geometry, Van Nostrand Reinhold (1969).

[24] I.R. Porteous, Clifford Algebras and the Classical Groups, Cambridge University Press (1995).

[25] H. Sato, Algebraic Topology: An Intuitive Approach, Amer. Math. Soc. Trans. of Mathematical Monographs **183** (1996).

[26] J.-P. Serre, Complex Semisimple Lie Algebras, Springer-Verlag (1987).

[27] S. Sternberg, Group Theory and Physics, Cambridge University Press (1994).

[28] G. Strang, Linear Algebra and its Applications, Second edition, Academic Press [Harcourt Brace Jovanovich] (1980).

[29] F.W. Warner, Foundations of Differentiable Manifolds and Lie Groups, Springer-Verlag (1983).

Additional entries added for the second printing

[30] J.C. Baez, The octonions, Bull. Amer. Math. Soc. **39** (2002), 145–205.

[31] T. Bröcker & T. tom Dieck, Representations of Compact Lie Groups, 3rd printing, Springer-Verlag (2003).

[32] J.J. Duistermaat & J.A.C. Kolk, Lie Groups, Springer-Verlag (2000).

[33] J. M. Lee, Introduction to Smooth Manifolds, Springer-Verlag (2003).

Index